geog.3

teacher's handbook

- ◆ starters and plenaries
- ◆ objectives and outcomes
- ◆ answers

< catherine hurst >< anna king >< john edwards >

< chris stevens >< jack mayhew >< david smith >

OXFORD
UNIVERSITY PRESS

Great Clarendon Street, Oxford OX2 6DP

Oxford University Press is a department of the University of Oxford.
It furthers the University's objective of excellence in research,
scholarship, and education by publishing worldwide in

Oxford   New York

Auckland   Cape Town   Dar es Salaam   Hong Kong   Karachi
Kuala Lumpur   Madrid   Melbourne   Mexico City   Nairobi
New Delhi   Shanghai   Taipei   Toronto

With offices in

Argentina   Austria   Brazil   Chile   Czech Republic   France   Greece
Guatemala   Hungary   Italy   Japan   Poland   Portugal   Singapore
South Korea   Switzerland   Thailand   Turkey   Ukraine   Vietnam

Oxford is a registered trade mark of Oxford University Press
in the UK and in certain other countries

Authors: Catherine Hurst, Anna King, John Edwards, Chris Stevens,
Jack Mayhew, David Smith

British Library Cataloguing in Publication Data

Data available

ISBN: 978-0-19-913473-1

10 9 8 7 6 5 4 3 2 1

Printed in Great Britain by Bell and Bain Ltd., Glasgow.

Contents

geog.SG is a complete course for Standard Grade geography. It offers excellent support for Assessment for Learning.

The course components

The course consists of the students' book, the teacher's handbook, and the teacher's resource file, which comes with a CD-ROM.

Find out more about the course components by looking at these panels.

The students' book

- A single book for the course
- Chapters divided into two-page units
- Chapter openers give the big picture – the big ideas behind the chapter – and the goals for the chapter
- Aims of unit given in student-friendly language at the start of each unit
- Activities at the end of each unit
- Glossary covering key vocabulary

The teacher's handbook

- Chapter overviews
- Help at a glance for each unit
- Ideas for starters and plenaries for each unit
- Outcomes for each unit
- Answers for Activities
- Glossary covering key vocabulary

The teacher's resource file

- Photocopiable decision-making exercises, enquiries, role-plays, and fieldwork, with assessment criteria and grade information in student-friendly language
- Photocopiable exam-style questions at General-Foundation and General-Credit level, with assessment criteria and grade information in student-friendly language
- Opportunities for teacher, self-, and peer assessment
- Photocopiable self-assessment forms – one for each chapter
- Outline maps
- Photocopiable glossary covering key vocabulary
- All material on CD-ROM, with all material provided as editable Word files

geog.SG provides a wide range of materials. The students' book is the core of the course. It combines a rigorous approach to content with an engaging style.

You can decide how to use the support materials. The whole package provides a comprehensive and flexible course for Standard Grade geography – which we hope you will enjoy using.

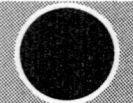

Using this teacher's handbook

This *geog.SG teacher's handbook* aims to save you time and effort! It offers full support for *geog.SG* students' book, and will help you prepare detailed course and lesson plans.

What it provides

For each chapter of the students' book, this teacher's handbook provides:

1 a chapter overview

2 help at a glance for each unit, including ideas for starters and plenaries and answers for Activities

This teacher's handbook also provides Standard Grade syllabus matching grids which show exactly how the students' book covers the syllabus (see pages 14-19).

Plus it has a glossary at the back, covering the geographical terms the students will meet.

Find out more about the two main components, below.

1 The chapter overview

This is your introduction to the corresponding students' chapter. Look at its sections.

Sets out the objectives and outcomes for the chapter, and the corresponding unit numbers.

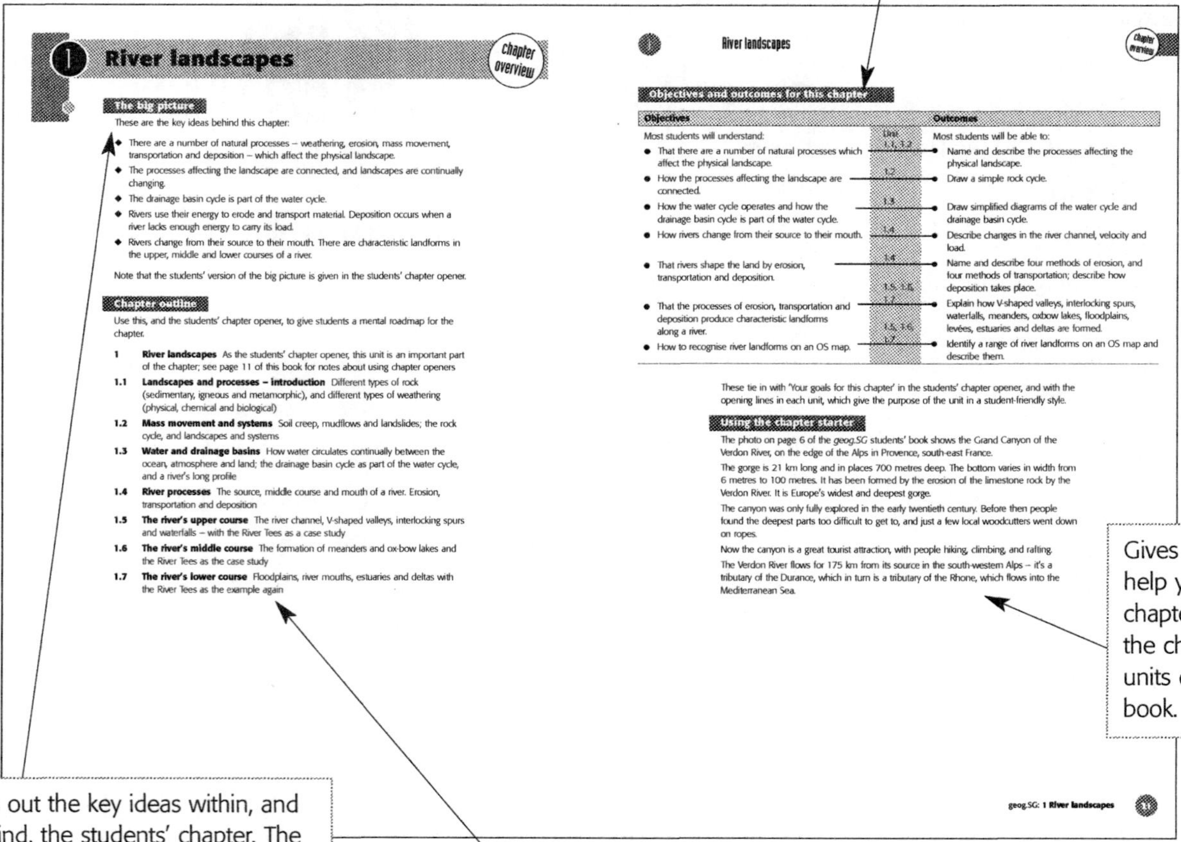
1 River landscapes

chapter overview

The big picture

These are the key ideas behind this chapter:

- There are a number of natural processes – weathering, erosion, mass movement, transportation and deposition – which affect the physical landscape.
- The processes affecting the landscape are connected, and landscapes are continually changing.
- The drainage basin cycle is part of the water cycle.
- Rivers use their energy to erode and transport material. Deposition occurs when a river lacks enough energy to carry its load.
- Rivers change from their source to their mouth. There are characteristic landforms in the upper, middle and lower courses of a river.

Note that the students' version of the big picture is given in the students' chapter opener.

Chapter outline

Use this, and the students' chapter opener, to give students a mental roadmap for the chapter.

1 **River landscapes** As the students' chapter opener, this unit is an important part of the chapter; see page 11 of this book for notes about using chapter openers

1.1 **Landscapes and processes – introduction** Different types of rock (sedimentary, igneous and metamorphic), and different types of weathering (physical, chemical and biological)

1.2 **Mass movement and systems** Soil creep, mudflows and landslides; the rock cycle, and landscapes and systems

1.3 **Water and drainage basins** How water circulates continually between the ocean, atmosphere and land; the drainage basin cycle as part of the water cycle, and a river's long profile

1.4 **River processes** The source, middle course and mouth of a river. Erosion, transportation and deposition

1.5 **The river's upper course** The river channel, V-shaped valleys, interlocking spurs and waterfalls – with the River Tees as a case study

1.6 **The river's middle course** The formation of meanders and ox-bow lakes and the River Tees as the case study

1.7 **The river's lower course** Floodplains, river mouths, estuaries and deltas with the River Tees as the example again

1 River landscapes

Objectives and outcomes for this chapter

Objectives	Unit	Outcomes
Most students will understand:		Most students will be able to:
That there are a number of natural processes which affect the physical landscape.	1.1, 1.2	Name and describe the processes affecting the physical landscape.
How the processes affecting the landscape are connected.	1.2	Draw a simple rock cycle.
How the water cycle operates and how the drainage basin cycle is part of the water cycle.	1.3	Draw simplified diagrams of the water cycle and drainage basin cycle.
How rivers change from their source to their mouth.	1.4	Describe changes in the river channel, velocity and load.
That rivers shape the land by erosion, transportation and deposition.	1.4	Name and describe four methods of erosion, and four methods of transportation; describe how deposition takes place.
That the processes of erosion, transportation and deposition produce characteristic landforms along a river.	1.5, 1.6, 1.7	Explain how V-shaped valleys, interlocking spurs, waterfalls, meanders, oxbow lakes, floodplains, levées, estuaries and deltas are formed.
How to recognise river landforms on an OS map.	1.5, 1.6, 1.7	Identify a range of river landforms on an OS map and describe them.

These tie in with 'Your goals for this chapter' in the students' chapter opener, and with the opening lines in each unit, which give the purpose of the unit in a student-friendly style.

Using the chapter starter

The photo on page 6 of the *geog.SG* students' book shows the Grand Canyon of the Verdon River, on the edge of the Alps in Provence, south-east France.

The gorge is 21 km long and in places 700 metres deep. The bottom varies in width from 6 metres to 100 metres. It has been formed by the erosion of the limestone rock by the Verdon River. It is Europe's widest and deepest gorge.

The canyon was only fully explored in the early twentieth century. Before then people found the deepest parts too difficult to get to, and just a few local woodcutters went down on ropes.

Now the canyon is a great tourist attraction, with people hiking, climbing and rafting.

The Verdon River flows for 175 km from its source in the south-western Alps – it's a tributary of the Durance, which in turn is a tributary of the Rhone, which flows into the Mediterranean Sea.

geog.SG: 1 River landscapes 11

Gives information to help you with the chapter starters, in the chapter opening units of the students' book.

Sets out the key ideas within, and behind, the students' chapter. The students' version of this is given in their chapter opening unit.

Gives a very brief summary of what's covered in the students' chapter. Together with the chapter opening unit in the students' book, it will help you give students a roadmap for the chapter.

2 Help at a glance for each unit

These pages give comprehensive help for each unit of *geog.SG* students' book.

Starts with a brief walk through the unit, to show you how it develops.

Summarises ideas covered in the unit, plus underlying ideas where appropriate.

Full answers to the Activities in the students' book, to save you time.

New vocabulary introduced in the unit. See the glossary at the back of this teacher's handbook.

A breakdown of the skills practised. It will help you identify where students may need extra support.

Expected outcomes for the unit. They tie in with the expected outcomes for the chapter.

Suggestions for a starter.

Suggests plenaries for use throughout the lesson, not just at the end.

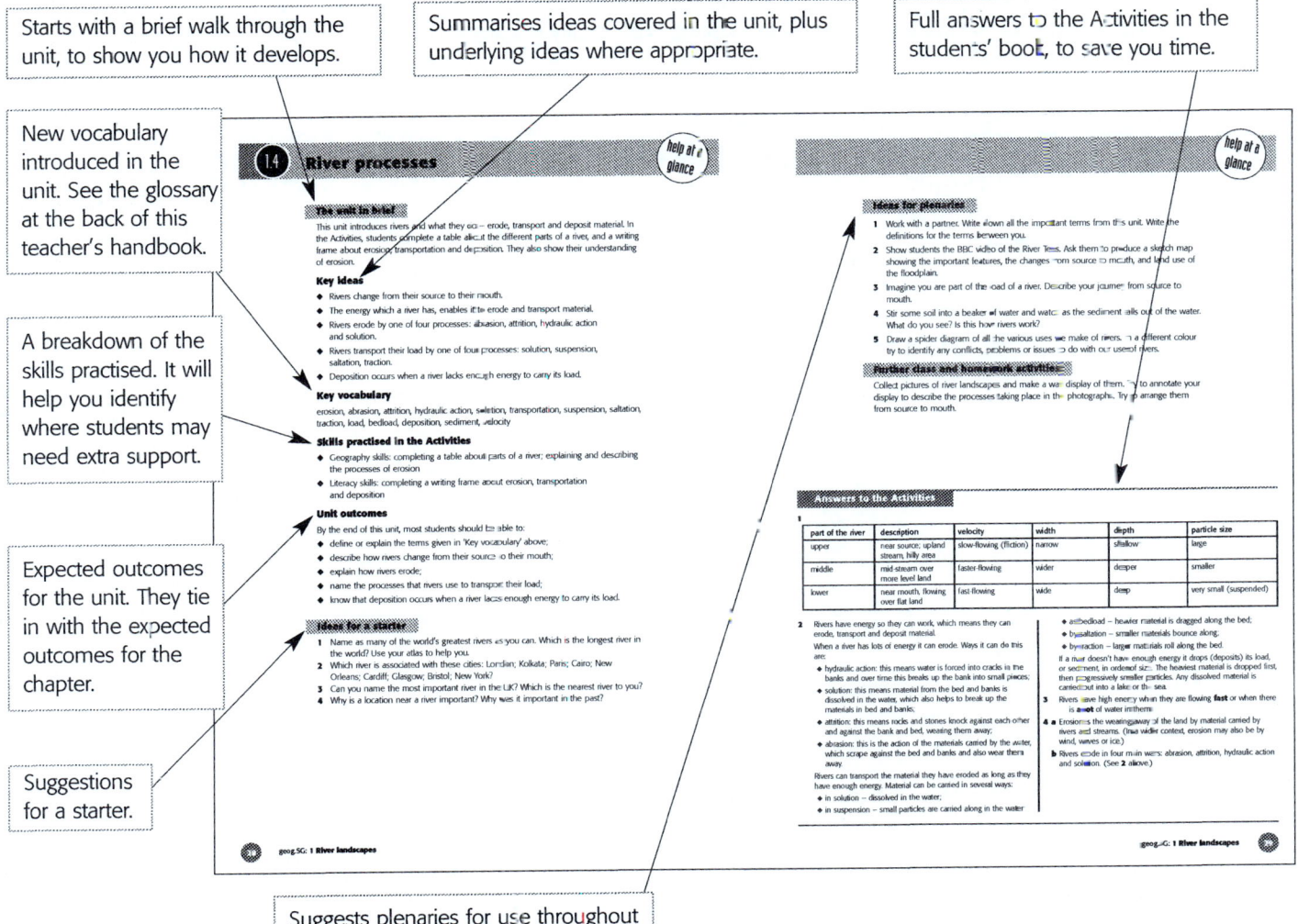

Planning for high-quality lessons

Well-planned and well-structured lessons are a key requirement, for delivering high-quality teaching and learning in any subject, at any level. The *geog.SG* course aims to make it easy to plan, structure, and deliver, high-quality lessons for Standard Grade geography.

Structure of a typical lesson

You will already be familiar with guidelines on structuring lessons. This shows a typical lesson structure:

STARTER

Purpose: To capture students' attention and focus the class. Use it as the lesson hook, or to find out what students know already about a new topic, or for quick revision of earlier work.

INTRODUCTION

Purpose: To prepare students for the activities ahead.

- If this is a new topic, tell students the topic objectives. Write these on the board.
- If it's a continuation of a topic, you can refer back to an objective as appropriate.

ACTIVITIES

This is the main body of the lesson.

Purpose: To achieve one or more of the topic objectives.

- Emphasis on exploration and investigation.
- Provide for practice in different types of skill: geographical, literacy, numeracy, thinking, listening, speaking, teamworking, and ICT skills.
- Choose from a variety of activities: reading, answering questions, enquiries, role play, game playing, fieldwork, and ICT.

Plenaries: note that plenaries can be used as staging posts throughout the activities, to gain feedback, check understanding, link to earlier work, and encourage reflection on what is being learnt, and how.

FINAL PLENARY

Purpose: To round off and review what has been done, and to assess what has been achieved against the topic objectives. This is where you help students to:

- check, and crystallise, their understanding
- generalise, for example from an individual case study
- set work in context, and make links to work already done, or to be done in the future
- reflect on how they have learned, as well as what they have learned
- check how well they have achieved the topic objectives (self-assessment).

HOMEWORK

Purpose: To confirm, give practice in, and extend, what has been learnt in the lesson.

- The homework can lead on from the final plenary, and be the basis for a starter for the next lesson.

Planning around *geog.SG*

Now see how the components of *geog.SG* provide material for each part of your lesson.

STARTERS

- The *Help at a glance* pages in this teacher's handbook have suggestions for lesson starters.
- See further notes about starters in this teacher's handbook.

OBJECTIVES

- The opening lines of each unit in the students' book give the purpose of the unit, in student-friendly language. The goals for each chapter are given in its opening unit.
- See also the objectives and outcomes given in this teacher's handbook.

ACTIVITIES

Using the students' book

- The text in the students' book provides the core information students need. Some lends itself to reading aloud, but try 'quiet time' too.
- You can let students work through the text uninterrupted, or break it up with Activities. (These generally follow the order of the text).
- The questions give practice in literacy, thinking, and geography skills.
- Some are ideal a whole-class questions with verbal response. Others can be worked through by students working alone, in pairs or in small groups. Some questions are open-ended questions that challenge students to show what they can do.
- For students who finish early, you could ask them to select and write definitions of key vocabulary from the unit, to make revision notes about the lesson, or to draft a question for an 'Ask the expert' session at the end of the lesson. You could also get them to start an activity from this file.

Using the teacher's resource file

- This file has two main types of activities you can base lessons on:
  - exam-style questions at General-Foundation and General-Credit level, with assessment criteria and grade information
  - decision-making exercises, enquiries, role-plays, and fieldwork, with assessment criteria and grade information.

Plenaries

- The *Help at a glance* pages in this teacher's handbook give suggestions for plenaries, for use throughout the lesson as well as at the end.
- See further notes about plenaries in this teacher's handbook.

Homework

- Some of the Activities in the students' book could be used for homework, particularly certain open-ended questions and those identified as research activities.
- You could use the exam-style questions or longer activities from the resource file.
- At the end of a chapter, students could complete the self-assessment form, and identify parts of the topic they need to re-visit or get extra help on.

More about starters and plenaries

Planning your starters and plenaries

Effective starters and plenaries need to be planned for. With planning, you can ensure that they'll help you to meet your lesson objectives, and that you won't have to rely on sudden inspiration in the classroom. But even where they are planned, you may want and need to modify them as you go along, in response to your students.

Our suggestions for starters and plenaries

The kinds of activities you feel comfortable with, for starters and plenaries, will depend on your teaching style, and the individual class. So the suggestions for starters and plenaries in this book are just that: suggestions! You may want to use some as described, or adapt them. Or they may provide inspiration for new ideas of your own.

The starters
*   Most of these are intended for use with the students' book closed, before students have looked at the new unit. But they lead seamlessly into the work in the students' book.
*   In some cases you may want to combine two starters to give a more extended one.
*   A number of starters require the use of an atlas, and can be an excellent way of giving your students atlas practice that's fun.
*   Other starters require both physical and mental activity – for example creating a graffiti wall on the board. This is a good way to get everyone involved.

The plenaries
*   There are suggestions for plenaries for use throughout the lesson, not just at the end.
*   They have been chosen for a variety of purposes: to encourage feedback; assess understanding; promote reflection; build bridges with material already covered (or still to be covered), with other subjects, and with the real world; help crystallise what has been learnt; and see whether it applies to other situations.
*   Some of the plenaries are single questions. You will find that you can readily combine some to make more extended plenaries.
*   Some need more preparation than others. You might not want to choose these for every class, but it's a good idea to ring the changes, and keep your students surprised.
*   Together with the Activities, the *Ideas for plenaries* section is a rich resource to help you deliver fresh, exciting and effective lessons.

Resources for starters and plenaries

Images
Many of the starters, and some plenaries, require images – mostly photos. These can be printed, on OHTs, or displayed from a computer via an interactive whiteboard or projector.

The Internet is an excellent source for geographical photos and other images. (Try a google image search, for example, with different sets of key words.) You can download the ones you need in advance, rather than hunt during class. Please check with the appropriate people in your school regarding copyright issues.

Building a resource library
Some resources, such as photos, can be used over and over. You may want to create your own resource library. Laminating printed photos, and other resources (such as the True/False cards) will extend their lives and save you time and effort in the future.

Using the chapter openers in the students' book

The chapter openers

The chapter openers in the students' book are in effect the starters for new topics – and you can return to them as an end-of-topic plenary.

Below is a typical chapter opener.

Large photo to hook your students' attention (we hope!). The opening photos usually relate to specific material within the chapter.

Gives the big underlying ideas for the chapter. These provide the context for new learning. At the end of the chapter they can be reviewed, to help crystallise the learning.

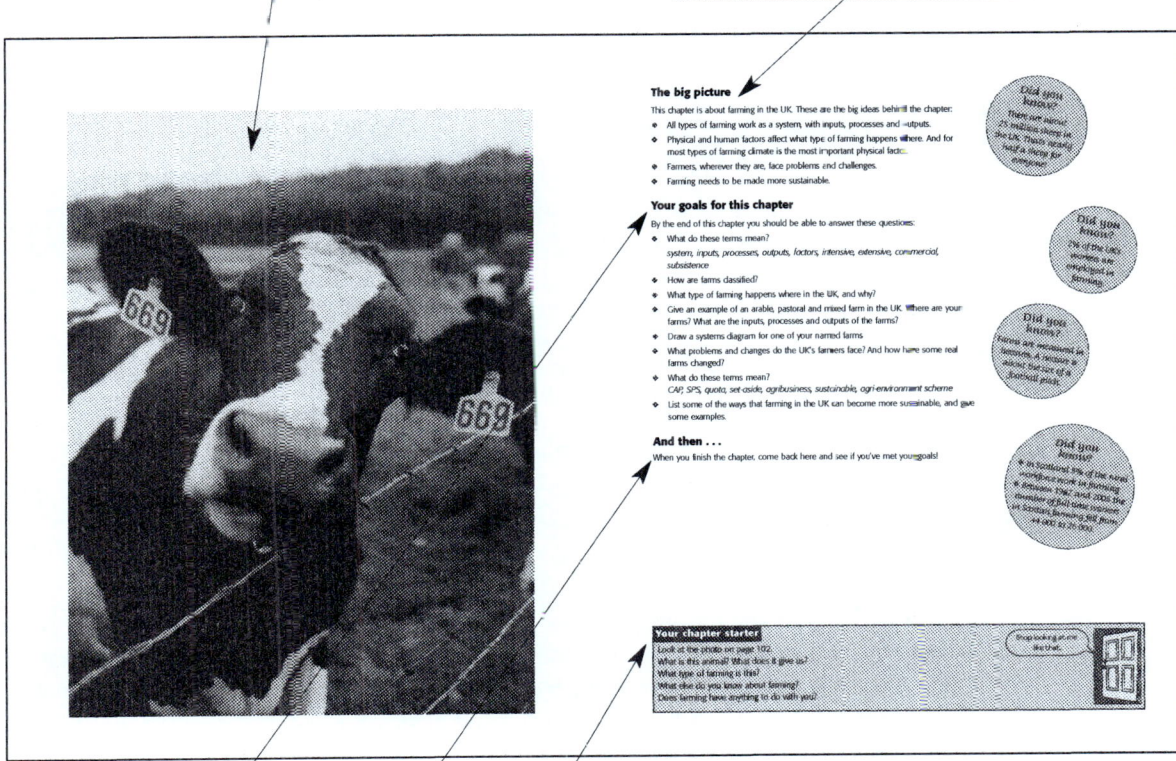

Sets goals for the students, in the form of questions they should be able to answer by the end of the chapter.

Chapter starter questions, to get your students thinking. The Chapter overviews in this book give information about the photos, and some background for the starter questions. (Look in the final sections.)

Invites students to revisit the goals at the end of the chapter. Note that *geog.SG teachers' resource file* has a students' self-assessment form for each chapter, which refers to these goals.

Using the chapter openers

As you can see, the chapter openers can do quite a lot of useful work, so it's worth spending some time on them.

- 'The big picture' can be read aloud, and discussed.
- You can work through 'Your goals for this chapter' in advance, to find out what students know already. Most will probably be able to answer at least a couple of questions.
- Then the next step is to give students a mental roadmap for the chapter, using the corresponding **Chapter overviews** in this book.

About the teacher's resource file

The *geog.SG teacher's resource file* contains vital support material for *geog.SG* students' book, to help you plan and deliver an effective Standard Grade geography course.

- It offers a wide range of photocopiable learning activities and exam-style questions.
- All materials are also provided on the accompanying CD-ROM, in pdf format.
- All materials and forms are also provided on the CD-ROM as Word files, so that you can adapt them.

Learning activities

There's a learning activity for each chapter – a decision-making exercise, enquiry, role-play, or piece of fieldwork.

Some of these activities will take several hours to complete. Some are for individual work, while others provide opportunities for students to work in pairs or small groups; some provide opportunities for whole-class work and feedback.

Most of these activities come with assessment criteria and grade information for all or part of the activity, written in student-friendly language. You can choose whether to show this to your students before or after the activity, or not at all. It can be used for teacher, self-, or peer assessment.

The assessment criteria form allows you to record the mark achieved, to comment, and to identify areas for improvement. It also allows the student to take part in this process.

The form can then form part of the student's assessment portfolio.

There are teacher's notes for each activity, giving the aims of the activity, and advice on how to set up and run it.

Exam-style questions

There are two exam-style questions for each chapter, and there's a General-Foundation and General-Credit level version of each question. The marks available for each part of the question are given. Assessment criteria and grade information is given for each question, written in student-friendly language. It can be used for teacher, self-, and peer assessment.

These questions could be used in class or for homework, or as end-of-topic 'tests'. You could put several together to create a mock exam.

The reflection activity allows you to record the mark achieved, to comment, and to identify areas for improvement. It also allows the student to take part in this process.

The form can then form part of the student's assessment portfolio.

Self-assessment forms

There's a self-assessment form for each chapter. Designed to be used at the end of the chapter, it allows individual students to review and analyse their own work.

The table relates to the text 'Your goals for this chapter' laid out at the start of each chapter in the students' book.

Glossary

A glossary covering key vocabulary is provided. You could use the Word file to create worksheets.

Outline maps

Maps of the British Isles, Europe (political), and the World (political) are provided.

Standard Grade syllabus matching grid

The physical environment

In this grid the Key Ideas, Explanatory comment and Areal/Thematic context for external assessment are taken from the Standard Grade syllabus. All other information relates to the *geog.SG* students' book.

Key idea	Explanatory comment	Areal/Thematic context for external assessment	geog.SG	Case studies
1 Physical landscapes are the product of natural processes and are always changing.	Physical landscapes are formed by a set of interdependent processes – weathering, mass movement, erosion, transportation, deposition – which can be viewed as systems.	UK/Western Europe **In external papers examples will be drawn from rivers and their valleys and glaciated areas**	River landscapes pages 6-21 Glacial landscapes pages 22-39	River Tees pages 14-21. Lake District pages 26-33; Cairngorms pages 34-39.
2 The elements of weather can be identified, observed, measured, recorded and classified. As a result, dynamic patterns can be identified and used for forecasting.	Weather is the state of a set of atmospheric variables – temperature, precipitation, humidity, cloud, wind, pressure. The most important weather patterns affecting Western Europe are associated with frontal conditions, depressions and anticyclones. The characteristics of weather associated with these patterns should be studied, including the use of synoptic charts and weather forecasting.	UK/Western Europe	Weather pages 40-53	British Isles pages 50-53.
3 The world can be divided into major climatic zones.	The emphasis should be on identifying and describing the characteristics of major climatic types through interpretation of, for example, climate graphs and maps.	Global. **In external papers examples will be drawn from Equatorial rainforests, Tropical deserts, Mediterranean and tundra regions.**	Climate zones pages 54-61, 64-65, 68-69, 72-73	Hot desert climate pages 60-61; Mediterranean climate pages 64-65; tundra regions pages 68-69; equatorial climate pages 72-73.

4 The physical environment offers a range of possibilities for, and limitations on, human activities.	Consideration should be given to those physical factors (e.g. water, soil, vegetation) which influence human activities, in particular land use and food production. Attention should focus on the relationship between physical factors and land uses in selected environments.	Global	Glacial landscapes pages 34-37 Climate zones pages 60-75 Farming pages 102-111	Cairngorms pages 34-37. Human activity in: hot deserts pages 60-63; Mediterranean regions pages 64-67; tundra regions pages 68-71; equatorial climate regions pages 72-75. Lynford House Farm, East Anglia (arable farm) page 108; Herdship Farm, Teesdale (pastoral farm) page 109; Penllan Farm, Herefordshire (mixed farm) page 110; Ardalanish Organic Farm, Isle of Mull (mixed farm) page 111.
5 There are many competing demands for the use of rural landscapes.	There is competition between the major existing land uses (e.g. farming, forestry, recreation). Conservation and urban expansion are added pressures in particular localities.	Scotland	Glacial landscapes pages 34-39 Settlement pages 98-101	Cairngorms pages 34-39; Edinburgh green belt developments pages 100-101.
6 The physical environment is a resource which has to be used with care and its management is a global issue.	Many current developments affect the environments of continents or even the whole globe. In fragile environments human activity can trigger short-term and long-term problems. There is a need to recognise the issues involved and the preventative and remedial measures required. Environmental deterioration can serve to stimulate international co-operation but may also cause disagreements.	Global. **In external papers examples will be drawn from the threat to tropical forests, the spread of tropical deserts and the use and misuse of oceans.**	Climate zones pages 62-63, 66-67, 70-71, 74-79	Human impact in hot deserts (desertification) pages 62-63; human impact in the Mediterranean (Benidorm and use and abuse of the Mediterranean Sea) pages 66-67; human impact in the tundra (oil exploitation) pages 70-71; human impact in the tropical rainforest (economic exploitation) pages 74-75; climate change and global warming pages 76-79.

Standard Grade syllabus matching grid

The human environment

In this grid the Key Ideas, Explanatory comment and Areal/Thematic context for external assessment are taken from the Standard Grade syllabus. All other information relates to the *geog.SG* students' book.

Key idea	Explanatory comment	Areal/Thematic context for external assessment	geog.SG	Case studies
7 Settlements have many common characteristics related to site, situation and function.	The initial location of a settlement may be closely related to site characteristics. Its sphere of influence will affect its growth and functions. These functions can change through time.	UK	Settlement pages 80-87	
8 Urban settlements have dynamic patterns relating to their size, form and function.	Most settlements undergo change which can create problems (e.g. congestion, loss of social cohesion within communities, loss of economic base). These changes and solutions to the problems can alter the internal structure of a settlement. Solutions (e.g. urban renewal, suburbanisation, new town development) provide opportunities to improve the urban environment. Patterns of development may be conditioned by features from the past.	UK	Settlement pages 88-101 Industry and economic change pages 124, 130, 132-133.	Brindley Place, Birmingham pages 94-95; traffic, Edinburgh pages 96-97; green belt developments, Edinburgh pages 100-101. South Wales pages 124 and 130; West Lothian pages 132-133.

9 Farming systems provide food supplies and raw materials.	Farms may be studied as systems, with inputs (physical, economic, technological, social, political), processes and outputs, in which the farmer is viewed as a decision-maker.	UK **In external papers examples will be drawn from arable, pastoral, and mixed farming systems in the UK.**	Farming pages 102-117	Lynford House Farm, East Anglia (arable farm) page 108; Herdship Farm, Teesdale (pastoral farm) page 109; Penllan Farm, Herefordshire (mixed farm) page 110; Ardalanish Organic Farm, Isle of Mull (mixed farm) page 111.
10 The viability of manufacturing industry is affected by a variety of factors.	Industrial location, expansion and decay result from the interaction of a number of factors, e.g. profitability, site, labour, raw materials (including power), markets, government policies. Of particular importance is technological change. The relative suitability of locations and the future of individual units and whole industries change over time.	UK/Western Europe	Industry and economic change pages 118-133	Toyota plant, Burnaston page 122; South Wales pages 124 and 130; the Ruhr pages 125 and 131; Heriot-Watt Research Park page 127; West Lothian pages 132-133.
11 Economic change has social and environmental consequences.	Economic change has positive and negative implications for employment patterns and opportunities, for the welfare of countries, communities and for local environments.	UK/Western Europe	Industry and economic change pages 124-125, 128-133	South Wales pages 124 and 130; the Ruhr pages 125 and 131; West Lothian pages 132-133.

Standard Grade syllabus matching grid

International issues

In this grid the Key Ideas, Explanatory comment and Areal/Thematic context for external assessment are taken from the Standard Grade syllabus. All other information relates to the *geog.SG* students' book.

Key idea	Explanatory comment	Areal/Thematic context for external assessment	geog.SG	Case studies
12 Population is unevenly distributed.	Distribution can be analysed in terms of population totals and densities, on local, national and international scales and can be related to economic, political and environmental factors.	Global	Population pages 136-137	
13 Populations have measurable social and economic characteristics.	Population facts are gathered in different ways with varying degrees of accuracy. Measured social characteristics include birth rates, death rates, infant mortality and life expectancy together with various indicators of living standards.	Global	Population pages 150-151 Development and international alliances pages 160-161, 163	UK census page 150; Sudan's census page 151.
14 In any area the size and structure of the population are subject to change.	Population change within regions and nations is created by gain factors (births and in-migration) and loss factors (deaths and out-migration). Variations in these factors can be related to influences in the physical environment and also to socio-economic and political influences. Changes are often related to pressures on agricultural resources and migration to find employment.	Global	Population pages 134-135, 138-149, 152-157	Migration and the UK pages 146-147; Darfur refugees pages 148-149; population change in the UK pages 154-155; population change in India pages 156-157.

15 International relations are dominated by a limited number of countries acting in conjunction with others.	Certain countries and alliances are dominant in international relations and trade because they control large resources. Their influence is a reflection of their location, historical development, size, population, resource base and level of technology.	Global **In external papers examples will be drawn from Europe, the USA and Japan.**	Development and international alliances pages 162, 172-175 Development, trade and aid pages 176-187	The EU, UNICEF and NATO pages 174-175; Kenya and Japan pages 178-179, 181; Trading blocs and WTO pages 182-183; TNCs and Ford in India pages 184-187.
16 Regions of the world are linked through trade.	Trade in resources and manufactured goods illustrates the inter-dependence of different parts of the world as producers and consumers.	Global	Development, trade and aid pages 176-187	Kenya and Japan pages 178-179, 181; TNCs and Ford in India pages 184-187.
17 Schemes of self-help, along with national and international aid, seek to encourage social and economic development.	In both the developing and the developed world, self-help and outside aid are required to meet a wide variety of needs. Self-help and aid can take different forms and can create different impacts. Schemes can be short-term (e.g. disaster relief) or long-term (e.g. agricultural and multi purpose development projects, development of trade and industry).	Global	Development, trade and aid pages 188-193	Self-help schemes (magic stones, afforestation, education and training, recycling goats) pages 192-193.

 # River landscapes

 *chapter overview*

The big picture

These are the key ideas behind this chapter:

◆ There are a number of natural processes – weathering, erosion, mass movement, transportation and deposition – which affect the physical landscape.

◆ The processes affecting the landscape are connected, and landscapes are continually changing.

◆ The drainage basin cycle is part of the water cycle.

◆ Rivers use their energy to erode and transport material. Deposition occurs when a river lacks enough energy to carry its load.

◆ Rivers change from their source to their mouth. There are characteristic landforms in the upper, middle and lower courses of a river.

Note that the students' version of the big picture is given in the students' chapter opener.

Chapter outline

Use this, and the students' chapter opener, to give students a mental roadmap for the chapter.

1   **River landscapes**   As the students' chapter opener, this unit is an important part of the chapter; see page 11 of this book for notes about using chapter openers

1.1   **Landscapes and processes – introduction**   Different types of rock (sedimentary, igneous and metamorphic), and different types of weathering (physical, chemical and biological)

1.2   **Mass movement and systems**   Soil creep, mudflows and landslides; the rock cycle, and landscapes and systems

1.3   **Water and drainage basins**   How water circulates continually between the ocean, atmosphere and land; the drainage basin cycle as part of the water cycle, and a river's long profile

1.4   **River processes**   The source, middle course and mouth of a river. Erosion, transportation and deposition

1.5   **The river's upper course**   The river channel, V-shaped valleys, interlocking spurs and waterfalls – with the River Tees as a case study

1.6   **The river's middle course**   The formation of meanders and ox-bow lakes and the River Tees as the case study

1.7   **The river's lower course**   Floodplains, river mouths, estuaries and deltas with the River Tees as the example again

Objectives and outcomes for this chapter

Objectives	Unit	Outcomes
Most students will understand:		Most students will be able to:
• That there are a number of natural processes which affect the physical landscape.	1.1, 1.2	• Name and describe the processes affecting the physical landscape.
• How the processes affecting the landscape are connected.	1.2	• Draw a simple rock cycle.
• How the water cycle operates and how the drainage basin cycle is part of the water cycle.	1.3	• Draw simplified diagrams of the water cycle and drainage basin cycle.
• How rivers change from their source to their mouth.	1.4	• Describe changes in the river channel, velocity and load.
• That rivers shape the land by erosion, transportation and deposition.	1.4	• Name and describe four methods of erosion, and four methods of transportation; describe how deposition takes place.
• That the processes of erosion, transportation and deposition produce characteristic landforms along a river.	1.5, 1.6, 1.7	• Explain how V-shaped valleys, interlocking spurs, waterfalls, meanders, oxbow lakes, floodplains, levées, estuaries and deltas are formed.
• How to recognise river landforms on an OS map.	1.5, 1.6, 1.7	• Identify a range of river landforms on an OS map and describe them.

These tie in with 'Your goals for this chapter' in the students' chapter opener, and with the opening lines in each unit, which give the purpose of the unit in a student-friendly style.

Using the chapter starter

The photo on page 6 of the *geog.SG* students' book shows the Grand Canyon of the Verdon River, on the edge of the Alps in Provence, south-east France.

The gorge is 21 km long and in places 700 metres deep. The bottom varies in width from 6 metres to 100 metres. It has been formed by the erosion of the limestone rock by the Verdon River. It is Europe's widest and deepest gorge.

The canyon was only fully explored in the early twentieth century. Before then people found the deepest parts too difficult to get to, and just a few local woodcutters went down on ropes.

Now the canyon is a great tourist attraction, with people hiking, climbing, and rafting.

The Verdon River flows for 175 km from its source in the south-western Alps – it's a tributary of the Durance, which in turn is a tributary of the Rhone, which flows into the Mediterranean Sea.

The unit in brief

This unit introduces students to the processes which affect our physical landscape. It initially looks at the three main types of rock that make up the landscape – sedimentary, metamorphic and igneous – and explains how they were formed. It then looks at three different types of weathering – physical, chemical and biological. In the Activities, students draw a mind map to pull together the information on rock types, and complete a table about weathering.

Key ideas

◆ There are a lot of different processes which affect the physical landscape.

◆ The processes affect the rock which the landscape is made up of.

◆ The three categories of rock are sedimentary, metamorphic and igneous. They are formed in different ways.

◆ Weathering is the disintegration and decomposition of rocks in situ.

◆ Weathering can be physical, chemical or biological.

Key vocabulary

physical landscape, human landscape, processes, sedimentary, igneous, metamorphic, weathering, physical, freeze-thaw, frost-shattering, onion-weathering, exfoliation, chemical, biological

Skills practised in the Activities

◆ Geography skills: completing a table on weathering

◆ Thinking skills: producing a mind map about different rock types

Unit outcomes

By the end of this unit, most students should be able to:

◆ define the terms given in 'Key vocabulary' above;

◆ understand that the processes affecting the physical landscape affect the rock the landscape is made up of;

◆ explain how sedimentary, metamorphic and igneous rocks are formed;

◆ give examples of sedimentary, metamorphic and igneous rocks;

◆ give examples of the three main types of weathering.

Ideas for a starter

1 Bring samples of different types of rock into the classroom and allow students to handle them and write down their observations. Give them a checklist – colour, hardness, layers, crystals, etc.

2 Discuss the geology of the local area. Use a geology map – there may be one in the atlas – to help. Are there any quarries in the area that the students know of? If so, what is the rock used for?

3 Ask students why geographers need to know about the characteristics of rocks. Try to name some areas of study where the geology is important to understanding the geography of the area.

4 Show students photos of different types/signs of weathering – it happens everywhere. Show photos of flaking paint, rotting wood, rusting metal, moss on walls, old gravestones. Ask students what's happening.

Ideas for plenaries

1 Ask students to try to work out rules to identify each type of rock (sedimentary, igneous, metamorphic).

2 Use photographs of a range of rocks to test out students' rules of identification.

3 Ask students to name some physical features which may be caused by the different hardness of rocks.

4 Play 'Just a minute' – the topic is 'Rocks and weathering'. Students have the chance to talk for a minute on rocks and weathering without repetition or hesitation. As soon as a student repeats an idea, or hesitates, the next student takes over until the minute is up.

Answers to the Activities

1 Mind map might include:

Sedimentary – how formed – sediments e.g. sand, shells, skeletons have fallen to the bottom of a lake or sea. Examples – limestone, chalk, clay, sandstone.

Metamorphic – how formed – rocks are altered by heat or pressure. Examples – schists, marble.

Igneous – how formed – lava erupting from a volcano, or magma cooling within the Earth. Examples – basalt, granite.

2

Type of weathering	Example	What happens
Physical	Freeze-thaw, frost-shattering	Water enters cracks, or joints, in the rock. If the temperature drops below 0 °C the water freezes and expands, widening the crack. Continued freezing and thawing weakens the rock and eventually pieces break off.
Physical	Onion-weathering, exfoliation	In warm climates the surface layers of rock are heated and expand during the day. At night they cool and contract. Repeated heating and cooling causes the outer layers to peel off, like those of an onion.
Chemical	Limestone solution	This occurs when carbonic acid (which occurs naturally as a weak solution in rainwater) reacts with rocks containing calcium carbonate, e.g. limestone. The limestone slowly dissolves and is removed by running water.
Biological	–	This occurs when tree roots grow into cracks in rocks and widen them. It can also occur as a result of acids released by decaying vegetation attacking the rock.

The unit in brief

In this unit students look at mass movement – soil creep, mudflows and landslides – and find out how the processes affecting the landscape are connected, investigating the rock cycle and landscapes as part of a system with inputs, processes and outputs. In the Activities, students describe the processes of mass movement, explain how the landscape is shaped, and find out about a landscape which has changed.

Key ideas

◆ Mass movement is the downhill movement of weathered material under the force of gravity.

◆ The speed of mass movement can vary from soil creep (barely noticeable) to mudflows and landslides where movement is increasingly rapid.

◆ The processes affecting the physical landscape are part of the rock cycle.

◆ Physical landscapes are part of a system with inputs, processes and outputs.

◆ Physical landscapes are continually changing.

Key vocabulary

mass movement, soil creep, mudflows, landslides, weathering, transportation, erosion, deposition, system, inputs, processes, outputs

Skills practised in the Activities

◆ Geography skills: describing how soil creep, mudflows and landslides happen; explaining and describing how the landscape is shaped

Unit outcomes

By the end of this unit, most students should be able to:

◆ define or explain the terms given in 'Key vocabulary' above;

◆ understand that mass movement is the downhill movement of weathered material under the force of gravity;

◆ describe how soil creep, mudflows and landslides happen;

◆ draw a simple rock cycle to show how the processes affecting the landscape are connected;

◆ understand that physical landscapes are part of a system with inputs, processes and outputs;

◆ understand that physical landscapes are continually changing.

Ideas for a starter

1 Brainstorm to find out what students know about mass movement.

2 Show photos of soil creep, landslides and mudflows. Ask students: What's happened? How has it happened? Has it affected people? How?

3 Draw an empty systems diagram with boxes for inputs, processes and outputs. Fill in one input, one process and the output (a changed landscape). Ask for suggestions to complete the other inputs and processes.

4 Show students photos of a changed landscape, e.g. Happisburgh or the Eiger (large cracks appeared on the side of the mountain, resulting in a large rockfall in July 2006). Ask students what is happening, or what has happened? How did it happen?

Ideas for plenaries

1 Make a graffiti wall of what students have learned today.

2 A quick-fire test: call out a student's name and a definition (e.g. for mass movement, weathering). The student has five seconds to give you the term.

3 Did you find anything difficult about the work in this unit? What? Why? What would help to make it less difficult?

4 If starter 4 has not been used, it could be used as a plenary – either as well as, or instead of, activity 3.

Answers to the Activities

1 Soil creep happens on very gentle slopes. Soil creeps very slowly downhill (it can be less than 1cm a year) under the force of gravity.

Mudflows and landslides happen much faster than soil creep. They may occur after periods of heavy rain when surface material becomes saturated. On steep slopes the extra weight and gravity makes the saturated material move downhill. Mudflows and landslides can happen as a result of earthquakes and volcanic eruptions.

2 Weather conditions – how hot, cold or wet it is – will affect the type and rate of weathering, e.g. chemical weathering is much faster in a hot, damp climate than in a cold, dry one.

Action of a glacier, river or the sea – can all shape the landscape by processes of erosion, transportation and deposition.

Volcanic activity – can create and alter rocks; can create and alter landforms and cause mudflows.

Gravity – causes weathered material to move downhill by the process of mass movement (soil creep, mudflows and landslides).

3 Answer will depend on the landscape chosen by students.

The unit in brief

This unit is about the water cycle, how water reaches rivers and ends up in the sea. In the Activities, pupils draw and label diagrams of the water cycle and the drainage basin cycle; and show that they know the difference between various key terms.

Key ideas

◆ The water (or hydrological) cycle is a continuous transfer of water from the oceans, into the atmosphere, onto land and finally back into oceans.

◆ The drainage basin cycle (or system) is part of the water cycle which operates on the land.

◆ A river, or drainage basin, is the area of land drained by a river and its tributaries.

◆ The watershed separates one drainage basin from the next.

◆ A river's long profile is a slice along its length – the slope of the river gets flatter, and the bed becomes smoother as you go downstream.

Key vocabulary

evaporation, transpiration, condensation, precipitation, stem-flow, drip-flow, interception, surface run-off, infiltration, throughflow, percolation, groundwater flow, permeable, impermeable, source, confluence, tributary, floodplain, mouth, watershed, drainage basin, long profile

Skills practised in the Activities

◆ Geography skills: copying and labelling a diagram of the water cycle; drawing and labelling a diagram of the drainage basin cycle; explaining the difference between key terms

Unit outcomes

By the end of this unit, most pupils should be able to:

◆ define or explain the terms given in 'Key vocabulary' above;

◆ understand that the water cycle is a continuous transfer of water from the oceans, into the atmosphere, onto land and finally back into oceans;

◆ draw a diagram of the drainage basin cycle;

◆ understand that the drainage basin cycle is part of the water cycle;

◆ describe how a river changes from its source to its mouth (long profile).

Ideas for a starter

1 Draw the hot water system in your home. Describe the water's journey around the home.

2 It's raining today (or yesterday or …). Where does the water come from? Where is it going? Build up the water cycle on the board.

3 Who can remember the water cycle? Can you draw a simple version on the board? How does the drainage basin cycle fit into the water cycle?

Ideas for plenaries

1 Have a set of cards with definitions and different terms to do with the water cycle and drainage basin cycle. Ask one student to read out the definition on the card they have and this is then paired with the correct term from someone else's card. This continues until all the definitions have been paired up. Alternatively, this can be played as a card game in which cards are placed face down with the definition on one and the term on another. Students then try to turn over the matching pair. The student with most sets of pairs wins!

2 Describe the journey of a water molecule through the water cycle. Make it an adventure story!

3 Let's look at the global water cycle. How much water is stored in the oceans? How much is stored in the ice caps and in glaciers? Present this as a flow diagram drawn to scale.

4 Tell your neighbour two key things you have learned today about drainage basins.

Answers to the Activities

1

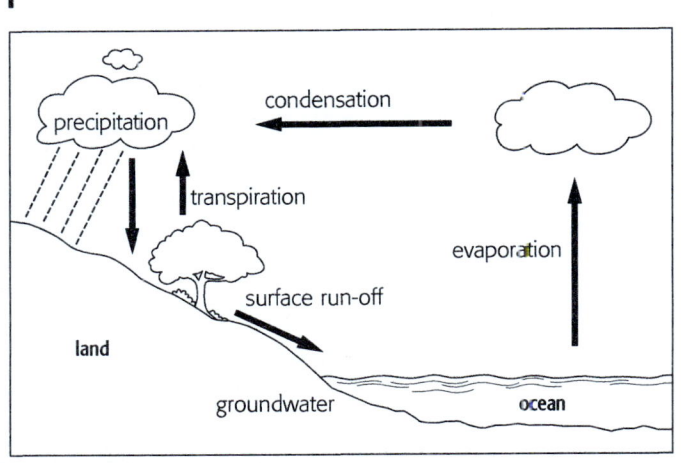

2 Students can draw a simple outline copy of the diagram on page 12 of the students' book, replacing the numbers with the key terms. However, they should be able to explain what each of these terms means, if they are challenged to do so. They will find it useful to create a list of these definitions.

3

a The source is the starting point of the river. The mouth of the river is where it flows into a lake, the sea or ocean.

b The watershed is an imaginary line separating one drainage basin from the next. The drainage basin is the area of land drained by a river and its tributaries.

c A tributary is a smaller river joining a bigger river. The confluence is the point where the two rivers join.

The unit in brief

This unit introduces rivers and what they do – erode, transport and deposit material. In the Activities, students complete a table about the different parts of a river, and a writing frame about erosion, transportation and deposition. They also show their understanding of erosion.

Key ideas

◆ Rivers change from their source to their mouth.

◆ The energy which a river has, enables it to erode and transport material.

◆ Rivers erode by one of four processes: abrasion, attrition, hydraulic action and solution.

◆ Rivers transport their load by one of four processes: solution, suspension, saltation, traction.

◆ Deposition occurs when a river lacks enough energy to carry its load.

Key vocabulary

erosion, abrasion, attrition, hydraulic action, solution, transportation, suspension, saltation, traction, load, bedload, deposition, sediment, velocity

Skills practised in the Activities

◆ Geography skills: completing a table about parts of a river; explaining and describing the processes of erosion

◆ Literacy skills: completing a writing frame about erosion, transportation and deposition

Unit outcomes

By the end of this unit, most students should be able to:

◆ define or explain the terms given in 'Key vocabulary' above;

◆ describe how rivers change from their source to their mouth;

◆ explain how rivers erode;

◆ name the processes that rivers use to transport their load;

◆ know that deposition occurs when a river lacks enough energy to carry its load.

Ideas for a starter

1 Name as many of the world's greatest rivers as you can. Which is the longest river in the world? Use your atlas to help you.

2 Which river is associated with these cities: London; Kolkata; Paris; Cairo; New Orleans; Cardiff; Glasgow; Bristol; New York?

3 Can you name the most important river in the UK? Which is the nearest river to you?

4 Why is a location near a river important? Why was it important in the past?

Ideas for plenaries

1 Work with a partner. Write down all the important terms from this unit. Write the definitions for the terms between you.

2 Show students the BBC video of the River Tees. Ask them to produce a sketch map showing the important features, the changes from source to mouth, and land use of the floodplain.

3 Imagine you are part of the load of a river. Describe your journey from source to mouth.

4 Stir some soil into a beaker of water and watch as the sediment falls out of the water. What do you see? Is this how rivers work?

5 Draw a spider diagram of all the various uses we make of rivers. In a different colour try to identify any conflicts, problems or issues to do with our use of rivers.

Further class and homework activities

Collect pictures of river landscapes and make a wall display of them. Try to annotate your display to describe the processes taking place in the photographs. Try to arrange them from source to mouth.

Answers to the Activities

1

part of the river	description	velocity	width	depth	particle size
upper	near source; upland stream, hilly area	slow-flowing (friction)	narrow	shallow	large
middle	mid-stream over more level land	faster-flowing	wider	deeper	smaller
lower	near mouth, flowing over flat land	fast-flowing	wide	deep	very small (suspended)

2 Rivers have energy so they can work, which means they can erode, transport and deposit material.

When a river has lots of energy it can erode. Ways it can do this are:

◆ hydraulic action: this means water is forced into cracks in the banks and over time this breaks up the bank into small pieces;

◆ solution: this means material from the bed and banks is dissolved in the water, which also helps to break up the materials in bed and banks;

◆ attrition: this means rocks and stones knock against each other and against the bank and bed, wearing them away;

◆ abrasion: this is the action of the materials carried by the water, which scrape against the bed and banks and also wear them away.

Rivers can transport the material they have eroded as long as they have enough energy. Material can be carried in several ways:

◆ in solution – dissolved in the water;

◆ in suspension – small particles are carried along in the water;

◆ as bedload – heavier material is dragged along the bed;

◆ by saltation – smaller materials bounce along;

◆ by traction – larger materials roll along the bed.

If a river doesn't have enough energy it drops (deposits) its load, or sediment, in order of size. The heaviest material is dropped first, then progressively smaller particles. Any dissolved material is carried out into a lake or the sea.

3 Rivers have high energy when they are flowing **fast** or when there is **a lot** of water in them.

4 a Erosion is the wearing away of the land by material carried by rivers and streams. (In a wider context, erosion may also be by wind, waves or ice.)

b Rivers erode in four main ways: abrasion, attrition, hydraulic action and solution. (See **2** above.)

The unit in brief

This unit looks at the river near to its source, and uses the River Tees (introduced in Unit 1.4) as a case study. The river channel, V-shaped valleys, interlocking spurs and waterfalls are covered in this unit. In the Activities, students use an extract from an OS map to draw a cross-section of a V-shaped valley; produce a flow diagram to explain how High Force developed; draw and label a sketch diagram of High Force from a photo and complete a paragraph about the formation of waterfalls.

Key ideas

In their upper course, rivers:

◆ flow slowly as they carry a large bedload, and the channel is narrow and shallow;

◆ cut downwards by vertical erosion to create steep-sided V-shaped valleys;

◆ flow around protruding hillsides, creating interlocking spurs;

◆ may have waterfalls. These develop when they meet a band of softer rock after flowing over a harder, more-resistant one.

Key vocabulary

river channel, bedload, V-shaped valley, interlocking spurs, waterfall, gorge

Skills practised in the Activities

◆ Geography skills: drawing and annotating a cross-section of the Upper Tees valley from an OS map extract; drawing and labelling a sketch diagram of High Force from a photo

◆ Literacy skills: completing a paragraph on waterfalls

◆ Thinking skills: explaining interlocking spurs; producing a flow diagram to explain how High Force developed

Unit outcomes

By the end of this unit, most students should be able to:

◆ define or explain the terms given in 'Key vocabulary' above;

◆ draw a cross-section from an OS map and explain how V-shaped valleys develop;

◆ explain how waterfalls develop;

◆ draw a sketch diagram from a photo.

Ideas for a starter

1 Draw a generalized long profile of a river from source to mouth on the board. Ask students to identify the upper course, the middle course and the lower course. What are the main processes – erosion, deposition, transport – taking place at each of the stages?

2 The amount of water in a river channel will vary from time to time throughout the year. Why? How does the amount of erosion, deposition and transportation vary with the amount of water in the channel?

3 Show photos of a V-shaped valley, interlocking spurs, waterfalls and gorges. Ask the class to name them and suggest how they were formed.

Ideas for plenaries

1 Give students photos of waterfalls. They have five minutes to sketch one with labels and annotations. Students swap and check each other's sketches.

2 Provide students with a suitable OS map. Ask them to follow a river from its source and try to find examples of steep-sided V-shaped valleys, waterfalls and gorges.

3 Draw accurate cross-sections across the river valley using the OS map. Remember to use the same vertical and horizontal scales.

4 Provide a student with labelled samples of hard and soft rock (e.g. granite and shale). The student has to demonstrate how a waterfall forms.

Further class and homework activities

1 Investigate other well-known waterfalls and see if they have formed gorges.

2 Find examples of river flowing over different rock types. Compare the shapes of the valleys and find out about the relative resistance of the rocks concerned.

Answers to the Activities

1 Students should draw an annotated cross-section of line A-B to explain V-shaped valleys.

2 Interlocking spurs are the alternating hills that stick out into a river's path. A river in its upper course doesn't have the power to erode these hills, so it has to flow around them.

3 Students should produce a flow diagram with sketches similar to the diagrams at the top of page 17 of the students' book, applied to High Force.

4 a An annotated sketch diagram of the photo at the bottom of page 16 of the students' book.

b Missing words: hydraulic action, plunge pool, retreat.

The unit in brief

This unit looks at the middle course of the river, and again uses the River Tees as a case study. Meanders and oxbow lakes are covered in this unit. In the Activities, students use an extract from an OS map to draw a cross-section of a meander; draw a sketch map of a meander now and when an oxbow lake has been created; draw a cross-section of the river channel at a meander and explain how the river is eroding the river cliff.

Key ideas

In their middle courses:

◆ rivers erode laterally as well as vertically;

◆ rivers develop large bends, known as meanders;

◆ differences in velocity mean that there is more erosion on one side of the river meander, and more deposition on the other side;

◆ oxbow lakes form when a river erodes a new channel across the neck of a meander.

Key vocabulary

meanders, thalweg, river cliff, point bar, slip-off slope, oxbow lake, swan's neck meander

Skills practised in the Activities

◆ Geography skills: drawing and annotating a cross-section of the Middle Tees valley from an OS map extract; drawing and annotating sketch maps of meanders; drawing a cross-section of a river channel at a meander, and explaining how rivers erode river cliffs

Unit outcomes

By the end of this unit, most students should be able to:

◆ define or explain the terms given in 'Key vocabulary' above;

◆ draw a cross-section from an OS map, and explain how meanders develop;

◆ explain how oxbow lakes form.

Ideas for a starter

1 Recap the development of a waterfall, and key terms so far in this chapter.

2 Show photos of a river's middle course. Ask students to identify the features they can see and suggest how they were formed.

Ideas for plenaries

1 Provide students with a suitable OS map. Ask them to identify the middle course of a river and features typical of the middle course.

2 Use OS maps to look at the middle section of rivers. Rivers are often used to mark boundaries. Try to find examples of this and you may find instances of where the river has changed its channel position while the boundary is in the former position. What does this tell you about the processes operating in the river?

3 What sort of land use should happen on the floodplain? Give reasons for your answer. What sort of land use should not happen on the flood plain?

4 Tell me two important things you have learned today about meanders, and two important things you have learned today about oxbow lakes. Now tell me another two things about meanders and oxbow lakes.

Answers to the Activities

1 Students should draw an annotated cross-section of line A-B to explain the meander.

2 Students draw two sketches to explain the likely changes to the river meander in the map extract over time, culminating in the creation of an oxbow lake.

3 a Students draw a labelled sketch of the cross-section of a river channel at a meander, similar to the diagram in the bottom left of page 18 of the students' book.

  b The river is eroding at the river cliff because on this side of the river bend the water is flowing at a greater velocity. As a result, it has greater power to erode the bank and bed on this side, which is deeper.

The unit in brief

This unit is about the lower course of a river, with the River Tees still used as the case study. Floodplains, estuaries and deltas are covered in this unit. In the Activities, students complete a paragraph about a river's lower course; draw a labelled diagram of a floodplain and explain how terraces and levées form; and draw and label sketch maps of an estuary and delta from resources in the students' book.

Key ideas

◆ In their lower courses rivers flow in a wide, deep channel.

◆ Floodplains are the land on either side of the river.

◆ When rivers flood, the coarsest material is deposited first – forming a levée.

◆ Estuaries are wide, deep river mouths.

◆ Deltas form when a river deposits its load and the sea currents aren't strong enough to remove it.

Key vocabulary

mudflats, floodplain, alluvium, terraces, levées, estuaries, deltas, distributaries, bird's foot delta, arcuate delta

Skills practised in the Activities

◆ Geography skills: drawing and labelling a diagram of a floodplain; drawing and labelling a sketch map of Tees estuary from an OS map or photo; drawing and annotating a sketch map of the Rhône delta from a photo

◆ Literacy skills: completing a paragraph about a river's lower course

◆ Thinking skills: explaining floodplains, alluvium and the formation of terraces and levées

Unit outcomes

By the end of this unit, most students should be able to:

◆ define or explain the terms given in 'Key vocabulary' above;

◆ draw a diagram of a floodplain and explain how terraces and levées form;

◆ draw a sketch map of the Tees estuary from an OS map or photo;

◆ draw a sketch map of the Rhône delta from a photo and explain how it has formed.

Ideas for a starter

1 Create a graffiti wall of what students have learned so far about rivers, the work of rivers and landforms.

2 When the river deposits its load does it dump it all in one go? How does it deposit its load? What landforms are created?

3 Ask what do students know about floodplains, estuaries and deltas? Can they give examples?

Ideas for plenaries

1 Which other rivers have deltas? What type are they? How have they formed?

2 Draw a diagram showing how the following change from source to mouth: channel width; channel depth; channel smoothness; size of load; amount of water (discharge); pollution; main types of transport; amount of load, etc.

3 Do rivers slow down, or speed up as they move from source to mouth? Explain your answer.

4 Prepare an odd-one-out to try on a partner using what you have learnt about rivers and landforms so far.

5 Choose a river landform feature and draw an annotated diagram to show how it was formed.

Further class and homework activities

Investigate the use of the River Nile and the current concern about the delta of the River Nile.

Answers to the Activities

1 Missing words: flat, industry, marshland, deep, fast, small, mudflats.

2 a A floodplain is the flat area on either side of a river at the lower end of its course, which is often flooded.

b Alluvium is fine material (silt) that is left behind after a river floods. It is carried down by a river when it has sufficient energy to transport it. Alluvium usually makes good soil for farming.

c Terraces form in some places because at one time the sea level was higher than it is today, so rivers did not have to fall so far to reach the sea. When the sea level dropped, rivers had to cut down further to reach the sea, and they left behind the terraces of alluvium that were originally formed in those earlier times.

d Levées are formed when a river floods over its usual channel. As it does so, its flow slows down and it drops (deposits) its load. It deposits the heaviest material first, leaving mounds or levées at the side of the main river channel.

e Students should be able to re-create the drawing on page 20 of the students' book, and label it correctly.

3 Students should outline the main course of the river, the coastline, and general areas of industry along the river, adding the labels and title specified in the activity. Drawing sketch maps of this type is useful practice in selecting material from a detailed map. Sketch maps should not be accurate to the last detail, but nor should they be over-generalised. It is useful, too, to show just a few important named locations. The map can be saved as a useful revision case study exemplar.

4 Labels for a sketch map of the delta photo should provide explanations, for example:
   ◆ The fast-flowing river carries a large amount of sediment.
   ◆ The river slows as it flows towards the sea.
   ◆ It begins to deposit the sediment (load) as it slows down.
   ◆ The sea here is fairly calm and the land is quite flat, so the load builds up in the sea to form a delta.
   ◆ Distributaries form where sediment has blocked river channels.

5 a If sea level rises, the sea would flood over the low-lying delta. The river would have less far to travel to the sea, it would drop its load further upstream, and the delta would retreat.

b If a dam is built upriver, it will mean that less water is flowing in the river bed below it, so it will have less energy to carry a load. Some of the load will also have been trapped by the dam. If there is less load, there is less material to be deposited in the delta. The sea may also begin to erode the materials in the delta, and if there is not enough load to replace it, the delta will retreat.

 # Glacial landscapes

The big picture

These are the key ideas behind this chapter:

◆ Glaciers are like frozen rivers which flow very slowly.

◆ Like rivers, glaciers erode, transport and deposit material. They create a range of different landforms as a result of erosion and deposition.

◆ Glacial landforms can be recognised on OS maps.

◆ There are many competing demands for the use of upland landscapes.

◆ Tourism creates problems which have to be managed.

Note that the students' version of the big picture is given in the students' chapter opener.

Chapter outline

Use this, and the students' chapter opener, to give students a mental roadmap for the chapter.

2 **Glacial landscapes** As the students' chapter opener, this unit is an important part of the chapter; see page 11 of this book for notes about using chapter openers

2.1 **Introducing glaciers** Where glaciers are found and how they shape the landscape

2.2 **Glacial landforms – erosional features** Corries, arêtes and pyramidal peaks, and how they are formed

2.3 **Erosional features continued** The formation of glacial troughs, truncated spurs and hanging valleys

2.4 **Glacial landforms – depositional features** Terminal moraine, ribbon lakes, drumlins, eskers and kames, and how they are formed

2.5 **Glacial landforms on maps** Using an extract from an OS 1:50 000 scale map to recognise glacial landforms

2.6 **Competing demands in the Cairngorms** Forestry, farming and wind power in an upland glaciated area

2.7 **Tourism in the Cairngorms** The importance of tourism to the local economy, and the problems it brings

2.8 **Managing tourists in the Cairngorms** Solutions to some of the problems tourists cause, and the Cairngorm Mountain Funicular Railway

Objectives and outcomes for this chapter

Objectives	Unit	Outcomes
Most students will understand:		Most students will be able to:
● What a glacier is.	2.1	● Explain that a glacier is like a river made of ice, and that it flows much more slowly.
● How glaciers change the landscape as a result of erosion, transportation and deposition, and that freeze-thaw weathering plays a part.	2.1, 2.2 2.3, 2.4	● Describe how glaciers erode and transport material; name the glacial landforms covered in the chapter; explain how they are formed; describe the process of freeze-thaw weathering
● How to recognise glacial landforms on OS maps.	2.5	● Identify a range of glacial landforms on an OS map and describe them.
● That there are competing demands for the use of upland landscapes.	2.6, 2.7	● Describe how people use the land in an area that was glaciated.
● That tourism is important to the local economy, but can cause conflict.	2.7, 2.8	● Give figures to explain how important tourism is to the local economy; identify the problems that tourists cause.
● That there are solutions to the problems that tourists cause, and that people have different views about some of the solutions.	2.8	● Give at least six solutions to the problems tourists cause; explain why people have different views about some of the solutions.

These tie in with 'Your goals for this chapter' in the students' chapter opener, and with the opening lines in each unit, which give the purpose of the unit in a student-friendly style.

Using the chapter starter

The photo on page 22 of the *geog.SG* students' book shows the Aletsch Glacier, Switzerland. The dark stripes are medial moraines, and give a clear impression of the glacier 'flowing' down the valley.

The Aletsch is the largest glacier in the Alps. It is 24 km long; in places it's 1.5 km wide and more than 900 metres deep.

Its start point is at 4000 metres. It covers an area of 120 sq km (45 sq miles). It descends into the valley of the Upper Rhone. At its fastest it moves at about 180 metres a year, or about 50 cm a day.

The weight of the ice has been calculated at 27 billion tonnes. If we melted the glacier, it would give everyone on Earth 1 litre of water a day for six years.

Scientists study the world's ice sheets and glaciers because they can tell us a lot about the history of climate change and the current trend towards global warming.

Like many glaciers, the Aletsch Glacier is retreating under the influence of global warming. In the middle of the 19th century it extended 3 km further down the valley than it does today, and since then its thickness has been reduced by about 100 metres. It is now retreating by up to 50 metres a year.

The unit in brief

This unit is about glaciers. It explains what glaciers are, where they are found, and how they shape the landscape. In the Activities, students produce a mind map about glaciers based on one in the students' book. They can add extra information to it based on their own research. They also explain the process of freeze-thaw weathering and describe the effects of this process on the landscape.

Key ideas

◆ A glacier is a body of ice which moves downhill.

◆ Glaciers are found in big mountain ranges, and wherever it is cold enough. Ice sheets are found in places like Greenland, Iceland and Antarctica.

◆ Glaciers erode by plucking and abrasion.

◆ Moraine is material that is transported and later deposited by glaciers.

◆ Much of the surface moraine results from freeze-thaw weathering.

Key vocabulary

glacier, ice sheet, ice age, melt-water, crevasse, plucking, abrasion, striation, moraine (ground, lateral, medial), freeze-thaw weathering

Skills practised in the Activities

◆ Geography skills: describing the effect of freeze-thaw weathering on the landscape

◆ Thinking skills: producing a mind map about glaciers; explaining freeze-thaw weathering

Unit outcomes

By the end of this unit, most students should be able to:

◆ define or explain the terms given in 'Key vocabulary' above;

◆ understand that a glacier is a body of ice which moves downhill;

◆ know where glaciers and ice sheets are found;

◆ explain how glaciers erode;

◆ name different types of moraine;

◆ explain the process of freeze-thaw weathering.

Ideas for a starter

1   Brainstorm glaciation to find out what students know about the topic already. What do they know about ice ages, glaciers and the effect that glaciation has had on the landscape?

2   Show photos of glaciated landscapes and point out the features created by, or altered by, glaciation. Ask the students to list the features and suggest how they might have been formed.

3   Draw a U-shaped valley on the board. Superimpose a V-shaped river valley over it. Ask students to estimate the depth of ice in a valley glacier, and consider the erosive powers of a glacier as well as how much material a glacier erodes.

4   Mind movie time! You are being lowered into a crevasse in a glacier. What does it feel like? What can you see and hear? Tell us. If you have seen 'Touching the Void', this might help!

Ideas for plenaries

1   Use the map on page 24 in the students' book. Do you live in an area which was once covered in ice or in an area which was unglaciated? Do you know any features which could be of glacial origin? Make a class list.

2   If you live in an unglaciated area do you think your landscape would have been unaffected totally by the effects of ice? What might have affected your area?

3   If you did not use starter 4 ask students to imagine what it would be like being lowered into a crevasse in a glacier. Ask them to think of 10 key words which sum up what it might be like. If they have seen 'Touching the Void', this might help.

4   What might be happening to glaciers as a consequence of global warming? What will the consequences be of this?

5   Write down as many words as you can relating to today's work.

Further class and homework activities

Investigate Ice Ages and be ready to report back to the class at the beginning of the next lesson.

Answers to the Activities

1   At this stage students should just re-create, on a large sheet of paper (A3) the basic diagram, with plenty of space around it to add features as they work through the chapter – for example all the landforms they encounter in Units 2.2–2.5, and any more that they research in the course of their work on glaciation.

Glaciation can be a large topic; make sure in this unit that students understand the basic processes (how glaciers move), and that they are clear about the differences between the various types of moraine. Through research, they should select one particular area on which there is plenty of information so that in an exam answer they can give named examples. For this reason it is probably best to select an area in the UK (Wales, Lake District, Scotland), which is well covered by Ordnance Survey maps on which the various features created by the processes of glaciation can easily be seen today.

2 a   Freeze-thaw weathering is a weathering process that occurs in cold climates where the temperature fluctuates above and below 0 °C, allowing ice to form and melt repeatedly.

b   Water fills cracks in the rock. When the temperature falls below 0 °C, the water freezes and expands by about 9%, making the cracks wider. When the ice melts, more water enters the crack and freezes in turn, forcing the crack even wider. Over time, the repetition of this process forces pieces of rock to break off. These pieces of rock can then be carried away by ice or water.

c   Freeze-thaw is an important process in cold climates, especially upland regions in many parts of the world, because it breaks up the bedrock and enables further weathering and transportation of material – and eventually the creation of soils in which plants may grow. Therefore, over time, it allows a change in the landscape from solid rock to a place where vegetation can take a hold. In the long term, it provides an environment for other plants and for animals that feed on those plants.

Glacial landforms – erosional features

The unit in brief

This unit looks at some of the erosional features created by glaciers, namely corries, arêtes and pyramidal peaks. In the Activities, students produce a flow diagram to show how corries form, make notes on arêtes and pyramidal peaks, and draw and annotate a diagram of one of the photos on the spread in the students' book to show how either corries, arêtes or pyramidal peaks are formed.

Key ideas

◆ Glaciers create landforms as a result of erosion and deposition.

◆ Corries (cirques or cwms) are formed when ice moves out of a hollow. When the ice melts, a corrie-lake or tarn may be left behind.

◆ If two corries form back-to-back, or side by side, a ridge called an arête forms between them.

◆ When three or more corries cut backwards into a mountain, a pyramidal peak, or horn, develops.

Key vocabulary

corries, cirques, cwms, back wall, lip, corrie-lake, tarn, arête, pyramidal peak, horn

Skills practised in the Activities

◆ Geography skills: drawing a sketch diagram of a photo and adding annotations

◆ Literacy skills: making notes on arêtes and pyramidal peaks

◆ Thinking skills: producing a flow diagram to show how corries form

Unit outcomes

By the end of this unit, most students should be able to:

◆ define or explain the terms given in 'Key vocabulary' above;

◆ understand that glaciers create landforms as a result of both erosion and deposition;

◆ explain how corries, arêtes and pyramidal peaks form;

◆ draw and annotate a diagram to show how corries, arêtes or pyramidal peaks form.

Ideas for a starter

1 Recap: Ask students to describe the processes of plucking, abrasion and freeze-thaw weathering.

2 Ask students to name features formed by glacial erosion. What do they know so far about how they were formed?

3 Show a video of glacial landforms. Ask students to describe what they see.

4 Ask a few students to report back on their investigation into Ice Ages.

Ideas for plenaries

1 Work in pairs. Each pair writes the definition for the key vocabulary in this unit. Their partner has to work out what is being described.

2 Provide photos of upland glaciated areas in the UK. Ask students to annotate them with any glacial features they can identify.

3 Make a graffiti wall of what students have learned today.

4 Sum up what you have learned today in less than 40 words.

Further class and homework activities

Work on a model of erosion. Start with a high upland area with some small indentations and show how corries, arêtes and pyramidal peaks develop. Draw diagrams to support your understanding.

Answers to the Activities

1

Snow collects in a sheltered hollow at the top of a mountain …

↓

It turns to ice and then begins to move downhill, pulled by gravity.

↓

As it moves, it curves or rotates (rotational slippage).

↓

The material in the hollow is loosened by freeze-thaw weathering, and the moving ice plucks and abrades this material away from the sides of the hollow, making the hollow wider and deeper.

↓

Most material is eroded from the back wall, which becomes very steep. Less is eroded at the front so a lip forms here.

↓

When the ice melts, a lake may form in the hollow behind this lip.

2 An alternative to the sentence beginnings in the students' book would be to ask students to create a four-column table, like the one below. They could eventually make similar notes for all the features described in this chapter. The format of this table could also provide the basis for revision cards on which the information is set out systematically (just one example is included here).

Feature	Looks like …	Formation	Example/s
Arête	A sharp ridge	Two corries form next to each other, leaving a sharp ridge between them	Striding Edge, the Lake District

3 Suitable additional images for drawing and annotating can be found on a number of Internet sites, including a search of Google Images – insert the names of known glaciated locations.

Students might be challenged to see how many different glacial features they can accurately label on a single such image. For example, on the top photograph on page 27 they could also identify and add labels for the corrie lake on the left (ask them to find out its name – the answer, Red Tarn, can be found on the map on page 32), and the steep scree slopes on either side of the arête which have been scoured by glaciers.

The unit in brief

This unit looks at some of the other erosional features created by glaciers, i.e. glacial troughs, truncated spurs and hanging valleys. In the Activities, students draw and annotate a sketch of a photo in the students' book; match beginnings and endings of sentences about erosional features, and consider why there's a lake at Wast Water in the Lake District.

Key ideas

◆ As glaciers move, they erode the sides and floor of the old river valley to form a glacial trough.

◆ The ends of interlocking spurs are eroded, leaving truncated spurs.

◆ When the glacier melts, tributary valleys are left hanging on the valley sides, creating hanging valleys.

Key vocabulary

abrasion, plucking, glacial trough, U-shaped valley, interlocking spurs, truncated spurs, hanging valleys, misfit stream

Skills practised in the Activities

◆ Geography skills: drawing and annotating a field sketch from a photo
◆ Thinking skills: matching beginnings and endings of sentences about erosional features; thinking about the formation of a lake in a glacial trough

Unit outcomes

By the end of this unit, most students should be able to:

◆ define or explain the terms given in 'Key vocabulary' above;

◆ understand and explain how glacial troughs, truncated spurs and hanging valleys are formed;

◆ draw and annotate a field sketch from a photo.

Ideas for a starter

1 Who can remind me how corries, arêtes and pyramidal peaks are formed? Who can draw a diagram on the board to show their formation?

2 If you did not show a video of glacial landforms as a starter for Unit 2.2 you could show it for this unit. Ask students to describe what they see.

Ideas for plenaries

1 Provide students with a word search which includes all the key vocabulary covered so far in this chapter.

2 Students write definitions for the key vocabulary found in the word search.

3 Provide photos of upland glaciated areas. Ask students to identify glacial features. For each feature they identify, describe how it is formed.

4 Make 10-15 statements about glaciers and erosional features based on what students have learned so far, some true, some false. Students hold up **True** or **False** cards. Where statements are false, ask students to correct them.

Answers to the Activities

1 Students must draw an appropriately annotated field sketch.

2 A glacial trough is … the eroded valley left by a glacier.

A hanging valley is … a tributary valley that enters a main valley part way up the valley side.

Truncated spurs are … the ends of interlocking spurs that have been cut off by the glacier.

This activity can be extended by asking students to prepare similar definitions for other features they study in this chapter, and then exchanging them with each other. They might also be asked to comment on each other's definitions – it's surprising how many variations there can be in describing just one feature! (All definitions must, of course, be accurate.)

3 Misfit streams are small streams that flow in the bottom of a glacial trough.

[These streams seem small in comparison with the size of the valley that they occupy, and the concept of misfit streams is not always an easy one to grasp. It requires some leap of the imagination to envisage what the valleys would have looked like when they were filled with ice. One way of demonstrating this is to create a three-dimensional model to show how the ice scraped away huge quantities of material both from the sides and the base of previously much smaller valleys.]

4 It will help students to answer this question if they have an Ordnance Survey map of the area (1:50 000 sheet 89, or use the Internet), which shows the contours in the area. Wast Water is the deepest lake in the Lake District (70 metres). During the Ice Age the ice moved down from the high land to the north and west and scoured out this deep, rocky trough which is now filled by a long, narrow ribbon lake. It has a lip at one end and this holds the water within the trough. The south-east side of Wast Water is exceptionally steep, rising from the 70 metres' depth of the lake to 609 metres above sea level, in a distance of about a kilometre.

The unit in brief

This unit is about glacial landforms created by deposition. It looks at terminal moraine, ribbon lakes, drumlins, eskers and kames. In the Activities, students explain what terminal moraines and ribbon lakes are, and try to make the connection between terminal moraine and global warming. They suggest reasons why Keswick developed by Derwent Water; write out true and false statements based on their knowledge of glaciation so far, read them out to the rest of the class for them to decide which are true and false; and make revision cards for all the features they have learnt about in this chapter.

Key ideas

- ◆ The material deposited by a glacier is called till.
- ◆ Terminal moraine marks the furthest point a glacier reached.
- ◆ Where water is trapped behind a terminal moraine in a glacial trough, a ribbon lake will form.
- ◆ Drumlins are shaped by the movement of the ice.
- ◆ Eskers are long ridges of material deposited by streams which flow under glaciers.
- ◆ Kames are mounds of debris which accumulated in crevasses in the ice.

Key vocabulary

till, terminal moraine, ribbon lake, drumlin, esker, kame

Skills practised in the Activities

- ◆ Geography skills: making revision cards for all the features learnt about in the chapter
- ◆ Thinking skills: explaining terminal moraines and ribbon lakes; making connections; suggesting reasons why Keswick developed by Derwent Water; writing and identifying true and false statements

Unit outcomes

By the end of this unit, most students should be able to:

- ◆ define or explain the terms given in 'Key vocabulary' above;
- ◆ explain how terminal moraine and ribbon lakes are formed;
- ◆ understand how drumlins, eskers and kames are formed;
- ◆ suggest why Keswick developed next to Derwent Water.

Ideas for a starter

1 Look out of the window. Until 10 000 years ago the ground was buried under ice. What features did the ice create as it deposited material?
2 Produce a diagram of a valley glacier, then remove the glacier.
   - ◆ What features are left behind after the ice has melted?
   - ◆ What effect will the melted water have on the remaining features in the landscape?

Ideas for plenaries

1 Provide photos of glaciated areas which have a number of depositional features. Ask students to identify as many as possible.

2 How do humans use these depositional features?

3 Create an acrostic. Write GLACIAL LANDFORMS down the side of a page. Make each letter the first letter of a word, phrase or sentence about erosional or depositional landforms.

4 Did you find anything difficult about the work on glacial landforms? What? Why? What would help to make it less difficult?

Further class and homework activities

Find out about well-known, or significant, examples of glacial depositional features.

Answers to the Activities

1 a The terminal moraine is the material that is left behind as a ridge when the ice melts. It marks the furthest point reached by a glacier.

Students should be clear about the distinctions between a terminal moraine, a ground moraine, a lateral moraine and a medial moraine (see Unit 2.1).

b A ribbon lake forms when water is trapped in the glacial trough after the ice has retreated, usually because the outlet is dammed by a terminal moraine. (An example is Wast Water – see activity 4 in Unit 2.3.)

c Currently, many glaciers around the world are in retreat, almost certainly as a result of global warming. This is clear to scientists and other observers, because the glacier's terminal moraine is visible some distance ahead of the glacier's front edge, or 'snout'. This means that the ice is no longer moving forward or even at a standstill, but is in clear retreat – it is melting. Terminal moraines seen in landscapes that are no longer covered by ice represent the point at which the climate changed (got warmer) in the past.

2 Students should consider all the factors involved in choosing an early settlement site, and its situation. Initially these would have been related to the lie of the land and access to life needs (food, water, shelter, defence, materials).

Selecting a site for a settlement on or near a lake is usually a good choice because it provides constant water and a means of defence. Keswick is on a low, relatively level site surrounded by hills, so it would have been easier to build here, and sheltered from extremes of weather. The lower land (which is obviously moraine material) was probably good farming land (moraines often provide good soils), with the hills offering a supply of building materials (wood, stone) and an environment suitable for grazing sheep (and possibly cattle lower down). The trees on the lower slopes of the hills would also have provided wood for fuel. The rising land would provide a dry site for settlement if it was too wet very close to the lake. The fact that this area is on lower land means that it would also have been a focus of routeways through the neighbouring hills.

3 These statements will depend on students' own ideas. Where students recognise that false statements have been read out, they should be encouraged to write out the correct version, rather than just say 'false'.

A couple of examples of possible statements:

A terminal moraine is material that collects at the side of the glacier. (False: this is the definition of a lateral moraine.)

An esker is a long wiggly ridge that formed under the ice. It is made up of material that was carried by a stream running at the base of the glacier. (True.)

Statements should cover not only individual features but also the processes that take place in glaciated areas, e.g. freeze-thaw weathering, plucking and abrasion.

4 This activity also ties in with the work started in Unit 2.2 Activity 2. The following is a list of individual features that should be covered by students who have worked through the chapter on glaciation. They are all visible today in a landscape that was once glaciated, i.e. after the ice has retreated.

arête

corrie/cwm/cirque (and terms back wall, lip, tarn)

drumlin

esker

glacial trough/U-shaped valley

hanging valley

kame

lateral moraine

medial moraine

misfit stream

pyramidal peak (horn)

ribbon lake

striation

terminal moraine

till

truncated spur

The unit in brief

This unit includes a full page extract from an Ordnance Survey map (1:50 000 scale) of the Lake District (the area around Thirlmere, Helvellyn and Grasmere). In the Activities, students use the map to identify glacial landforms and practise a wide range of skills, as detailed below.

Key ideas

This spread is different to most others in the students' book, in that there is no content as such. It includes activities based on the OS map and the large photo of Thirlmere.

Key vocabulary

relief

Skills practised in the Activities

◆ Geography skills: describing how height and steepness are shown on OS maps, and describing the relief of the area shown on the map in the students' book; using six-figure grid references and reading the map to match locations, places and landforms; drawing a cross-section; explaining the location of Grasmere and the A591 using map evidence; using a map and photo together; using the map to give examples of landforms, describing the landforms and explaining how they were formed.

Unit outcomes

By the end of this unit, most students should be able to:

◆ explain the term given in 'Key vocabulary' above;
◆ demonstrate a range of map skills, including:
  – describing relief;
  – using six-figure grid references;
  – identifying landforms;
  – drawing a cross-section;
  – using a map in association with a photograph;
◆ describe landforms and explain how they are formed.

Ideas for a starter

1 Who can name the glacial erosion features we have looked at? Who can name the depositional features we have looked at?
2 Call out glacial landforms and ask students to classify them according to whether they are erosional or depositional.
3 Use Activity 1a and 1b as a starter. You could also ask students – What colour are contours? What is the interval between contours? What is a spot height?

Ideas for plenaries

1 Provide photos of the glacial features shown on the OS map. Ask students to provide grid references for the features shown in the photos.
2 Pick three glacial features on the OS map. Describe a walk between the three to a partner. Use grid references to explain your route and describe the features you see.
3 Provide students with a table of glacial features and ask them to draw the contour patterns.
4 Tell your neighbour the two key things you have learned today.

Further class and homework activities

1 Design a tourist brochure for the Lake District or another upland glaciated area and include information about the glacial features found there. Include diagrams and sketches of the features.

2 Investigate Snowdonia. Collect information from a variety of sources and produce a wall display. Contact the National Park Office and look up websites on the internet.

Answers to the Activities

The map on page 32 of the students' book is a useful resource and reference for many of the activities in this chapter. By giving four- and six-figure grid references, individual glacial features can be identified and pinpointed for students to locate; for example, Red Tarn, a corrie lake/tarn, at 3415.

1 a Height is shown in two ways on OS 1:50 000 maps:

By contours: these are brown lines drawn at 10-metre vertical intervals, numbered at 50-metre intervals. Note also that lake depths are marked by blue contours (as in Thirlmere).

By spot heights: small black dots, numbered to the nearest metre above mean sea level. Note that these are not always at high points – ask students to find some 'low' spot heights, along roads for example. Can they suggest why these are included on a map?

b It is possible to see on a map where land is steep because this is where the contours are very close together, forming a dense brown mass (as around Helvellyn, for example).

c Note that relief describes both the rise and the fall of the landscape. Descriptions should not only cover the higher areas shown on the map, but should also consider the valleys between them. They should also cover the direction in which streams and valleys are running (to find this out students may need to look closely at the contours), and the relative steepness of slopes (again referring to contours – around Grasmere the land is much less steep than around Thirlmere, for example). Plenty of named examples and specific figures (maximum heights of hills, depths of valleys) should be included in the description.

2 342151 = spot height 949 on Helvellyn

321131 = car park

349079 = Alcock Tarn

This activity can be extended to provide useful practice in map reading and identification of specific features on a map – see note above.

3 Brown Cove = corrie

Thirlmere = ribbon lake

Striding Edge = arête

along the eastern edge of Thirlmere = truncated spurs

Red Tarn = corrie lake

Grisedale = glacial trough

along the eastern edge of Thirlmere = hanging valleys

4 Students draw an accurate cross-section of line A-B

5 The part of Grasmere shown in square 3307 is an area of low, flat land crossed by the River Rothay, which would have been a good site for an early settlement in this otherwise very hilly area: flat land, good water supply, a crossing-point of the river, on a routeway at a point where several routes meet probably also good soils here for growing crops.

6 The A591 is a relatively direct north-south route which follows the valleys where it can do so, and skirts the eastern shore of Thirlmere. It crosses the line of hills between Grasmere and Thirlmere at a point named Dunmail Raise (238 metres), a relatively low pass between two lines of higher hills. Students should notice that it is also labelled 'Roman Road', which means that the original road was built by the Romans, so the line of the route has more or less been maintained for about 2000 years. Building and maintaining roads in this type of environment is always difficult, and they are generally at as low a level as possible, avoiding steep gradients and extreme climatic conditions.

7 All of the trees planted around Thirlmere are conifers. Thirlmere is in fact a reservoir supplying water to Manchester. It makes use of the existing ribbon lake, and it is important to maintain the flow of clean water into the lake. Trees were planted to stabilise the ground around it and to prevent surrounding rocks and soil being washed into the lake. They also help to prevent evaporation of water from the surrounding land, as more water is trapped in the soil by the tree roots. The slopes here are so steep that the land cannot be used for farming, but planting trees may bring in a small income.

Following the Second World War, there was a deliberate policy to plant fast-growing conifers in the upland regions of Britain, including several large areas of the Lake District. Many such areas are now gradually being replaced by mixed native woodland species.

8 a From the south of square 3315.

b North-west.

9 This activity is a continuation of Unit 2.3 activity 2 and Unit 2.4 activities 3 and 4. If they have been updating their revision notes, students should have no difficulty in completing and even adding to this table. Emphasise again how important it is that they memorise at least one named example for each feature (except drumlins, eskers and kames), so that they can quote these in exam questions referring to features and processes of glaciation. The Lake District has numerous examples of most of these features.

The unit in brief

This unit is about how people use the land in an upland glaciated area. The Cairngorms is used as the example here. People have to make a living – but competing demands over land use can cause conflict. In the Activities, students think about the effect of climate on farming and produce a mind map to show how different land uses in the Cairngorms might conflict.

Key ideas

◆ The Cairngorms is an upland glaciated area in north-east Scotland.

◆ 12% of the land in the Cairngorms is forested – mainly with coniferous trees.

◆ Balliefurth Farm is a pastoral farm on the banks of the River Spey.

◆ Farming is limited by physical factors – the climate is too cold and the soil too thin, for crops to grow.

◆ Windfarms provide a clean form of renewable energy – but people object to them in, or near, National Parks.

◆ Competing demands over land use can cause conflict.

Key vocabulary

visual pollution

Skills practised in the Activities

◆ Geography skills: describing the physical factors which make the Cairngorms unsuitable for arable farming; explaining biodiversity

◆ Thinking skills: producing a mind map of different land uses in the Cairngorms and explaining the links

Unit outcomes

By the end of this unit, most students should be able to:

◆ define the term given in 'Key vocabulary' above;

◆ give an example of an upland glaciated area;

◆ describe how people use the land in an upland glaciated area;

◆ understand why different land uses might conflict.

Ideas for a starter

1 Show photos of the Cairngorms. Brainstorm the type of activities people want to do in this area.

2 Add to the activities produced in starter **1** any others which might take place in an area like the Cairngorms and which students might not have thought of, such as: hill walking, mountain biking, farming, sight-seeing, climbing, army training, quarrying, etc. Ask students which activities could exist together and which would need to be apart.

3 Ask students to locate the Cairngorms on a blank outline map of the UK.

Ideas for plenaries

1 Use an OS map extract of part of the Cairngorms and find glacial features and examples of economic activity.

2 How should the Cairngorms be managed? Discuss various options: increase the forested area; encourage farm diversification; allow the development of windfarms. What are the consequences of each one?

3 Should upland areas be left unmanaged so that people are discouraged from visiting? This may only cater for people who will care for the area in a sustainable way. What do you think?

4 Take two minutes with a partner to think up one interesting question about how people use the land in an upland glaciated area that we have not covered today.

Answers to the Activities

1 a In upland glaciated areas such as the Cairngorms physical factors such as relief, poor thin soils combined with the cold climate, length of growing season, etc. means the area is unsuitable for arable farming.

b Biodiversity means biological diversity. In the context of Balliefurth Farm, it relates to the increase in species of plants and animals found on the farm.

2 The mind map should include forestry, farming; tourism; windpower and could include any others students have come up with if starters **1** and **2** have been used. Links between the different land uses should explain how they are connected and how different uses might cause conflict.

The unit in brief

In this unit students find out more about economic activity in an upland glaciated area. The focus here is on tourism in the Cairngorms, the benefits and disadvantages that it brings and the conflicts it can cause. In the Activities, students create a spider diagram to show problems that tourism causes in the Cairngorms; write a role play about conflict between local people and tourists; and find out about Aviemore or Braemar – two of the honeypot villages in the Cairngorms.

Key ideas

◆ Tourism is vital to the economy of the Cairngorms.
   – There are over 1 million visitors a year.
   – 3000 people have jobs related to tourism.
   – Tourism brings in approximately £500 million a year for the Highlands of Scotland.
   – Tourism and tourist-related businesses make up about 80% of the economy.
◆ Tourism brings problems too.
   – Skiing damages vegetation, chairlifts are unsightly.
   – Footpaths are eroded.
   – Salaries are low, house prices are high.
   – Jobs may be seasonal.
   – There is a high volume of traffic.
   – Litter causes visual pollution and is dangerous for wildlife.
   – Honeypot villages attract too many tourists.
   – Dogs worry sheep and disturb wildlife.

Key vocabulary

There is no key vocabulary in this unit.

Skills practised in the Activities

◆ Geography skills: finding out about Aviemore or Braemar
◆ Literacy skills: writing a role play of conflict between tourists and local people
◆ Thinking skills: making a spider diagram about the problems tourism causes

Unit outcomes

By the end of this unit, most students should be able to:

◆ give three reasons why tourism is important to the economy of the Cairngorms;
◆ describe some of the attractions of the Cairngorms;
◆ describe some of the problems tourists cause;
◆ understand why there is conflict between tourists and local people.

Ideas for a starter

1 Make a graffiti wall of tourist activities in an upland area such as the Cairngorms, and the impacts they might have.

2 Show photos of a variety of the effects tourism can have on an area such as the Cairngorms.

Ideas for plenaries

1 Provide students with an OS map extract of the Cairngorms. Ask them to identify glacial features as well as other tourist attractions (and provide grid references for all).

2 Ask students to create a thirty second soundbite for a local radio programme on the future of the Cairngorms. Different students can take on roles of people with different views. Ask several students to read this to the rest of the class.

3 Ask students to draw up a table listing the advantages and disadvantages (problems) that tourism brings to the Cairngorms.

4 Question time! Think back over the lesson and write down three questions related to what you have learned. The teacher will ask a member of the class to try to answer.

Further class and homework activities

1 Use an OS map extract of the Cairngorms to plan a touring holiday in which the visitors would stop off at various points for short walks and to visit various features. Plan opportunities for taking photographs.

2 Design a poster advertising the attractions of the Cairngorms.

3 Choose another National Park and investigate it. Compare the problems of the Cairngorms with the other area.

4 Investigate tourism within the UK – use this website to start off: http://www.staruk.org.uk/ Find out how important the Cairngorms is to Scotland's tourist industry.

Answers to the Activities

1 The spider diagram should include as many of the points listed on page 37 of the students' book as possible. Any other problems that students are aware of, and which may have been mentioned if starter 1 or 2 have been used, can be added to the spider diagram. Students should be aware that many of the problems that tourists cause in the Cairngorms are not unique to this area – they can be found in many popular tourist areas.

2 This can be an individual, group or whole-class activity. Decide first what the exact point of conflict is – some examples:

◆ scramble-bikers want to use a public bridleway that goes across a local sheep-farmer's land;

◆ ramblers are accused by a local landowner of leaving gates open;

◆ car drivers have been caught speeding through a local village.

Make sure that both sides of the argument are fully covered – it shouldn't be difficult for most students to see the problem from both points of view! Can they offer any solutions to the problems?

3 Presentation is important here. The final document will provide a useful case study for a student's revision portfolio, which they can refer to in appropriate exam answers. The research will also be useful practice in distilling information from the internet, making useful selection of material and re-presenting it at an appropriate level. Remind them that a map or diagram can often provide as much information as written text. It may be a useful exercise for students to compare their case studies, to share ideas on how

The unit in brief

This unit looks at how some of the problems caused by tourists in the Cairngorms can be managed, and in particular considers the Cairngorm Mountain Funicular Railway – built to replace some of the original skiing infrastructure. It caused conflict when it was proposed and is still creating problems. In the Activities, students draw a consequence map for the Cairngorm Mountain Funicular Railway. They also consider whether tourism in the Cairngorms is sustainable.

Key ideas

◆ There are a variety of management solutions to the problems caused by tourism in the Cairngorms. They revolve around:
  – traffic and transport;
  – protecting fragile parts of the National Park;
  – housing.

◆ Other solutions tackle footpath erosion, upgrading paths, attracting tourists out-of-season, etc.

◆ The Cairngorm Mountain Funicular Railway is important for the tourist industry, but some people were opposed to it:
  – a 'closed system' of visitor management is operated;
  – some groups are still campaigning to allow walkers to leave the railway at the top station.

Key vocabulary

There is no key vocabulary in this unit.

Skills practised in the Activities

◆ Thinking skills: drawing a consequence map for the Cairngorm Mountain Funicular Railway; considering whether tourism in the Cairngorms is sustainable

Unit outcomes

By the end of this unit, most pupils should be able to:

◆ explain how different solutions can be used to manage the problems created by tourists in the Cairngorms, and give at least six examples.

◆ explain why the Cairngorm Mountain Funicular Railway is important for the tourist industry; why some people were opposed to it; how visitors are managed and what might happen in the future.

Ideas for a starter

1 To check students' understanding of the key terms covered so far in this chapter devise a 'pairs' game where key terms and definitions have to be turned over together and matched.

2 Draw a sketch map of the Cairngorms on the board. Ask students to identify the glacial features, human uses and tourist attractions in the Cairngorms.

Ideas for plenaries

1 Provide students with a cross-word, including the key terms from this chapter.

2 Choose a student to be in the hot seat. Another student asks him or her a question about managing tourism in the Cairngorms. Then nominate different students (4-6 pairs in total). There's one golden rule – questions cannot be repeated.

3 Go back through the work from this chapter and highlight all the key topics. Take two minutes to tell your neighbour the key things you have learned in this chapter.

Further class and homework activities

1 Write a short report entitled 'The management of the Cairngorms', using the following structure:

  – An introduction to the area.
  – Competing land uses in the Cairngorms.
  – The importance of tourism in the Cairngorms.
  – The problems and conflicts tourists cause.
  – An evaluation of the solutions to the problems.

2 Produce a brochure for the Cairngorms combining information about the physical and human landscapes.

Answers to the Activities

1 The consequence map for the Cairngorm Mountain Funicular Railway should include the following:

The problem was … the original skiing infrastructure was becoming difficult to maintain (the original chairlift was built in 1961). Chair lifts and tows were susceptible to weather and could not operate in high winds.

The solution was … to replace the old tows with a funicular railway.

Conflicts … The Chief Executive of the Highlands and Islands Enterprise was for the funicular railway and claimed it was of national significance for Scotland's tourist industry, for improving facilities for visitors, and for the environment. WWF and the RSPB were amongst those opposing the funicular railway – they were concerned that it would quadruple or quintuple the number of visitors – both going up the mountain and going to the car park at the bottom – and the effect this would have on the environment and wildlife. Groups were also concerned that the extra visitors to the Cairngorms would mean fewer visitors elsewhere and that those areas would suffer.

What's happening now … Since the funicular railway was built, milder weather and poor snow conditions has meant fewer skiers so the railway is more dependent on other users. Groups are campaigning to allow walkers to leave at the top station. This had not been allowed previously in order to protect the fragile mountain environment. What will happen in the future?

2 Students need to pull together the information in Units 2.7 and 2 8 to answer the question 'Is tourism in the Cairngorms sustainable?' They could use the spider diagram (Activity 1 Unit 2 7) of problems that tourists cause and draw another to show some of the solutions. They then need to consider the issue of sustainability – can tourism meet people's needs now and in the future without harming the environment?

The big picture

These are the key ideas behind this chapter:

◆ Weather is the day-to-day state of the atmosphere and can change very quickly.

◆ Weather consists of a number of elements which can be measured.

◆ Air moves around the world in huge 'blocks' of air called air masses. (They move because of the temperature differential from equator to pole.)

◆ Two large-scale weather systems control the weather in the UK – depressions and anticyclones.

◆ Synoptic charts and satellite images are used for weather forecasting.

◆ Weather and climate affect all of us.

Note that the students' version of the big picture is given in the students' chapter opener.

Chapter outline

Use this, and the students' chapter opener, to give students a mental roadmap for the chapter.

3 **Weather** As the students' chapter opener, this unit is an important part of the chapter; see page 11 of this book for notes about using chapter openers

3.1 **Measuring the weather** The elements of weather, and how they are measured

3.2 **Air masses and anticyclones** The effect different air masses have on our weather, and winter and summer anticyclones

3.3 **The depression** How depressions form, and the weather they bring

3.4 **Synoptic charts and satellite images** What synoptic charts and satellite images show, and how to read them

3.5 **The climate of the British Isles** The temperate, maritime climate of the British Isles, and the formation of relief and convectional rainfall

3.6 **Climate and human activity in the UK** How the weather and climate in the UK affects sources of energy, farming, flooding, tourism and sport and leisure

Objectives and outcomes for this chapter

Objectives	Unit	Outcomes
Most students will understand:		Most students will be able to:
What weather is, and that it consists of different elements which can be measured.	3.1	Define weather; identify at least six elements of weather; say what instruments are used to measure them and give the units of measurement.
What air masses are, and how they affect our weather.	3.2	Explain that air moves around the world in huge blocks called air masses and that many different air masses cross Britain which explains why our weather can change so quickly. Name three different air masses and say what kind of weather they bring to the UK.
That anticyclones are high pressure weather systems and that there are differences between winter and summer anticyclones.	3.2	Describe the differences in weather associated with winter and summer anticyclones.
That depressions are low pressure weather systems; how they form; and how they affect our weather.	3.3	Describe how depressions form and the weather associated with them.
That three types of rainfall are responsible for the precipitation in the British Isles.	3.3, 3.5	Describe the formation of frontal, relief and convectional rainfall.
That synoptic charts and satellite images help us to forecast the weather.	3.4	Explain what synoptic charts and satellite images are; describe weather conditions using a synoptic chart.
That Britain has a temperate maritime climate, but climate varies across the British Isles.	3.5	Explain why climate varies across the British Isles.
That weather and climate affect what we do.	3.6	Give five examples of how climate affects our lives.

These tie in with 'Your goals for this chapter' in the students' chapter opener, and with the opening lines in each unit, which give the purpose of the unit in a student-friendly style

Using the chapter starter

The photo on page 40 of the *geog.SG* students' book shows a tornado in Kansas, USA. It narrowly missed the house in the picture, but continued on to wreck a house not far away. A tornado is a spinning column of air, shaped like a funnel. They're usually associated with thunderstorms and are known for being destructive – tornado winds can be over 300 mph. Tornado damage-paths can be up to 1.5 km (1 mile) wide and 75 km (50 miles) long. It's impossible to predict exactly where a tornado will hit.

Tornadoes form in storms all over the world. In the UK a tornado in Birmingham in the summer of 2005 left a thousand buildings damaged. But tornadoes are mostly associated with the American Mid-West. In the USA, in an average year about 1000 tornadoes are reported. Tornadoes cause an average of 80 deaths and 1500 injuries a year in the USA.

In the film *The Wizard of Oz* a Kansas girl and her dog are picked up by a tornado and transported to a fantasy land.

Although tornadoes aren't covered in this chapter, the photo was chosen because it's a stunning weather and climate image that should spark interest in the topic.

The unit in brief

This unit is about weather and how it can be measured. Weather can be divided into a number of elements, each of which can be observed and measured. In the Activities, students look at the single most important weather element – air (or atmospheric) pressure – and explain how it controls the other elements of weather; they describe different weather elements and say how they're recorded.

Key ideas

◆ Weather is the day-to-day condition of the atmosphere.

◆ Weather is made up of different elements which can be measured.

◆ Air pressure is the most important weather element.

◆ Weather stations are where all the weather instruments are located together. Automatic weather stations are linked to a computer to measure and record the weather.

Key vocabulary

weather, precipitation, rain gauge, temperature, maximum and minimum thermometer, Stevenson Screen, air pressure, humidity, hygrometer, wind direction, compass rose, wind vane, wind strength/speed, anemometer, visibility, cloud (type and cover), sunshine, sunshine recorder/heliograph, weather station

Skills practised in the Activities

◆ Geography skills: describing weather elements and how they are recorded

◆ Thinking skills: explaining air pressure

Unit outcomes

By the end of this unit, most students should be able to:

◆ define or explain the terms given in 'Key vocabulary' above;

◆ understand that weather is the day-to-day condition of the atmosphere;

◆ identify the different weather elements;

◆ explain how air pressure controls other weather elements;

◆ describe how different weather elements are measured.

Ideas for a starter

1 Ask: Who can tell me what weather is? What's the difference between weather and climate?

2 How do we measure weather? If you have thermometers, gauges, etc. use these as props to ask: What are they? What are they used to measure?

3 Show photos of different weather conditions, e.g. a sunny cloudless scene, another photo showing cloud, rain, obvious wind, etc. Ask students to identify as many weather elements as possible.

Ideas for plenaries

1 Why is weather so important that people measure it all the time, all around the world?

2 Ask some 'Why' questions, e.g.:
Why is the sky cloudy all over on some days, and clear on others?
Why is the weather different in different parts of the country today?
Ask students to think up other 'Why' questions about the weather. The class chooses

the five they think best, and write them down. You can revisit these nearer the end of the topic to see which have been answered. Some could become enquiry questions.

3   Have a set of cards, each with one term from this unit (e.g. rain gauge, wind speed), face down on the desk. A student chooses a card and silently draws clues on the board. The student who guesses the term has to explain it, then chooses the next card.

4   Prepare an odd-one-out for your partner on the elements of weather.

Further class and homework activities

Record the weather for a week, for your local area. Include temperature, precipitation, wind speed and direction, cloud cover. (Take measurements using the school weather station, or look up local weather reports.) Show your records as a table, and also plot graphs or do drawings where appropriate. (For example, a simple wind rose.)

Answers to the Activities

1 a Atmospheric pressure is effectively the weight of the air pushing down on the Earth's surface. Pressure is the application of force per unit area, so air pressure is the force of the atmosphere on the ground surface. It is measured in millibars or millimetres of mercury using a barometer. Pressure is generally high if over 1000 mb and low if less then 1000 mb.

b Air pressure indicates whether air is warm and rising, which results in low pressure, or cold and sinking which results in higher pressure. In this way, the air pressure controls many other elements of the weather. Low pressure is associated with clouds and wet, windy weather. As warm air rises, moisture cools and clouds develop giving precipitation; visibility becomes poor. High pressure is associated with clear skies, dry and calm conditions. Cool air sinks and warms as it moves towards the ground. It warms up and evaporates moisture, giving dry conditions with generally very clear visibility, with the exception of fog and pollution.

2   Students can choose any five weather elements. All those included in this unit are listed in the table below.

Weather element	Description	How recorded
Precipitation	All forms of moisture from clouds – rain, hail, sleet and snow.	Measured in a rain gauge in millimetres.
Temperature	How hot or cold it is – the heat in the air.	A maximum and minimum thermometer is used to measure temperature in degrees Centigrade.
Air (or atmospheric) pressure	The force the atmosphere exerts on the Earth's surface.	Measured by a barometer in millibars.
Humidity	The percentage of water vapour in the air.	Measured with a hygrometer.
Wind direction	The direction the wind is coming from.	Recorded using the eight points of the compass rose and a wind vane.
Wind strength (speed)	The strength or speed of the wind.	Measured with an anemometer, in knots, kilometres or miles per hour, or by the Beaufort Scale (an observation scale).
Visibility	How far we can see.	Measured by a visibility meter in metres or kilometres.
Cloud type	There are five main types of cloud: stratus (layered), cumulus, nimbus (rain bearing), cumulonimbus, cirrus (ice).	Identified by observation.
Cloud cover	How much of the sky is covered in cloud.	Measured in eighths, or oktas, by observation.
Sunshine	The length of time the sun shines.	Measured in hours using a sunshine recorder or heliograph.

The unit in brief

This unit is about air masses and the effect they have on our weather, and about anticyclones and the different types of weather associated with winter and summer anticyclones. In the Activities, students explain why different air masses bring different weather and suggest reasons for the weather associated with summer anticyclones.

Key ideas

◆ Air moves around the world in large blocks called air masses. (They move because of the temperature differential from equator to pole.)

◆ Air masses differ from each other in temperature and moisture content – they assume the temperature and humidity of the area where they originated.

◆ Different air masses cross Britain which is why our weather changes so quickly.

◆ Anticyclones and depressions are large-scale weather systems which affect the weather in the British Isles.

◆ There are differences in the weather associated with winter and summer anticyclones.

Key vocabulary

air masses, Polar maritime, Arctic maritime, Polar continental, Tropical continental, Tropical maritime, anticyclone, depression, isobars

Skills practised in the Activities

◆ Thinking skills: explaining air masses and why they bring different weather; suggesting reasons for summer anticyclonic weather

Unit outcomes

By the end of this unit, most students should be able to:

◆ define or explain the terms given in 'Key vocabulary' above;

◆ understand that air moves around the world in large blocks called air masses;

◆ explain why different air masses bring us different types of weather;

◆ name the two large-scale weather systems which control the weather in the British Isles;

◆ suggest reasons for the weather associated with summer and winter anticyclones.

Ideas for a starter

1   What would air over the Sahara desert be like? Hot or cold? Damp or dry? How would it affect our weather, if it drifted over the UK? What might air over the North Pole be like? How would it affect our weather, if it drifted over the UK? Explain that air masses from other places are always coming our way, causing changes in the weather. The map on page 44 of the students' book shows the directions they come from.

2   Think about the air above us. Is it the same air that was above us last week? If it's different, where did it come from? Where did the other air go? Explain that the air around the Earth is continually on the move. It moves in huge blocks that can be millions of sq km in area.

3   Show photos of typical summer and winter anticyclone conditions. Ask students to describe the weather.

Ideas for plenaries

1   Draw a rough map on the board like the one at the bottom of page 44 of the students' book. Write *Summer* above it. Then mark in the arrows from that map. These show the five common air masses that affect the UK.

Ask students to come up and shade in the arrows, red for warm and blue for cold. Then they write on each arrow *damp* or *dry*.

Ask the class how each air mass would affect the UK weather in summer. Then cross out *Summer* and write *Winter*, and repeat the exercise.

Only the Polar continental air mass changes its characteristics, from warm and dry to cold and dry. (But note the Tropical continental air mass does not usually come our way in winter.)

2   Provide students with a map of Western Europe showing isobars (you can download one from the Met Office website). Ask students to mark on the areas of highest pressure and the wind directions.

3   Anticyclones can be associated with high levels of pollution, especially over urban areas. Why?

4   Provide students with satellite images of anticyclones. Ask them to annotate them with information about the weather conditions.

Answers to the Activities

1 a   Air mass: a huge block of air, perhaps thousands of km across, with a particular set of characteristics (for example warm and dry, or cold and wet) depending on where it came from.

b   Air masses assume the temperature and humidity of the area where they originated, so for example an air mass originating over the North Pole will be cold and dry and will bring cold dry weather, one coming from the tropics will be warm and damp and will bring warm, damp weather.

2   The weather associated with a summer anticyclone includes clear skies and dry conditions. The winds are very calm. Daytime temperatures can be very high in the UK at over 25 °C, although evenings can be much cooler. On occasions, anticyclones can bring heat-waves and convectional thunderstorm activity.

The high summer temperatures are due to the higher angle of the sun in the sky during the summer months. The heating from the sun is more concentrated during the day. Temperatures can cool rapidly at night, though, as the daytime heat escapes into space. This cooling leaves dew on the ground in the morning. Sometimes temperatures become so hot that warm moist air rises locally and builds up towering cumulonimbus clouds that give thunderstorms and periods of torrential rain.

Students could extend their understanding of anticyclones by also explaining the winter contrasts.

The unit in brief

This unit is about depressions – how they form, and how they affect our weather. In the Activities, students produce a weather forecast for Birmingham as a depression passes over the UK, and describe the differences between the weather associated with a depression and that associated with a summer anticyclone.

Key ideas

◆ A depression is a low pressure weather system.

◆ In a depression the warm front has warm air behind it, the cold front has cold air behind it.

◆ There are distinct stages in the development of a depression.

◆ Depressions produce a distinct weather pattern.

Key vocabulary

depression, warm front, warm sector, cold front, frontal rainfall cold sector, pressure, temperature, cloud cover, wind speed and direction, precipitation

Skills practised in the Activities

◆ Geography skills: describing the differences between weather associated with depressions and summer anticyclones

◆ Literacy skills: producing a weather forecast

Unit outcomes

By the end of this unit, most students should be able to:

◆ define or explain the terms given in 'Key vocabulary' above;

◆ understand that a depression is a low pressure weather system;

◆ remember that in a depression the warm front has warm air behind it, the cold front has cold air behind it;

◆ understand that there are distinct stages in the development of a depression;

◆ describe the weather associated with a depression.

Ideas for a starter

1 Show a satellite image of a depression from the Met Office website on the whiteboard or give students copies of the satellite image. Ask the students to describe what they can see. What type of weather feature is this? What type of weather might it bring?

2 Have a diary prepared which records changes in weather as a depression passes over. Ask students to read it out.

3 Ask students to look out of the window. Describe the weather. What was it like yesterday? What will it be like tomorrow? What weather feature is giving us this weather?

Ideas for plenaries

1 Provide students with satellite images from a depression passing over the British Isles. Ask students to annotate them to explain what is happening.

2 What do you think the weather conditions along an occluded front will be?

3 Show a video of a weather report from the television. Ask: How effective do you think it is? Can you think of ways to improve it?

4 Describe the weather associated with the passage of a depression to your neighbour.

5 Did you find anything difficult about the work in this unit? What? Why? What would help to make it less difficult?

Further class and homework activities

1 Log the progression of an actual depression as it moves across the British Isles. Use the Met Office website to help you.

2 Investigate the sequence of cloud type and cover as the depression advances and passes overhead.

3 Look up the different ways the weather is presented on a variety of websites.

Answers to the Activities

1 An effective way of answering this question would be to divide up the depression chart into 6-hour periods and to encourage students to describe what is happening in each section, structuring their forecast more carefully. Use the table below to identify the correct information for Birmingham.

Time period	Passage of depression	Weather features
0-6 hours	Ahead of warm front	Air pressure starts to fall rapidly with high-level cirrus clouds developing. Conditions are dry with a south to south east wind direction.
6-12 hours	Passage of warm front	Air pressure continues to fall and there is a noticeable increase in temperature. Stratus cloud is low and thick, giving steady and continuous precipitation. The wind direction changes to south westerly.
12-18 hours	Warm sector	The air pressure remains steady and the temperatures are mild in the warm air mass. Cloud development becomes patchy and rain may stop or there could be light drizzle.
18-24 hours	Cold front to cold sector	During the passage of the cold front the cloud will be thick cumulonimbus with heavy precipitation. The winds will change to come from the colder north west. There is a sudden drop in temperature and the air pressure begins to rise. Following the cold front, squally showers persist in cool conditions, the cloud cover becoming increasingly patchy.

2 Five differences between the weather associated with a depression and a summer anticyclone are shown and explained in the table below:

Weather feature	Depression	Summer anticyclone
Average temperature and cloud cover	Mild due to warm sector air from the tropics. Cloud cover in a depression prevents the temperatures from getting very warm in the summer.	Very high temperatures due to clear skies and high angle of the sun, with no cloud to block the penetrating heat.
Precipitation	Very wet conditions at both the warm and cold fronts, particularly the latter. Frontal rainfall means that warm air is rising over cold air in the case of the warm front, or being undercut in the case of the cold front. Cooling leads to condensation and cloud development.	Sinking air warms as it moves towards the ground surface. This encourages the evaporation of moisture, limiting the development of clouds. Consequently the skies are clear and there is no precipitation. The exception comes from convectional activity on very hot summer days, which can result in thunderstorms.
Wind speed	Higher wind speed as the isobars will be much closer together, indicating a greater change in air pressure.	Calm wind speed as the isobars will be further apart, indicating a more gradual change in air pressure.
Air pressure	Low as air is rising to form frontal rainfall.	High as air is descending to give clear and dry conditions.

The unit in brief

This unit introduces students to synoptic charts and satellite images. In the Activities, students study synoptic charts to write a weather forecast, describe weather conditions at specific places, and suggest how the weather will change.

Key ideas

◆ Synoptic charts are maps that show weather conditions.

◆ Satellite images are photos taken from space.

◆ Satellite images show cloud cover and are used for weather forecasting.

◆ Weather symbols are used to show conditions at specific places.

Key vocabulary

synoptic chart, depression, occlusion, anticyclone, high-pressure ridge, precipitation, satellite image, weather symbols

Skills practised in the Activities

◆ Geography skills: using synoptic charts to describe weather conditions

◆ Literacy skills: writing a weather forecast

◆ Thinking skills: analysing synoptic charts; suggesting reasons for changes in weather

Unit outcomes

By the end of this unit, most students should be able to:

◆ define or explain the terms given in 'Key vocabulary' above;

◆ describe weather using a synoptic chart;

◆ know that satellite images are photos taken from space;

◆ understand that satellite images show cloud cover and are used for weather forecasting;

◆ understand that weather symbols are used to show conditions at specific places.

Ideas for a starter

1 Recap: the precipitation process (cooling → condensation → cloud formation → precipitation); frontal rainfall; weather conditions associated with the passage of a depression.

2 Show a satellite image and synoptic chart for the same time period. What does each show?

3 Brainstorm to find out what students know about synoptic charts and satellite images. What are they? What are they used for?

4 How do weather forecasters find out about the weather? (Satellites provide much of the information. They can show weather systems some distance away from the UK that may be heading our way. This helps forecasters to make predictions.)

Ideas for plenaries

1 Show students flashcards of weather symbols and ask them to say what they are in a quick-fire quiz.

2 What is the really important thing on satellite images that give us clues about the weather? (The clouds. The cloud patterns give meteorologists a great deal of information).

3 Provide students with a synoptic chart and satellite image for the same time period. Ask students to draw the likely cloud formation on the synoptic chart and then check the actual formation against the satellite image.

4 Weather forecasters don't like to predict the weather more than five days in advance. Why do you think this is?

5 Make a graffiti wall of a range of jobs that would find the weather forecast useful.

Answers to the Activities

1 The synoptic chart for 11 August indicates a large area of high pressure to the west of the British Isles. This is an anticyclone. The isobars are far apart indicating a gradual change in air pressure so winds would be expected to be light or calm. The winds would be circulating in a clockwise direction approaching the UK from the North West. For most of the country skies would be clear giving rise to higher temperatures as the intense sun heats the land during the day, although these temperatures may fall quite readily during the evening. In all areas conditions will be dry.

2 a The second synoptic chart shows a classic depression over the UK.

Weather feature	Place A	Place B
temperature	4 °C	5 °C
precipitation	dry	rain showers
cloud cover	Overcast, 8 oktas	partial cloud cover, 4 oktas
wind direction	south to south west	westerly
wind strength	28-32 knots	18-22 knots
air pressure	low, declining	increasing

b There is a trick to showing how the weather conditions at place A might change over a 24-hour period. As the depression moves across the UK, the conditions described at place B will be heading towards place A, and will eventually reach the area over the time period indicated. The reason for the change is that the weather system is tracking across the UK from west to east and the winds are circulating in an anti-clockwise direction. This means that after 24 hours, place A will be behind the cold front and the winds will have veered to the west-north west.

The unit in brief

This unit is about the climate of the British Isles, and explains the formation of relief and convectional rainfall. In the Activities, students draw their own sketch map of the British Isles, divide it into climate zones and annotate it. They compare climate areas, and using a map provided describe and suggest reasons for the distribution of rainfall across the British Isles.

Key ideas

◆ The British Isles has a temperate maritime climate.

◆ The British Isles can be divided into four climate areas.

◆ Three types of rainfall are responsible for the precipitation in the British Isles: relief, convectional and frontal.

Key vocabulary

temperate maritime climate, relief rainfall, convectional rainfall, frontal rainfall, rain shadow, convection currents

Skills practised in the Activities

◆ Geography skills: drawing a sketch map of the British Isles, dividing it into climate areas and annotating it; comparing climate areas; describing the distribution of rainfall

◆ Numeracy skills: using climate data to annotate a map of climate areas

◆ Thinking skills: explaining differences; suggesting reasons

Unit outcomes

By the end of this unit, most students should be able to:

◆ define or explain the terms given in 'Key vocabulary' above;

◆ understand that the British Isles has a temperate, maritime climate;

◆ divide the British Isles into four climate areas;

◆ understand that three types of rainfall are responsible for the precipitation in the British Isles;

◆ explain the formation of relief and convectional rainfall.

Ideas for a starter

1 Make a graffiti wall of what students have learned so far about weather and climate.

2 Provide temperature and rainfall figures for places in the four different climate areas of the UK. Ask groups of students to draw climate graphs for the different places. What do the different graphs show us about climate in the UK?

3 Find the British Isles in an atlas. What would you expect the climate to be like given its latitude? What is the North Atlantic Drift? How does it affect our weather and climate? How does relief affect our weather and climate?

Ideas for plenaries

1 How could the pattern of climate across the British Isles influence economic activity? How could it affect farming for example?

2 How does the North Atlantic Drift affect the climate in the summer and the winter?

3 Make up 10-15 statements about the climate of the British Isles based on what students have learned so far, some true, some false. Students hold up True or False cards. Where statements are false ask students to correct them.

Futher class and homework activities

Collect data on temperature and precipitation for a variety of weather stations over the UK and produce a series of isotherm and isohyet maps to show the temperature and precipitation distribution. Do they look like the maps in the Atlas? Make a wall display of these maps and annotate them to explain the pattern.

Answers to the Activities

1 a Using an outline map of the British Isles, students should design their climate map using the example on page 50 as a model.

b Use the table below to help students to identify the main climate descriptive annotations that are required.

UK region	January temp	July temp	Annual range	Total precipitation	Monthly max	Monthly min
North west	3 °C	14.5 °C	11.5 °C	1979 mm	December 235 mm	May 90 mm
North East	3 °C	16 °C	13 °C	761 mm	July 90 mm	April 45 mm
South West	5 °C	16.5 °C	11.5 °C	1090 mm	December 130 mm	April 45 mm
South East	3 °C	18 °C	15 °C	558 mm	July 60 mm	February 35 mm

2 Students are required to compare the regions and not to write descriptive paragraphs for each separate region. They should be encouraged to calculate the differences for both temperature and precipitation, using the words 'more' or 'less' for the two chosen regions. The statistics given in the above table will make it simpler for these comparisons to be worked out.

3 The rainfall distribution map for the UK shows that most precipitation falls in the west of the country. Over 2400 mm falls in the north west of Scotland and the Welsh mountains, for example. Towards the east, there is less rainfall. Areas such as Cambridgeshire and London only receive between 6-800 mm each year.

There is much more frontal and relief rainfall in the west of Britain. The east is in a rain shadow so receives less.

Climate and human activity in the UK

The unit in brief

This unit looks at how weather and climate affects human activity in the UK – ranging from providing sources of energy through to its effect on sport. In the Activities, students look at farming – comparing maps of hours of sunshine and areas of wheat growing and explaining the patterns shown; and produce a mind map which shows the links between climate and human activity.

Key ideas

◆ Climate and weather has a direct impact on human activity.

◆ The UK's temperate maritime climate provides two sources of alternative energy – HEP and wind power.

◆ There is a direct link between farming and climate.

◆ The threat from flooding determines land use.

◆ Weather and climate affect tourism and sport.

Key vocabulary

temperate, maritime, hydro-electric power, wind power

Skills practised in the Activities

◆ Geography skills: comparing maps and explaining patterns

◆ Thinking skills: producing a mind map and explaining the links between climate and human activity

Unit outcomes

By the end of this unit, most students should be able to:

◆ define or explain the terms given in 'Key vocabulary' above;

◆ produce a mind map to show that climate and weather has a direct impact on human activity;

◆ understand how the UK's temperate maritime climate provides two sources of alternative energy – HEP and wind power;

◆ explain the link between farming and climate;

◆ realise that the threat from flooding determines land use;

◆ understand that weather and climate affect tourism and sport.

Ideas for a starter

1 Ask: Who can remind me of the climate we have in the British Isles? What type of rainfall and weather systems do we have?

2 Recap: Show a map with the British Isles divided into 4 climate areas. Show students 4 climate graphs (one for each area). Ask them to match the correct graph with the correct area.

3 Ask: How does the climate in the British Isles affect you? Record answers on a spider diagram on the board. Then ask: How does climate affect other people, or other activities in the British Isles? Record answers on the spider diagram in a different colour.

Ideas for plenaries

1 The construction industry uses the services from the Met Office. Can you suggest why?

2 Test your neighbour! Spend 3 minutes testing your neighbour on the key vocabulary and definitions you have covered in this chapter.

3 Write 'climate and human activity in the UK' in the middle of a page. Create a mind map around the phrase. How many ideas can you come up with in 2 minutes?

Further class and homework activities

Contact a local farmer and find out what information is required for the effective running of the farm. You may find that the farmer buys in specialist services from the Met Office. Try to find out about what these are and how important the weather forecast is.

Answers to the Activities

1 The maps show a very close relationship between the climate in terms of sunshine hours and the major areas for the growth and ripening of wheat crops. The map for wheat production clearly shows that over 30% of this crop in the UK is grown in eastern counties. This is due to the drier conditions in the summer and the average 4-4.5 hours of daily sunshine needed for the ripening process. Since the west of Britain is much wetter, there is a greater dominance of cloud. This is less ideal for arable wheat farmers and, therefore, only around 10% of production is based in western counties, where sunshine is typically around 3 hours per day or less.

2 Further examples of how the climate of the UK has a direct impact on human activity could include:

◆ Drought and hosepipe bans/lawn sprinklers. A problem for gardeners who want to show their home lawns off to their best.

◆ Retail sales are very strongly influenced by the British climate. Ice cream sales rise dramatically in the summer months. As do the sales of summer clothing and other fashion items such as hats and sun glasses.

◆ Attractions based on landscaped gardens, such as National Trust properties, showcase places with various plants in flower at different times of the year. Snow drops in February, Bluebells in May, etc. The number of visitors in each month is dependent upon the flowers on show, which is ultimately controlled by the climate.

◆ Retirement homes along the south coast have increased in number, due to more elderly members of the population preferring the milder climate found in the south. Examples include settlements such as Bournemouth and Weymouth.

3 Students should be encouraged to design a mind map using climate features as the main branches. For example:

◆ Strong winds
◆ High sunshine hours
◆ Low sunshine hours
◆ Heavy rainfall
◆ Low rainfall
◆ High temperatures
◆ Low temperatures

They can then add human activities that are influenced by each of these climatic features. Students could use the example from activity 1 about wheat growing as a starting example. They can then map out the ideas from the text pages and their own ideas. The more able should be able to spot some interrelationships between human activities. For example, wheat growth and ice cream sales are largely controlled by sunshine and higher temperatures. Sailing and wind power are connected by the need for stronger wind speeds. Flooding and reservoirs for drinking water are related by the requirement for heavy rainfall.

There should be no shortage of ideas for activities 2 and 3!

Climate zones

chapter overview

The big picture

These are the key ideas behind this chapter:

◆ Climate is the average weather in a place. It is worked out by taking measurements over a long period (usually 30 years), and calculating the average.

◆ The world can be divided into climate zones – large areas with a similar climate.

◆ Climate depends on a range of factors – but the main one is the effect of latitude.

◆ The world can be divided into eight major ecosystems or biomes. Individual biomes are mainly determined by climate.

◆ Climate affects people, but people have a major impact on the world's ecosystems.

◆ We are learning to use and manage ecosystems in a sustainable way.

◆ Global warming and climate change will affect us all.

Note that the students' version of the big picture is given in the students' chapter opener.

Chapter outline

Use this, and the students' chapter opener, to give students a mental roadmap for the chapter.

4 **Climate zones** As the students' chapter opener, this unit is an important part of the chapter; see page 11 of this book for notes about using chapter openers

4.1 **Global climate explained** The global distribution of climate and factors affecting climate

4.2 **Climate and ecosystems** What ecosystems are, how they're related to climate and how humans fit in

4.3 **Hot desert climate** What the climate is like, how it affects people living there and how vegetation has adapted to the climate

4.4 **Human impact in hot deserts** Desertification, the causes, and sustainable land management

4.5 **Mediterranean climate** What the climate is like, how it affects people living there, vegetation and land use

4.6 **Human impact in the Mediterranean** How tourism has affected Benidorm, and pollution in the Mediterranean Sea

4.7 **Tundra regions** Climate in the Arctic tundra, coping with the climate, and how vegetation has adapted to the climate

4.8 **Human impact in the tundra** The issue of oil exploitation in the Arctic National Wildlife Refuge

4.9 **Equatorial climate** What the climate is like, traditional farming and how the vegetation has adapted to the climate

4.10 **Human impact in the tropical rainforest** How people have cleared the rainforest for mining, logging, cattle ranching and peasant farming, and sustainable management strategies

4.11 **Climate change** Causes and possible effects

4.12 **What can we do about global warming?** What we can do to tackle global warming on an international, national and individual level

Objectives and outcomes for this chapter

Objectives	Unit	Outcomes
Most students will understand:		Most students will be able to:
• How climate varies around the world, and that climate is the result of a range of factors.	4.1	Describe the global distribution of climate; give five factors that affect climate and describe their effect.
• What an ecosystem is.	4.2	Define ecosystem, and give examples of ecosystems.
• That climate is the main factor affecting the global distribution of ecosystems, or biomes.	4.2	Explain why climate is the main factor affecting the global distribution of ecosystems; describe the global distribution of hot deserts, the Mediterranean climate and ecosystem, tundra, and tropical rainforest.
• How climate affects people's lives in different ways.	4.3, 4.5, 4.7, 4.9	Describe how people cope with hot desert, Mediterranean, Arctic tundra and equatorial climates; describe a hot desert, Mediterranean, Arctic tundra and equatorial climate.
• That vegetation has adapted to the climate in different ecosystems	4.3, 4.5, 4.7, 4.9	Give examples of how the vegetation has adapted to the climate in hot deserts, the Mediterranean, tundra and tropical rainforests.
• The impact that people have had on different ecosystems.	4.2, 4.4, 4.6, 4.8, 4.10	Give examples of human activity in hot deserts, the Mediterranean, tundra and tropical rainforests and describe the consequences.
• That it is possible to use and manage ecosystems in a sustainable way.	4.4, 4.6, 4.8, 4.10	Give examples of sustainable management in hot deserts, the Mediterranean, and tropical rainforests.
• What causes global warming; the possible effects of global warming.	4.11	Explain the greenhouse effect; list the possible effects of global warming.
• That we can do something about global warming at an international, national and individual level.	4.12	Explain why countries need to cooperate to reduce greenhouse gas emissions; explain what Scotland could do to reduce its production of greenhouse gases; list how individuals could reduce their energy use.

These tie in with 'Your goals for this chapter' in the students' chapter opener, and with the opening lines in each unit, which give the purpose of the unit in a student-friendly style.

Using the chapter starter

The photo on page 54 of the geog.SG students' book shows a polar bear in the Archipel de Svalbard. Svalbard is the group of islands within the Arctic Circle, directly north of Scandinavia, approximately 1000 km from the North Pole.

Svalbard has a relatively mild climate, with average temperatures ranging from -14°C in winter to 6°C in summer. Svalbard could be described as an 'Arctic desert' with annual rain- and snowfall at only 200-300 mm.

In Svalbard the polar bears are a protected species, and there are perhaps 500 on the island of Svalbard itself. Polar bears rely on the thin ice on the edge of the polar region for hunting. But the glaciers that make up two-thirds of the archipelago began an almost continual retreat in 1900, as temperatures rose due to global warming. And a lack of ice makes it difficult for the bears to reach their hunting grounds. In Manitoba, Canada, bears have to be tranquillized and air-lifted back to their natural habitat, or else hunger sends them to the human settlements to forage for food.

Some scientists claim that polar bears could be extinct by 2050 due to the effects of global warming. But wildlife in the Arctic is affected by humans in a number of other ways, such as oil drilling, and chemical pollutants absorbed into the food chain.

Global climate explained

The unit in brief

This unit introduces climate, and is about the global distribution of climate and the main factors affecting temperature (and therefore climate). In the Activities, students use a map of climate zones to describe the global distribution of certain types of climate; draw a climate graph, and compare climate zones.

Key ideas

◆ Climate is the average weather of a place taken over a long period of time.

◆ The world can be divided into climate zones.

◆ The main factors affecting temperature are: latitude; distance from the sea; the prevailing wind; altitude; ocean currents.

Key vocabulary

climate, latitude, distance from the sea, maritime climate, continental climate, prevailing wind, altitude, ocean currents

Skills practised in the Activities

◆ Geography skills: describing the global distribution of climate zones; drawing climate graph; comparing climate zones

◆ Numeracy skills: using statistical data from climate graphs

◆ Thinking skills: suggesting reasons for differences

Unit outcomes

By the end of this unit, most students should be able to:

◆ define or explain the terms given in 'Key vocabulary' above;

◆ understand that climate is the average weather of a place taken over a long period of time;

◆ describe the global distribution of climate zones;

◆ draw a climate graph;

◆ identify the main factors affecting temperature and climate.

Ideas for a starter

1 Ask students: Who can define weather and climate for me?

2 In the UK we have a temperate, maritime climate. What does this tell you?

3 Read out the weather forecast for today. How does this match up with the climate description?

4 Use a globe to introduce ideas about the variation of temperature and precipitation over the planet.

5 Tell me a country that's hotter than the UK? And one that's colder? Can you explain why? Elicit that latitude is the main reason.

Ideas for plenaries

1 Produce a spider diagram showing how different factors affect climate. Annotate the diagram.

2 Can you think of an example where altitude has a bigger effect than latitude on the climate? (On very high mountains. Some high mountains in the tropics have glaciers!)

3 How will climate affect human activity in each of the climates shown?

4 How can people influence climate?

5 What is a microclimate? Research the school's microclimate.

Answers to the Activities

1 a The equatorial climate is generally located in a zone around the equator from 0°-5° in latitude, including regions such as the Amazon rainforest, Central Africa and much of Indonesia.

b The Mediterranean climate is found on the west coast of continents roughly between latitudes 30 and 40°, and around the Mediterranean Sea. The Mediterranean region has a Mediterranean climate, of course, as does the southern tip of South Africa.

c The desert climate is located in broad zones around the Tropics of Cancer and Capricorn. The largest of the deserts is the Sahara in North Africa, although other famous deserts include Nevada in the western USA and the Arabian desert.

d Polar climates are found in a broad zone around the Arctic Circle and south of the Antarctic Circle. Northern Canada, Iceland and Greenland have polar climates.

In all cases, it's important that students identify the global latitude bands where the climate is located and then exemplify it with some named areas.

2 a Models of the climate graphs can be found on page 56 of the students' book. Students should use this as a guide and be encouraged to use graph paper to promote accuracy. They should be reminded that the blue bars represent precipitation (measured in mm) and that the red line represents temperature (measured in °C).

b When comparing climate data students are tempted to write simple descriptive paragraphs for each graph all too easily. This is not a comparison. They should be encouraged to calculate the differences between the temperate climate and the other identified climate types.

They should give January and July temperatures, monthly maximum and minimum precipitation figures and could work out total precipitation figures.

3 The table on the right provides a summary of the main reasons for the differences in climate students will have described in activity 2. Most of the reasons are provided on page 57, although more-able students should be able to identify additional explanations.

Climate type	Temperature	Rainfall
Temperate	Due to the curvature of the Earth, there are contrasts between summer and winter. The angle of the sun is higher in summer, pushing up temperatures. A lower angle in the winter means cooler temperatures.	The mountains found towards the west of the UK encourage relief rainfall and, as the prevailing wind comes from the Atlantic (south west), rain falls reliably all year.
Desert	With clear skies and a very high angle of the sun in summer, temperatures can get very hot. Less intense sunlight in the winter means temperatures dip by about 10 °C.	The prevailing winds blow across the desert land and remain dry. The continental winds pick up no moisture, preventing cloud and rain formation. Convectional thunderstorms in the summer months provide more rain but only up to 35 mm in one month.
Equatorial	Constant temperatures around 27 °C, as the intense sun shines all year round with no seasonal contrast. The sunlight is direct so temperatures are high.	High precipitation totals are due to the process of convectional rainfall, page 51. The strong intense sunlight means higher temperatures and rapid convection to produce towering cumulonimbus clouds and torrential rainfall
Polar	Summers are short but relatively warm due to the long hours of daylight. The sun is always at a low angle in the sky. Winters are very long and cold due to the northerly latitude – for a period the sun never rises above the horizon. Strong winds can lower temperatures.	Precipitation is light throughout the year. The air is too cold to hold much moisture and most places are a long way from any rain-bearing winds from the sea.
Mediterr-anean	Summers are hot, partly because the sun is at a high angle in the sky (though never directly over-head) and prevailing winds blow from the warm land. Winters are warm, as, although the sun is lower in the sky, it is still higher than places further from the equator (e.g. the UK).	Summers are dry. Winds blow across a dry land surface unable to pick up much moisture. There may be occasional thunderstorms. In winter the prevailing winds blow from the sea, bringing warm moist air, which gives large amounts of relief rainfall as the air is forced to rise over coastal mountains.

The unit in brief

This unit looks at ecosystems – what they are, how they are related to climate, the global distribution of ecosystems, and how we affect them. In the Activities, students use the map in the students' book to describe the distribution of one of the world's main biomes, and consider the effect humans have on ecosystems.

Key ideas

◆ Ecosystems are made up of two parts; living things, and the non-living environment.

◆ Ecosystems vary in size, e.g. from a pond to a tropical rainforest.

◆ The world has eight major ecosystems or biomes.

◆ Climate is the main factor affecting the distribution of ecosystems.

◆ The human impact on ecosystems can be measured in terms of our ecological footprint.

◆ Ecosystems need to be treated in a sustainable way.

Key vocabulary

ecosystem, environment, biomes, ecological footprint, sustainable

Skills practised in the Activities

◆ Geography skills: defining ecosystem and ecological footprint; describing the distribution of biomes

◆ Thinking skills: explaining; thinking; justifying answers

Unit outcomes

By the end of this unit, most students should be able to:

◆ define or explain the terms given in 'Key vocabulary' above;

◆ understand that ecosystems are made up of two parts;

◆ understand that ecosystems vary in size;

◆ describe the distribution of one of the world's main biomes;

◆ understand that climate is the main factor affecting the distribution of ecosystems;

◆ explain how humans affect ecosystems.

Ideas for a starter

1 Show photos of the tropical rainforest, a hot desert, a Mediterranean area and the tundra. Ask: Why are they different? What is the vegetation like? What is the climate like? How are the vegetation, climate and people linked in each ecosystem? What would happen if there was a major change in the ecosystem?

2 Ask students to describe a journey from the equator to the North Pole. They can use atlases to trace their journey. They should concentrate on the change of natural vegetation as they move northwards.

3 Brainstorm to find out what students know about the destruction or misuse of ecosystems. Why should we be concerned about this?

Ideas for plenaries

1 Show a graph of world population growth. How do you think this growth has affected the Earth's ecosystems? What if it continues at this rate?

2 Write the words ecosystem, environment, biomes, ecological footprint and sustainable and their meanings on separate sheets of paper. Ask ten students to hold up the ten sheets. The rest of the class have to match them.

3 Investigate the hot desert climate – its characteristics, vegetation and how people can cope with the climate. Prepare a short presentation for the next lesson.

4 Write down the five key things you have learned from today's lesson.

Further class and homework activities

1 a Find a website on ecological footprints. Calculate your own footprint. Now do this for the class, year group and school. Plot these areas on a map. What does this tell you about our use of resources? How sustainable is this?

b What can we do to live in a more sustainable way?

Answers to the Activities

1 Ecosystem – this is an interactive environment where living things depend upon their non-living surroundings. The Earth has some very large ecosystems, where the living organisms depend upon and interact with the climate, vegetation and soils.

Ecological footprint – this refers to the human impact on ecosystems. Human activities have spread across the globe and it is suggested that, in one lifetime, each human individual 'uses' two hectares of land for food, water, transport, shelter, waste disposal and so on.

2 Students' answers will depend on which biome they have chosen. They may need an atlas to help with this activity as students can then refer to global latitude bands where the biome is found. They could also exemplify their answers with named areas. Students should be encouraged to refer to the map showing the global distribution of climate zones in Unit 4.1.

3 The answer to this question requires some justification and not a simple 'yes' response. A good answer would mention greater respect and appreciation for plants and animals, whatever their size, as they are all important and sustain our existence. There could be reference to appreciating the interdependence between the living and non-living environment. For example, fostering

respect for earthworms in the soil to protect the habitats for owls. The best answers would be where a student could independently give an example(s) about how studying ecosystems will affect how they think about the world around them. These examples could be from the local scale, such as the preservation of hedges, to the global scale and helping to save on energy consumption (thereby continuing to reduce global warming).

4 The response to the notion of humans being the greatest global 'pest' requires a balanced response, even though the conclusion will be the student's own individual opinion. It is important that their view is justified. There are clearly many negative examples of the human impact on ecosystems which justify the term 'pest'. Mass deforestation, soil erosion, harnessing energy and global warming plus the exploitation of other natural resources, provide clear examples of the harm human civilisation has caused to the planet. However, on the positive side, there are sustainable human activities that actively protect the Earth's ecosystems. Conserving core areas of virgin rainforest, reducing greenhouse gas emissions and preventing soil erosion are good examples of this. The question for the future is whether the negative impacts will continue to outweigh the positive things that people can do.

The unit in brief

This unit is about hot deserts (sometimes called tropical deserts) – the climate, and how the climate affects the people living in hot desert areas. It also looks at how the vegetation has adapted to survive in the harsh climate. In the Activities, students complete a mind map for the hot desert climate and describe this climate type.

Key ideas

◆ Hot desert climates are found around the Tropics of Cancer and Capricorn, and on the western side of continents.

◆ Hot deserts have a high temperature range – daytime temperatures can rise to over 40° during the summer but plunge rapidly at night due to the lack of cloud cover.

◆ Rainfall in hot deserts is limited – less than 250 mm pa, and it may not rain for months, or even years.

◆ Traditionally desert inhabitants were nomadic – travelling in search of food and water.

◆ Water is in scarce supply – large dams provide water for irrigation and domestic use in some countries, in others desalination plants have been built.

◆ Vegetation has adapted to survive in the harsh climate.

Key vocabulary

wadi, nomadic, oasis, irrigation, desalination, succulents

Skills practised in the Activities

◆ Geography skills: describing a hot desert climate

◆ Thinking skills: producing a mind map of the hot desert climate

Unit outcomes

By the end of this unit, most students should be able to:

◆ define the terms given in 'Key vocabulary' above;

◆ describe where hot desert climates are found;

◆ describe the hot desert climate;

◆ understand why traditionally desert inhabitants were nomadic;

◆ say how some countries have increased their water supply;

◆ give three ways that vegetation has adapted to the hot desert climate.

Ideas for a starter

1 Ask: What do you know about hot deserts? Tell me five things about them.
2 Show a climate graph for a place with a hot desert climate – you could use the graph for Khartoum on page 60 of the student book. Ask questions about the temperature and rainfall patterns.
3 Show photos of a variety of desert vegetation. Ask students to describe what they see.
4 Mind movie time! You are in a hot desert – all alone. What can you see, hear and smell? What do you feel? Tell us.

Ideas for plenaries

1 Provide students with temperature and rainfall figures of a place with a hot desert climate. Ask them to draw a climate graph using the figures.
2 Produce a table on the hot desert climate. Include information on the climate, coping with the climate and vegetation.
3 Write a two-minute soundbite for a radio programme on tourism in hot deserts. What activities can people do? What will the weather be like? What clothes will they need for a desert holiday?
4 Tell your neighbour the three most important things you learned today.

Answers to the Activities

1 The mind map should include the following:

Location - found around the Tropics of Cancer and Capricorn, and on the western side of continents.

Main features – Hot and dry. Daytime temperatures can rise to over 40 °C during the summer, but they can get very cold at night. Hot deserts may have less than 250 mm of rain a year. It may not rain for many months (even years).

Vegetation – Roots are either long (to tap into groundwater sources) or horizontal, lying just below the surface. Plants may store water in their roots, stems, leaves or fruit (these are called succulents). They have small leaves or spines, glossy and waxy leaves to reduce water loss. Seeds can be dormant for years, but germinate quickly when it rains.

Traditional life – Traditionally most inhabitants were nomadic (e.g. the Bedouin). They travelled in search of food and water for themselves and their animals. In the 1950s and 60s many Bedouin started leaving their traditional nomadic lives to live and work in the cities of the Middle East as their grazing areas shrank and population increased.

2 The hot desert climate is characterised by being exceptionally hot and dry.

Daytime temperatures can rise to over 40 °C during the summer. But, hot deserts can get very cold at night as there is no cloud cover to keep the heat in. So there can be a high temperature range. Hot deserts have less than 250 mm of rain a year. It may not rain in a hot desert for many months (even years). When it does rain it is often a torrential downpour. A heavy desert storm can bring up to 1 mm of rainfall per minute.

The unit in brief

This unit is about the spreading of deserts – how people have helped to cause this problem and how we might be able to stop, or even reverse, the process of desertification using sustainable land management practices. In the Activities, students focus on desertification – writing a definition of the term, looking at the causes, and explaining the 'spiral of desertification'.

Key ideas

◆ The Sahel runs across Africa south of the Sahara desert. It is about 500 km wide.

◆ The Sahel is under intense pressure from human activity.

◆ Desertification is a process of land degradation. It happens when human and climatic processes combine so that the land can't support vegetation.

◆ Causes of desertification include overcultivation, overgrazing, deforestation, and climate change.

◆ There are a number of sustainable land management practices that if used well can stop, or even reverse the process of desertification.

Key vocabulary

the Sahel, desertification, overcultivation, overgrazing, deforestation, climate change, sustainable land management

Skills practised in the Activities

◆ Geography skills: defining desertification

◆ Thinking skills: considering the causes of desertification; justifying answer; explaining the 'spiral of desertification'

Unit outcomes

By the end of this unit, most pupils should be able to:

◆ define or explain the terms given in 'Key vocabulary' above;

◆ know where the Sahel is;

◆ understand that the Sahel is under intense pressure from human activity;

◆ understand that desertification is a process of land degradation;

◆ explain the causes of desertification;

◆ give examples of sustainable land management practices.

Ideas for a starter

1 Draw a diagram on the board to show how climate, natural vegetation and soils are linked in the Sahel. Ask students what happens if the soil is damaged?

2 Ask: Who can tell me what desertification is? Do you know what causes it?

3 Show 'before' and 'after' photos – of land before desertification, and one of an area suffering desertification. Ask: What has happened? What problems does this cause? What can be done about it?

Ideas for plenaries

1 Draw a spider diagram showing human activity in hot deserts. How is this human activity damaging the ecosystem?

2 Draw up a table showing different ways of managing the Sahel. Include a column for the possible impacts – economic, social and environmental. Use your table to evaluate ways of managing the Sahel.

3 Write a letter to the government of Niger explaining the need to use sustainable land management practices.

4 Make a graffiti wall of what students have learned today.

Further class and homework activities

1 Investigate how much of the problem of desertification is physical – to do with climate, and how much is human – to do with population and economic change.

2 Write a role play to do with managing issues in areas suffering from desertification. Include local people, farmers, tourists, etc.

Answers to the Activities

1 Desertification refers to the spreading of a desert into areas that were once productive and fertile in terms of vegetation. The outcome of desertification is a dry and barren landscape where plant life finds conditions too difficult to re-establish itself.

2 Desertification is partly due to natural climate change, although it could be strongly argued that even this is a direct result of global human activity, especially climate change related to global warming. The rapidly increasing population pressure of many African states in the Sahel requires ever more intense use of the land and deforestation for fuel wood. Therefore it could be considered that the issue of desertification is very much a problem of human creation, both regionally and globally.

Ensure that answers are justified.

3 Students should make reference to the problems of climate change, overgrazing, overcultivation and deforestation to comprehensively explain how the process of desertification can commence and drastically change the characteristics of places in the Sahel. In this context, climate change can be regarded as the natural element, whilst the other reasons are all linked to the pressure of rapidly growing populations in LEDC countries. Once the process of desertification becomes established, it is very difficult to stop. As the population of an area increases and people continue to need more food and wood as a fuel source, the cycle of human impact becomes ever more severe. If the rainy season fails, the spiral into desertification worsens.

The unit in brief

This unit is about the Mediterranean climate – where it's found, what it's like, how people cope with the climate, and land use and vegetation in the Mediterranean. In the Activities, students complete a mind map for the Mediterranean climate, describe the climate and compare it with one of the other climates studied.

Key ideas

◆ Mediterranean climates are found on the west coast of continents, roughly between latitudes 30° and 40° north and south of the equator, and around the Mediterranean Sea.

◆ The Mediterranean climate has two distinct seasons – hot, dry summers, and warm, wet winters.

◆ Water can be in short supply in the summer – reservoirs provide water and crops need to be irrigated.

◆ Long hot summers attract tourists.

◆ Land use changes with altitude.

◆ Mediterranean regions have two natural types of vegetation – woodland and scrub. All the vegetation has adapted to the summer drought.

Key vocabulary

Mediterranean, irrigation, shelter belt, Meltemi, siesta, maquis, garigue

Skills practised in the Activities

◆ Geography skills: describing a Mediterranean climate; comparing climate graphs

◆ Thinking skills: producing a mind map of the Mediterranean climate

Unit outcomes

By the end of this unit, most students should be able to:

◆ define or explain the terms given in 'Key vocabulary' above;

◆ describe where the Mediterranean climate is found;

◆ describe the Mediterranean climate;

◆ give two examples of how the Mediterranean climate affects people;

◆ describe how land use changes with altitude in a Mediterranean region;

◆ name two types of Mediterranean vegetation, and say how vegetation has adapted to the summer drought.

Ideas for a starter

1   Show photos of typical Mediterranean landscapes from a variety of known locations. Ask students: What can you tell me about the vegetation and landscape of these areas? What can you work out about the climate? Where are these places (look them up in an atlas)?

2   Show a photo of Mediterranean vegetation – either maquis or garigue. Ask: Why is the vegetation like this? How has it adapted to the climate? Where do you find this type of vegetation?

3   Show a climate graph for a place with a Mediterranean climate – e.g. Palermo (but don't tell the students where it is). Give students a set of statements and ask them to put them in the correct place on the graph, e.g.:

   ◆   The local fire brigade is put on standby due to the increased risk of forest fires.

   ◆   Waiters and waitresses can have their own holiday.

   ◆   Too hot to work – time to start taking a siesta!

   Ask for suggestions about where the climate graph could be for.

Ideas for plenaries

1   Draw a concept map to show how climate, people, vegetation and land use are linked in areas with a Mediterranean climate. Explain the links.

2   If starter **3** was not used, it could be used as a plenary.

3   Close your book. Describe the distribution of areas with a Mediterranean climate to your neighbour.

4   Prepare an odd-one-out for your neighbour on what you have learned today.

5   Sum up what you have learned today in 35 words (or less).

Answers to the Activities

1   The mind map should include the following:

Location – Mediterranean climates are found on the west coasts of continents, roughly between latitudes 30° and 40° north and south of the equator, and around the Mediterranean Sea.

Main features – Hot dry summers (up to nearly 30 °C, little or no rain – any rain comes in short, heavy thunderstorms), warm wet winters (average winter temperature is 10 °C, annual precipitation is less than 1000 mm pa – rain mostly falls in the winter).

Vegetation – two natural types of vegetation – woodland and scrub. In Europe there are two types of scrub – maquis and garigue. Vegetation has to adapt to the summer drought – leaves are small, waxy or glossy, or plants have thorns and thick bark to reduce water loss. Long taproots reach groundwater supplies. Plants germinate in winter rain, flower in the spring and are dormant in summer.

Land use – changes with altitude. Coastal plain is used for tourist resorts, transport links, etc. Lower slopes and foothills are traditional agricultural areas, as these are more fertile than higher areas. Vines and olives are grown here but land is also now used for tourist facilities, e.g. golf courses, water parks, luxury villas, etc.

Sierra and hilly areas – citrus groves are found on hillsides with old villages on hilltops.

2 a   A Mediterranean climate has two distinct seasons – hot dry summers, and warm wet winters. Temperatures in the summer can be high – up to nearly 30 °C. Winters are mild and temperatures rarely reach freezing (except in areas with a high altitude).The average winter temperature is 10 °C.

On average the annual precipitation is less than 1000 mm per year. Summer months can be very dry with little or no rain, (though if it does rain it comes in short, heavy thunderstorms) and places with a Mediterranean climate can experience drought conditions. Winters though, can be very wet.

Winds can be strong in Mediterranean climate regions, particularly in coastal areas.

b   Answer will depend on which climate graph is used for comparison.

The unit in brief

This unit includes two case studies – one of Benidorm in Spain to show how tourism has affected the landscape there, and one of pollution in the Mediterranean Sea. Both case studies include examples of sustainable management – attempts to overcome some of the problems that humans cause. In the Activities, students draw spider diagrams to show the problems that tourists have caused in and around Benidorm, and how those problems are being tackled. They look at the issue of pollution in the Mediterranean Sea and consider how the sea and surrounding landscape can be protected from increasing numbers of tourists.

Key ideas

◆ Mass tourism transformed Benidorm from a quiet fishing town in 1960, to a large, noisy, tourist resort filled with poor-quality hotels by the late 1980s.

◆ The Spanish government took more control of development in the 1990s, improving hotels.

◆ Tourism puts pressure on the coastline and creates water shortages.

◆ The Mediterranean Sea is polluted with vast amounts of untreated sewage, oil, mercury, lead and phosphates.

◆ Most of the pollution reaches the Mediterranean via rivers.

◆ There are sustainable management plans in place to protect and improve both Benidorm and the Mediterranean Sea – but the region faces further problems due to increased numbers of tourists.

Key vocabulary

package tour, pollution (visual and noise), GDP

Skills practised in the Activities

◆ Thinking skills: drawing spider diagrams to show the problems tourism causes, and how the problems are being tackled; identifying consequences of pollution; suggesting how the Mediterranean landscape and sea can be protected from increasing tourism

Unit outcomes

By the end of this unit, most students should be able to:

◆ define the terms given in 'Key vocabulary' above;

◆ explain why Benidorm grew into a major tourist resort;

◆ identify the problems tourism has caused in and around Benidorm;

◆ identify the problems pollution causes in the Mediterranean;

◆ say how the Spanish and local governments have been tackling Benidorm's problems;

◆ describe how the Mediterranean Sea is being protected.

Ideas for a starter

1 Tell me ten ways that people have affected the Mediterranean landscape and the Mediterranean Sea.

2 Give out tourist brochures showing package holidays on the east coast of Spain. Ask students to suggest what would attract people to an area like this. Then ask them what effect package holidays might have on the area.

3 Show the photo of the monk seal on page 67 in the students' book, or a similar photo. Tell students that the seal is threatened by pollution in the Mediterranean. Ask: Where does the pollution come from? What type of pollution is it? How does it get there? What can be done about it? Will it get better or worse?

Ideas for plenaries

1 Compare the two photos at the top of page 66 of the students' book. What differences do you notice? How has tourism affected Benidorm?

2 Show photos of the effects of pollution in the Mediterranean. Ask: Is it going to get better or worse? Why?

3 Create an acrostic. Write MEDITERRANEAN down one side of the page. Make each letter the first letter of a word, phrase or sentence about the Mediterranean climate or the Mediterranean region.

4 Take two minutes with a partner and think up one interesting question about human impact in the Mediterranean that we have not covered today. This could produce a good enquiry question which the class could follow through.

Answers to the Activities

1 a Spider diagram should include: poor-quality buildings (visual pollution); bars, clubs and restaurants contribute to noise pollution; waste (sewage systems unable to cope, sewage washed into sea); erosion and damage to beaches; over-use of water (has led to aquifer levels falling, farmland drying out, sea water seeping into aquifers).

b Spider diagram should include: grants provided for clean-up operations; sustainable tourism policies developed; nightclubs to be quieter and have fewer people allowed in; high-quality beaches awarded EU blue flags; bad hotels improved and new ones built.

2 Pollution is a threat to the natural habitat of the entire Mediterranean region – endangering wildlife both in the sea, and around its shores. Tourism both creates pollution and is likely to be adversely affected by it. Pollution threatens the livelihoods of those who make their living from the sea, or from tourism.

3 Students could argue that all Mediterranean countries should agree to the Barcelona Convention to reduce/prevent pollution. Water and energy resources are being overstretched in the Mediterranean and need to be conserved. Tour operators need to start taking sustainability seriously and give tourists information about how the Mediterranean environment can be protected. Sustainable tourism should be developed.

The unit in brief

This unit is about the Arctic tundra – where it's found, what the climate is like, how people cope with the climate and how vegetation has adapted to survive in the harsh climate. In the Activities, students complete a mind map for the tundra, describe the climate at Barrow, Alaska, and compare it with one of the other climates studied.

Key ideas

◆ Arctic tundra regions are found in the Northern Hemisphere in areas with a polar climate, mostly to the north of the Arctic Circle.

◆ Arctic tundra regions are very cold, have low precipitation and have strong, cold winds.

◆ Building techniques, services, cars, etc. have to be adapted to the permafrost and extreme cold.

◆ Indigenous peoples such as the Inuit and Athapaskan Indians may now have a more modern lifestyle, but they still have strong traditions and values.

◆ Vegetation has adapted to survive in the harsh climate.

Key vocabulary

tundra, indigenous, nomadic

Skills practised in the Activities

◆ Geography skills: describing the climate at Barrow; comparing climate graphs

◆ Thinking skills: producing a mind map of the tundra

Unit outcomes

By the end of this unit, most students should be able to:

◆ define the terms given in 'Key vocabulary' above;

◆ describe where Arctic tundra regions are found;

◆ describe the climate at Barrow;

◆ describe the problems that the extreme climate causes for people living in the Arctic tundra;

◆ understand that indigenous peoples such as the Inuit and Athapaskan Indians may now have a more modern lifestyle, but they still have strong traditions and values;

◆ give three ways that vegetation has adapted to survive in the harsh climate.

Ideas for a starter

1 Brainstorm to find out what students know about the tundra already.

2 Show a photo of a typical tundra landscape. Ask students: What can you tell me about the vegetation of this place? What can you work out about the climate? Where is this place? Who can find it in an atlas?

3 Show the climate graph for Barrow. Ask questions about the temperature and rainfall patterns.

4 Mind movie time! You are in the tundra, in Alaska, in the middle of winter – all alone. What can you see, hear and smell? What do you feel? Tell us.

Ideas for plenaries

1 Activity 1 could be used as a plenary.

2 Describe a journey through the tundra. What people, animals and vegetation would you see? What would the weather/climate be like?

3 Close your book. Describe the distribution of tundra regions to your neighbour.

4 Provide students with temperature and rainfall figures of a place in the tundra. Ask them to draw a climate graph using the figures.

5 Make 10-15 statements about climate zones, based on what students have learned so far, some true, some false. Students hold up True or False cards. Where statements are false, ask students to correct them.

Answers to the Activities

1 The mind map should include the following:

Location – Arctic tundra regions are found in the Northern Hemisphere in areas with a polar climate, mostly to the north of the Arctic Circle.

Main features – Very cold. In winter temperatures often drop to at least -25 °C and in summer temperatures rarely go above 10 °C. There's a big temperature range. Precipitation is low. Tundra regions receive less than 250 mm of rainfall a year and tend to get a small amount of precipitation every month. Anywhere north of the Arctic Circle experiences at least one day a year when the sun doesn't set, and one day when the sun doesn't rise. The tundra has strong cold winds blowing at over 48–97 km/h.

Vegetation – plants are low-growing to avoid the strong, cold winds. They have short roots (ground is often frozen) and a short life cycle adapted to the short growing season. In poorly drained areas, mosses and lichen grow.

Traditional life – The Inuit (**indigenous** people living in Arctic coastal areas) traditionally relied on fish, sea mammals, and land animals for food, heat, light, clothing, tools, and shelter. In winter some Inuit chose to live in igloos, while others used snow to insulate their houses made from whalebone and caribou hides. Traditionally Inuit are **nomadic**. Other indigenous peoples live inland, e.g. the Athapaskan Indians of northern Alaska. One group of Athapaskans, the Gwich'in Indians, supplement their diet by hunting caribou.

Now igloos and skin tents have been replaced with a more modern lifestyle and permanent housing.

2 a Very cold. In Barrow the winter temperatures often drop to at least -25 °C and in summer temperatures rarely go above 5 °C. There's a 30 °C temperature range. There's little precipitation – tundra regions receive less than 250 mm of rainfall a year. Barrow tends to get a small amount of precipitation every month. It is very windy, with winds blowing over 48–97 km/h. The winds are very cold when they blow off the Arctic ice cap. Barrow, like Prudhoe Bay, has two months of the year when the sun doesn't set and two months when it doesn't rise.

b Answer will depend on which climate graph is used for comparison.

The unit in brief

This unit is about the effect of resource exploitation in the tundra. Oil production began in Prudhoe Bay in 1977, and the government of the USA is now keen to exploit the significant oil reserves in the Arctic National Wildlife Refuge (ANWR) east of Prudhoe Bay. In the Activities, students consider whether drilling for oil should be allowed in the ANWR and what the effects of drilling would be. They also think about how the development of oil reserves could be managed sustainably.

Key ideas

◆ Oil production began in Prudhoe Bay in 1977, and it is estimated that all the oil will be extracted by 2015.

◆ Building the pipeline which transports the oil from Prudhoe Bay to Valdez (the closest ice-free port) presented special challenges.

◆ The government of the USA wants to exploit oil reserves in the ANWR.

◆ There are strong arguments for and against drilling for oil in the ANWR.

Key vocabulary

There is no key vocabulary in this Unit.

Skills practised in the Activities

◆ Thinking skills: analysing arguments for and against drilling for oil; justifying answer; identifying impacts of drilling for oil; suggesting how oil reserves could be managed sustainably

Unit outcomes

By the end of this unit, most students should be able to:

◆ locate Prudhoe Bay and Valdez on an atlas map;

◆ explain why building the oil pipeline was difficult;

◆ understand why the government of the USA wants to exploit the oil reserves in the ANWR;

◆ understand the arguments for and against drilling for oil in the ANWR;

◆ identify the likely impacts of drilling for oil.

Ideas for a starter

1 Where is Prudhoe Bay? Where is Valdez? What have these two places got to do with each other? Give students five minutes to find out.

2 Show students photos of the American president, the wilderness of the ANWR, an oil drilling platform and an Athapaskan Indian or Inupiat Eskimo. Ask: What's the connection between these photos?

3 What do you know about oil? Come up with ten things you know about it.

Ideas for plenaries

1 Activity 3 could be used as a plenary.

2 The ANWR consists of 19 million acres. The US government wants to use just 2000 acres for oil exploitation. Ask: Why shouldn't they?

3 Ask students: What other examples can you think of where resource exploitation is a major issue? Why is it an issue? What are the consequences of the exploitation? Think global and local scales.

4 Play 'Just a minute' – the topic is oil exploitation in the ANWR. Students have the chance to talk for a minute on oil exploitation without repetition or hesitation. As soon as a student repeats an idea, or hesitates, the next student takes over until the minute is up.

Further class and homework activities

Ask students to find out more about the issues surrounding oil exploitation in the ANWR. Hold a role-play where students take on the roles of those for and against drilling for oil. One student will need to act as chairperson.

Answers to the Activities

1 Students will need to carefully weigh up the arguments for and against drilling for oil in the ANWR. They should be aware that the arguments included in the students' book do not give them the whole picture – but they have to answer the question on the basis of the information available. Their answer will be based on their own personal view and should be justified.

2 Effects on people – The Gwich'in Indians believe that drilling for oil will have a negative impact on their culture and way of life because of the threat to the caribou birthing and nursing grounds (caribou are an important part of their culture).The Inupiat Eskimos believe that drilling will benefit them. They traditionally rely on the land and resources for their physical, cultural and economic well-being. The oil industry at Prudhoe Bay has brought them benefits including jobs, and they believe that drilling for oil in the ANWR will do the same.

Effects on the environment – evidence from Prudhoe Bay suggests that drilling for oil will have a negative effect on the environment. The oil complex at Prudhoe Bay has turned 1000 square miles of fragile tundra into a sprawling industrial zone with a landscape defaced with mountains of sewage sludge, scrap metal, rubbish and over 60 contaminated waste sites which contain, and often leak, acids, lead, pesticides, solvents and diesel fuel. Oil spills from tankers and leaking pipelines also threaten the environment.

Other effects – people argue that wildlife, e.g. polar bears, caribou and birds will be adversely affected – although evidence from Prudhoe Bay suggests that this has not been the case as far as polar bears and caribou are concerned.

3 Students may argue that it is debateable whether oil reserves can be developed and managed in a sustainable way. Whilst the exploitation of oil may help to improve some people's lives, e.g. by providing jobs and wealth, it has a negative effect on others' lives and on the environment – as the evidence from Prudhoe Bay suggests. Oil spills and leaks are bound to happen. Many environmentalists describe the ANWR as an important fragile ecosystem that would be irreparably damaged if opened up to oil companies. If lessons are learned from Prudhoe Bay then development may be managed more sustainably than in the past.

help at a glance

The unit in brief

This unit is about the equatorial climate and tropical rainforests – where they're found, what the climate is like, how the climate affects people living there and how the vegetation has adapted to the climate. In the Activities, students complete a mind map about the equatorial climate and describe the climate.

Key ideas

◆ Places with an equatorial climate are found in the tropics – 5° either side of the equator.

◆ Tropical rainforests are found in places with an equatorial climate.

◆ There are no real seasons in the equatorial climate, but the weather has a daily pattern.

◆ Indigenous inhabitants of the tropical rainforests farm traditionally using shifting cultivation.

◆ Rainforest vegetation has adapted to the climate.

Key vocabulary

equatorial, slash and burn, drip tips, buttress roots, lianas, epiphytes

Skills practised in the Activities

◆ Geography skills: describing the equatorial climate

◆ Thinking skills: producing a mind map of the equatorial climate

Unit outcomes

By the end of this unit, most students should be able to:

◆ define the terms given in 'Key vocabulary' above;

◆ describe where equatorial climates and tropical rainforests are found;

◆ describe the equatorial climate;

◆ understand how people who live in tropical rainforests farm;

◆ explain how rainforest vegetation has adapted to the climate.

Ideas for a starter

1 Show a photo of a tropical rainforest. Ask students to describe what they can see What would it sound like there? What would it smell like?

2 Read out the 'Climate control' text omitting 'in the rainforest' from the 'Daily rhythm' section. Ask students: What type of climate is this? Who can tell me the name of a place with this type of climate?

3 Show a climate graph for the equatorial climate on the board. Ask questions about the temperature and rainfall patterns. Ask why it's so humid in the rainforest and why most rain falls in the early evening.

Ideas for plenaries

1 Give students a photo of rainforest vegetation. Ask them to annotate it to show how the vegetation had adapted to the climate.

2 Close your book. Describe the equatorial climate to your partner. Include real figures (for temperature and precipitation) in your description.

3 You're a reporter for a holiday programme telling people about a holiday in the tropical rainforest. You need to tell people what they will see, what the weather/climate will be like and what kind of clothes they might need. You have five minutes to write your piece for the programme.

4 Indigenous inhabitants in the tropical rainforest farm using shifting cultivation clearing only small areas of rainforest. How else are tropical rainforests used? Draw a spider diagram to show how rainforests are used, in preparation for Unit 4.10.

5 Write down as many words as you can relating to today's work.

Answers to the Activities

1 The mind map should include the following:

Location – Places with an equatorial climate are found in the tropics – 5° either side of the equator.

Main features – no real seasons, but there is a daily pattern to the weather culminating in torrential downpours of rain. Average temperature is about 28 °C. There is a very small temperature range (it can be as low as 2°C). The climate is hot and humid – on average 2000mm of rain falls pa.

Vegetation – Trees are branchless, trunks are tall and thin (to reach the sunlight more easily).Trees look evergreen and shed leaves at any time. Leaves often have drip tips and are waxy to help shed heavy rain. Some trees have large buttress roots for support, others spread their roots over the surface. Lianas climb high into the canopy to reach the sunlight. Epiphytes grow in the branches of the trees

Traditional farming – Shifting cultivation called 'slash and burn'. A small area of rainforest is cleared and burnt. The area is planted with crops. After about 4-5 years the soil has lost its fertility and people move to a new area and begin again.

2 The equatorial climate is hot, wet and humid all year. The average daily temperature is usually about 28 °C. Only occasionally does it go above 35 °C. It never drops below 20 °C. The temperature range may be as little as 2 °C.

Around 2000mm rain falls pa, although it can be more. Precipitation is at its lowest from June to October.

Areas with an equatorial climate are unique in that there are no real seasons. The same weather is repeated on a daily basis:

◆ 3.00 a.m. – dry with intense heat from the sun

◆ 12.00 noon – high temperatures, up to 33 °C, and the formation of cumulus clouds.

◆ 6.00 p.m. – thick cumulonimbus clouds, thunderstorms, and torrential downpours of rain.

The unit in brief

This unit is about the economic exploitation of the tropical rainforest and considers the effects of mining, logging, cattle ranching and peasant farming in the Amazon rainforest. It introduces a variety of sustainable management strategies. In the Activities, students describe the causes of deforestation in the Amazon rainforest; consider which are the most important sustainable management strategies and identify the problems that could be caused by some of these strategies.

Key ideas

- ◆ Current rates of deforestation mean that the world's rainforests might all disappear by 2030.
- ◆ The Amazon rainforest is being destroyed by mining, logging, cattle ranching and peasant farming.
- ◆ Large-scale deforestation breaks the natural nutrient cycle, so rainforests can't regenerate.
- ◆ There are a range of sustainable management strategies which can be adopted to protect the rainforest ecosystem.

Key vocabulary

deforestation, sustainable management, agro-forestry, tree measuring, education, selective logging, afforestation, forest reserves

Skills practised in the Activities

- ◆ Geography skills: describing causes of deforestation
- ◆ Thinking skills: prioritising sustainable management strategies; justifying opinions; identifying problems caused by sustainable management strategies

Unit outcomes

By the end of this unit, most students should be able to:

- ◆ define or explain the terms given in 'Key vocabulary' above;
- ◆ recognise that current rates of deforestation mean that the world's rainforests might all disappear by 2030;
- ◆ describe how the Amazon rainforest is being destroyed;
- ◆ understand why large scale deforestation means rainforests can't regenerate;
- ◆ give examples of sustainable management strategies which can be adopted to protect the rainforest ecosystem.

Ideas for a starter

1. Show a video of deforestation in tropical rainforests. Ask: Why is deforestation happening? How quickly is it happening? What can be done about it?
2. If you used plenary **4** in Unit 4.9, go back to the spider diagram now and see what else the class can add to it. Otherwise, you could produce the spider diagram now.
3. Provide students with a quick wordsearch on the key terms to do with tropical rainforests.

Ideas for plenaries

1 You are an economic adviser to the Brazilian government. Argue the case for the Brazilians to develop the Amazon region as much as possible.

2 As a conservationist write a letter to the United Nations advising them of the problems caused by the loss of habitat in the tropical rainforest.

3 What influence might the loss of rainforest have on climate change?

4 How could countries in MEDCs help in conserving the rainforest? Consider sustainable farming methods and wood from renewable sources.

5 Draw up a table showing different ways of managing the rainforest from national park (no development), agroforestry, logging and replanting, ecotourism, zonation to deforestation (total redevelopment). Include a column for the possible impacts - economic, social and environmental including the global dimension. Use your table to evaluate the ways of managing the rainforest. How should the rainforest be managed so that it is sustainable?

Further class and homework activities

Role-play a meeting about developing the rainforest with different people who contribute their views – loggers, farmers, indigenous Indians, government officials, miners, conservationists, etc. Prepare speeches in advance and then argue the best way to develop the rainforest.

Answers to the Activities

1 Page 74 describes in some detail the range of human activities that are currently exploiting the natural resources of the rainforest. Students should be able to use these ideas but express them in their own words rather than copying. They may choose to design a mind map or other diagram to represent the information. The extension requires students to think of additional activities they have discovered. Some possibilities include:

- the development of transport links such as the Trans-Amazon Highway

- the development of settlements such as Manaus as a result of prospecting for mineral resources such as gold, the growth of the rubber trade and tourism. The citadel Carajas was specifically planned for the workers who operate the opencast mining operation shown in the photograph on page 74.

- construction of dams for reliable water and power supplies.

2 The best sustainable strategies are always integrated. So expect students to choose three ideas that link together very well. For example, forested reserves, selective logging and afforestation. However, students should also be prepared to justify their choices and explain the links. This part should focus on the need to protect the tree canopy and maintain the natural nutrient cycle to ensure the survival of the ecosystem. Forest reserves allow economic activity which facilitates Brazil's continued need to develop, whilst protecting vast areas of virgin forests at the same time.
This particularly applies to preventing the timber trade by illegal loggers. Selective logging is more sustainable, as only mature trees are carefully lumbered for their economic value without causing damage to 20 or 30 other trees in the process. It is a more controlled operation. Afforestation ensures that the tree roots bind the soil together, preventing erosion. The tree canopy

also feeds the nutrient cycle and prevents soil erosion by torrential rains. Expect agro-forestry to be a popular choice also for similar reasons. Students should emphasise the point that one strategy on its own will not be effective, and that the enforcement of sustainable strategies needs to be vigorously applied if they are to work. This is difficult in a rainforest as vast as the Amazon.

3 Slash and burn farming refers to a type of small-scale farming where trees are burnt to clear land for farming and also to provide ash as a source of nutrients. This means that crops can be grown but only for a limited period of time, as the nutrients soon become exhausted and the farmers then have to move to a new plot of land and repeat the process. Consequently, this form of farming practice is regarded as unsustainable as the nutrient cycle is broken, particularly if the scale of the slash and burn increases over larger areas of land. However, students could also argue that, if over small areas of land, the forest could naturally regenerate more easily and therefore be a sustainable practice. It is the scale of the slash and burn which is important in this case.

4 The problems associated with sustainable forest management could include:

- the difficulty of enforcing the strategies and combating the problems presented by groups of illegal loggers, peasant farmers or cattle ranchers seeking to increase their plots of land.

- the time taken for the sustainable strategies to have a positive impact on preserving the forest. For example, tropical hardwoods such as mahogany will take decades to mature before they can be harvested. Society has a tendency to be too impatient for natural regeneration processes.

- the success of programmes such education and ecotourism can never be guaranteed.

The unit in brief

This unit looks at the issue of climate change – what it is, what's causing it, and what the effects might be. In the Activities, students use a graph of global temperatures; explain the process of global warming; consider whether there is enough evidence to say that climate is changing and produce a diagram or table to show the impact of global warming.

Key ideas

◆ The process by which world temperatures are rising is known as global warming.

◆ The enhanced greenhouse effect is causing global warming – human activity is increasing the amount of greenhouse gases in the atmosphere.

◆ Most scientists predict a rise in global temperatures of between 2.5 °C and 10.5 °C over the next 100 years.

◆ The effects of global warming will be felt around the world.

Key vocabulary

climate change, global warming, greenhouse effect

Skills practised in the Activities

◆ Geography skills: producing a table or diagram to show the impacts of global warming

◆ Numeracy skills: reading a graph of global temperatures

◆ Thinking skills: explaining global warming; considering evidence

Unit outcomes

By the end of this unit, most students should be able to:

◆ define the terms given in 'Key vocabulary' above;

◆ explain the process of global warming;

◆ draw a simple diagram of the greenhouse effect;

◆ identify the effects of global warming on both MEDCs and LEDCs.

Ideas for a starter

1 Show students a photo similar to the one on the chapter opener page. Ask: what has this got to do with climate change? (Climate change is threatening polar bears with starvation. The bears' main food source – seals – is becoming less accessible. Seals live on ice but the Arctic sea ice is breaking up earlier, due to climate change, so the polar bears' hunting season is reduced.)

2 Ask students: What is the greenhouse effect? Why are plants grown in a greenhouse? How does it work?

3 You live here in the UK but the date is 2107. How has the climate changed? Tell us what the winter is like; what the summer is like. Is it hotter, colder, wetter or drier than now? How does that affect the way you live, and what you do?

Ideas for plenaries

1 Draw a diagram showing the greenhouse effect. Annotate it to show how human activity enhances the greenhouse effect.

2 Produce a graffiti wall display about the global effects of climate change.

3 Draw 2 speech bubbles on the board. One says: 'We're helping to cause global warming by burning so much fossil fuel.' The other says: 'Global warming is a natural change and nothing to do with us.' Should we:

◆ cut down on using fossil fuels?

◆ carry on as we are?

Point out that most experts now agree we are either causing global warming, or greatly accelerating a natural trend.

4 Choose a student to be in the hot seat. Another student asks him or her a question about climate change. Then nominate two different students (4-6 pairs in total). There's one golden rule – questions cannot be repeated.

5 Do an alphabet run from A-Z, with a word or phrase to do with climate for each letter.

Answers to the Activities

1 The average global temperature has risen about 1 °C since 1861.

2 Global warming is effectively the warming of the planet due to a rise in the average global temperature. It is based on the greenhouse effect, which is an essential process to sustain life on Earth. Incoming radiation from the sun warms the planet and heat escapes back into space. A layer of greenhouse gases, such as carbon dioxide and methane, traps some of the escaping heat - resulting in a warm planet capable of supporting life. Essentially it works like a greenhouse, hence the terminology used. However, the increased burning of fossil fuels and developments in agriculture has increased the atmospheric composition of the greenhouse gases, amplifying the process. Therefore, the planet is steadily warming. This is referred to as global warming and media reports suggest that the planet is 3 °C away from disaster as far as human society is concerned.

Commonly students mention the ozone layer in their responses to global warming and they need to be encouraged away from this confusion.

3 Climate needs to be monitored over sufficiently long periods of time before accurate claims about global warming can be made. The time period for this must be a minimum of 30 years. However, measurements taken on the Greenland ice sheet and Antarctica suggest that ice masses are melting at a faster rate than previously known. Accurate measurements showing an increase in atmospheric carbon dioxide have also been recorded. It would be appropriate for students to state that there is sufficient evidence to promote a cause for concern, only the future and continued careful monitoring will show whether climate is actually changing.

4 Some of the impacts of global warming on MEDC and LEDC nations are highlighted in the table below. The skill is in enabling students to automatically sort information out into appropriate categories.

Impact on MEDCs	Impact on LEDCs
• Decline in national forests due to heat and drought in places like Canada and Russia. • The location of wheat-farming areas changes in the USA to be further north. • Increased storm activity in places like Florida. • Loss of skiing industry in Alpine resorts in Europe. • Oil pipelines and supply damaged in regions like Alaska, USA. • Disappearance of beaches in tourist hot spots, due to sea level change. • Flooding in major economic centres, such as London.	• Extensive flooding in low-lying countries such as Bangladesh and loss of fertile land. The same areas could also be affected by increased storm activity. • Increased desertification in the Sahel. The Sahara desert could also expand further north into the Mediterranean. • Low-lying islands such as the Maldives could disappear. • Water shortages in the Middle East and Africa. • Amazon rainforest damaged by heat and drought conditions.

The unit in brief

This unit is about how we can tackle global warming, both from a government's point of view – which includes looking at the Kyoto Protocol – and from an individual's point of view. In the Activities, students consider why countries need to work together to reduce greenhouse gas emissions and how Scotland can reduce its emissions. They also produce a table to show the resources they used that day and suggest how they could use energy more carefully.

Key ideas

◆ In 1997, over 160 nations signed the Kyoto Protocol – an international agreement on climate change.

◆ The Protocol commits countries to reduce their greenhouse gas emissions to 5% below 1990 levels.

◆ Some countries objected to the Protocol, and the USA pulled out in 2001.

◆ The UK intends to increase production of energy from renewable sources – mainly using wind power.

◆ Individuals can help to tackle global warming by using less energy based on fossil fuels.

Key vocabulary

Kyoto Protocol

Skills practised in the Activities

◆ Geography skills: making a table of activities and resources used

◆ Thinking skills: identifying why countries need to co-operate and suggesting why this is difficult; suggesting how Scotland can reduce greenhouse gas emissions; suggesting ways of using resources more carefully

Unit outcomes

By the end of this unit, most students should be able to:

◆ explain the term given in 'Key vocabulary' above;

◆ explain the purpose of the Kyoto Protocol;

◆ recognise that it is difficult to get all countries to agree to the Kyoto Protocol;

◆ say how the UK intends to increase production of energy from renewable sources;

◆ suggest how individuals (including themselves) can help to tackle global warming.

Ideas for a starter

1 Recap: What is the natural greenhouse effect and why is this necessary for life on Earth? What is the enhanced greenhouse effect?

2 Show some of the best diagrams from plenary **1** in Unit 4.11 to remind students about how human activity increases global warming.

3 Ask: Can individuals act to tackle global warming, or is it just something that governments set targets about? Can you think of any examples of when individual actions made governments and TNCs adapt or change their policies?

Ideas for plenaries

1 Why is international agreement necessary for the reduction of greenhouse gases? Why is it sometimes difficult to get governments to agree targets for greenhouse gas emissions? Why don't they always meet their targets?

2 Why did the government of the USA pull out of the Kyoto Protocol? Do all Americans feel the same way as the government?

3 Provide students with information on alternative energy, such as the article at www.guardian.co.uk/renewable/0,2759,180749,00.html. Ask them to read it and make the case for having more alternative forms of energy in the UK.

4 Create an acrostic. Write GLOBAL WARMING down one side of the page. Make each letter the first letter of a word, phrase or sentence about global warming.

Further class and homework activities

1 Many MEDC governments are considering building more nuclear power stations. How would this help to reduce global warming? What do organisations such as Greenpeace and Friends of the Earth say about this? Find out.

2 Investigate advice that is given to householders by the British government and fuel supply companies about reducing energy use and improving insulation in the home. Make a wall display of the findings. Find out what is done in school to try to reduce energy consumption.

Answers to the Activities

1 a Climate change and global warming is a problem the whole world faces. Some countries currently produce far more greenhouse gas emissions than others, but the situation is likely to change in the future as emissions from developing countries increase. Countries cannot tackle the issue of global warming alone – it's a global problem – countries need to cooperate to find a global solution.

b Different countries have their own agendas, so it's difficult to get them to agree to a common goal, e.g. Australia said that agreeing to the Protocol would cost jobs and damage their industry. Canada wants credits for the 'clean energy' it exports to the USA, and won't agree to the Protocol until this is clarified. Russia argued that its forests absorbed 17 million metric tonnes of carbon a year and that this should be taken into account when considering its emissions reductions (Russia did agree to the Protocol in 2005).

2 Scotland could produce more of its energy from renewable sources e.g. using wind power, HEP, wave power, etc. Increased use of public transport will also help to reduce its production of greenhouse gases.

3 a Answers will depend on student activities and lifestyle.

b The text provides information on using energy more wisely. Individuals can do a number of things including:

* using less electricity in the home (e.g. only boiling as much water as you need instead of the whole kettle; turning computers and TVs off instead of leaving them on standby) and not wasting heat. In the UK, in 2005/06, a campaign started to encourage all home owners to reduce their energy bills by 20%, saving both money and carbon dioxide emissions.
* walking or cycling instead of using a car.
* sharing transport instead of using individual cars for all journeys.
* using more public rather than private transport

5 Settlement

chapter overview

The big picture

These are the key ideas behind this chapter:

◆ A settlement's site is the land it is built on. A range of factors were important for the original sites of settlements.

◆ All settlements have at least one function, most have several.

◆ A settlement hierarchy is a ranking of settlements with the largest at the top.

◆ Urbanisation is increasing.

◆ Land use changes across cities. The Burgess model shows land use in MEDC cities.

◆ The cost of land and accessibility create different urban zones.

◆ Cities suffer from problems of deprivation but these problems can be tackled.

◆ Other urban problems include traffic and urban sprawl.

Note that the students' version of the big picture is given in the students' chapter opener.

Chapter outline

Use this, and the students' chapter opener, to give students a mental roadmap for the chapter.

5   Settlement As the students' chapter opener, this unit is an important part of the chapter; see page 11 of this book for notes about using chapter openers

5.1   Site Factors important in choosing a site for a settlement

5.2   Settlement functions The main activities that happen in settlements, and Oxford's functions

5.3   Settlement hierarchies How settlements are ranked in order of size and importance, and sphere of influence

5.4   Urbanisation Increasing numbers of people are living in urban areas in MEDCs and LEDCs

5.5   You are entering the twilight zone How land use changes across a city, and the Burgess urban model

5.6   Urban zoning: why does it happen? Urban zones develop due to cost of land and accessibility

5.7   Urban problems in MEDCs Inner cities can have high levels of deprivation; redevelopment can bring a higher quality of life back to inner cities

5.8   Traffic – everybody's problem How Edinburgh is trying to solve its traffic congestion problems

5.9   Urban sprawl Counter-urbanisation; problems of building at the rural-urban fringe; green belts

5.10   Edinburgh – green belt developments The pressures on Edinburgh's green belt

chapter overview

Objectives and outcomes for this chapter

Objectives	Outcomes

Most students will understand:

- That there are different factors which were important in choosing the original site of a settlement.
- What settlement functions are; that most settlements have several functions.
- That settlements are ranked in order of size and importance, and that settlements have a sphere of influence.
- What urbanisation is, and the connection between urbanisation and industry.
- That land use changes across cities.

- That the cost of land and accessibility create different urban zones.
- That cities suffer from deprivation, but that these problems can be tackled.

- Why Edinburgh has a traffic congestion problem.
- That urban sprawl and counter-urbanisation create problems and put pressure on green belts.

Unit

5.1

5.2

5.3

5.4

5.5

5.6

5.7

5.8

5.9, 5.10

Most students will be able to:

- Give at least six factors that would have influenced a choice of site.
- Define settlement function; describe the functions settlements of different sizes might have.
- Describe a settlement hierarchy; explain how settlements are ranked; define sphere of influence.
- Define urbanisation; explain the connection between urbanisation and industry.
- Describe the pattern of land use in MEDC cities; draw the Burgess urban model.
- Explain how the cost of land and accessibility create different urban zones.
- Describe the deprivation in inner cities in the UK; identify the reasons for deprivation in the inner cities; describe how the inner cities can be improved.
- Explain why Edinburgh has a traffic problem; describe the measures being taken to solve the problem.
- Give reasons why people are moving out of cities; give reasons why some people don't want more building at the rural-urban fringe; give an alternative to building at the rural-urban fringe; give an example of development in an area of green belt.

These tie in with 'Your goals for this chapter' in the students' chapter opener, and with the opening lines in each unit, which give the purpose of the unit in a student-friendly style.

Using the chapter starter

The photo on page 80 of the *geog.SG* students' book shows a view of Haines (population 2800), in the Alaskan panhandle. The area was a route centre for native tribes. A mission and school were built in 1881. But Haines really got its start soon after as a gold mining and fishing town. The army built a fort in 1904. Logging became important.

The fort closed in 1946. The last fish cannery closed in 1972 due to declining fish stocks. Logging declined. Tourism is now the town's most important source of income. Wildlife and outdoor activities attract thousands of visitors. Many come in the autumn, when 4000 bald eagles gather to feast on spawning salmon. Haines is surrounded by mountains and is close to 20 million acres of protected wilderness.

Haines was named by 'Outside Magazine' as one the top twenty places to live. There's little crime, little unemployment, few social problems, and living standards are quite high. But Haines is quite remote, mosquitoes can be a nuisance in summer, and winters are long and cold.

The unit in brief

This unit is about the things people looked for when choosing a place to settle. In the Activities, students use the knowledge gained from the unit to describe the site of their own school. They decide which site factors apply to a range of different settlements shown in the photos in this unit; and, using the maps of London at different periods, think about why London's site was first chosen, and how London changed and adapted to become a site with many advantages.

Key ideas

◆ A settlement's site is the land it is built on.

◆ Situation describes where the settlement is located in relation to other settlements, communications, physical features, etc.

◆ Sites were chosen for different reasons.

◆ The sites where people chose to locate their settlement changed over time.

◆ Now settlements can develop almost anywhere, if there's a strong enough economic reason.

Key vocabulary

site, situation, resources, wet point, dry point, defence, shelter, aspect, gaps, trade, route centre, economic

Skills practised in the Activities

◆ Geography skills: describing the site of the school

◆ Thinking skills: deciding about site factors for different settlements; thinking about sites, change and adaptation

Unit outcomes

By the end of this unit, most students should be able to:

◆ define the terms given in 'Key vocabulary' above;

◆ understand the difference between site and situation;

◆ give six different location factors for choosing different sites;

◆ explain why the sites people chose to locate their settlement on might change over time;

◆ understand that now settlements can develop almost anywhere if there's a strong enough economic reason.

Ideas for a starter

1 Brainstorm: What do we mean by the terms settlement, site and situation? Why have settlements developed in different places?

2 Ask students: Think about our own settlement (or nearest large town/city). What things do we use there (e.g. shops, leisure facilities, etc.)?

3 Show a photograph of a city in the UK. Ask the students about their perception of urban areas. What are the three words that they think of when they see the photograph? Are these positive or negative views of the urban area?

Ideas for plenaries

1 Use an atlas and choose four towns and cities across the UK. Why did they develop in these places? What factors were important?

2 Provide students with an OS map of the local area. Ask them to choose four settlements (but not their own). What factors led to the development of these places? Why were these sites chosen? What evidence can you find on the map?

3 Use the OS map of your local area. Draw an annotated sketch map to show the advantages of the site of your own settlement.

4 You have 30 seconds. Tell your neighbour as many location factors (wet point, dry point, etc.) as you can for the site of a settlement.

Further class and homework activities

Investigate one major city and try to find out which factors were important in its development. Why was this site chosen?

Answers to the Activities

1 Answers will depend on individual schools. Ensure students are using the correct terms, and are clear about the difference between site and situation.

2 A is Ely – a dry point site in the middle of the fens, with good farmland created from the drained marshes.

B is Durham, a wet point site located within a river meander, so that the town is almost entirely surrounded by water; it's an excellent defensive site, since it's also on a steep hill. It's also a bridging point for trade, and the river provides water for households and small industries.

C is a coastal hilltop defensive site, with good all-round observation. It is also located on the Riviera, which attracts thousands of tourists each year.

D is a Dogon settlement in Senegal. The houses are sheltered under a cliff overhang from the sun, and are high up, again for defence.

E is a typical alpine valley, with an adret, or sunny side, usually south-facing.

F is another defensive site, located next to a road for trade at the end of a pass between hills.

3 a It's on a major river - good for water transport, which was much faster than roads at the time. It provided an easy route to the coast, and inland for trade; and a safe harbour. The Thames also supplied water, as did the streams to the west and centre of the site. It was also a defensive site, with one side protected by the river.

b By Tudor times, London was the lowest bridging point of the Thames (the one closest to the sea). Routes ran to the bridge, and London became a route centre. Victorian London was obviously much bigger, with seven bridges over the river. It also had railways radiating out from the centre.

The unit in brief

This unit is about settlement functions – the things that happen in settlements and the role that settlements play in our lives. In the Activities, students use the OS map extract of Oxford included in this unit to identify the city's main function and to find evidence of other functions. They describe the main function of the place where they live and of any nearby settlements.

Key ideas

◆ The term 'function' describes the main activities of a settlement.

◆ All settlements have one or more functions. Most have several.

◆ Functions can change over time.

◆ Land use in urban areas reflects the primary functions of the settlement.

Key vocabulary

function, route centre, hamlet, village, market town, primary function

Skills practised in the Activities

◆ Geography skills: explaining the term 'urban function'; using an OS map to provide evidence of functions and giving grid references; describing the functions of own settlement and nearby settlements

Unit outcomes

By the end of this unit, most students should be able to:

◆ define the terms given in 'Key vocabulary' above;

◆ explain what an urban function is;

◆ identify Oxford's functions using map evidence and give grid references;

◆ understand that functions can change over time.

◆ understand that land use in urban areas reflects the primary functions of the settlement.

Ideas for a starter

1 Recap: definitions of site and situation, and location factors for settlements from Unit 5.1.

2 Ask several students to report back on the city they investigated in Unit 5 1.

3 Brainstorm to find out what students know about settlements of different sizes. You are looking for terms such as hamlet, village, etc. Develop to ask what they know about 'functions'. Ask for examples of functions and draw a spider diagram on the board.

Ideas for plenaries

1 Provide students with a variety of OS maps. Ask them to look at a range of towns/cities and try to decide on the range of functions of these places. They should provide evidence of these functions and grid references.

2 Use an OS map of your local area. What is the current function of your settlement? Has it changed over time? What was its main function when it originally developed? What was it 100 years ago? 50 years ago?

3 Quick-fire test. Ask different students to give examples of functions; types of settlements; typical functions in settlements of different sizes; definition of function etc.

Further class and homework activities

Urban areas change their functions over time. Investigate London Docklands and find out how functions have changed here over the last 100 years. Produce a 'before' and 'after' picture/wall display of Docklands. There are a number of websites that could help.

Answers to the Activities

1 a A job or a process which happens in a town or city.

b Education. The university, shown as coll or colleges on the map. The university accounts for almost 15 000 of Oxford's 115 000 population.

c Encourage students to use headings:
Shopping, including a daily or weekly market (retail)
Business (commerce)
Religion
Entertainment and leisure
Industry
Residential
Route centre

d Students should give named examples of places with the functions mentioned in the Activity where possible.

2 The colleges in the city centre. [Students could give a number of appropriate grid references.]

3 For cultural centre – museums, overall size of city; for industrial centre – motor works; for county town – County Hall; for historic town – the colleges, various antiquities. e.g. Godstow Abbey or castle remains; for recreational centre – golf courses, country park, the river, museums; for route centre – major roads converging on ring road, railway; for tourist town – the colleges, riverside pubs; for residential settlement – areas of housing [Students could give a number of appropriate grid references.]

The unit in brief

This unit is about settlement hierarchies – how settlements are ranked in order of size and importance. In the Activities, students explain the terms hierarchy and sphere of influence and give examples of settlements of different sizes in the UK. They make and explain the connection between the size of towns and the urban percentage of a country's population. They analyse an extract from an atlas map in terms of settlement hierachy and sphere of influence.

Key ideas

◆ A settlement hierarchy is a ranking of settlements.

◆ Different countries define different settlement types in different ways.

◆ Settlement hierarchies are pyramid-shaped. The larger the settlement, the fewer there are.

◆ Settlements are ranked by population size, number and variety of functions, the distance between a settlement and the nearest one of similar size, and the sphere of influence.

◆ The sphere of influence is the area served by a particular settlement.

Key vocabulary

settlement hierarchy, ranking, sphere of influence, village, town, city

Skills practised in the Activities

◆ Geography skills: explaining the terms 'hierarchy' and 'sphere of influence'; naming UK settlements of different sizes; analysing an atlas style map

◆ Thinking skills: giving reasons; making and explaining connections between population sizes of towns and urban populations as a percentage of a country's total population

Unit outcomes

By the end of this unit, most students should be able to:

◆ define the terms given in 'Key vocabulary' above;

◆ understand that a settlement hierarchy is a ranking of settlements;

◆ recognise that different countries define different settlement types in different ways;

◆ explain why settlement hierarchies are pyramid-shaped;

◆ describe how settlements are ranked.

Ideas for a starter

1 Ask: Who can remind me about settlement functions? Who can give me five functions of a settlement?

2 Ask: What does the term hierarchy mean? Try to draw out the organisational framework of your school starting at the top with your headteacher and students at the bottom. Establish that this ranking is called a hierarchy and you can rank settlements in the same way based on size and function.

3 Ask: What's a village? What's a city? What size would they be? (Note that different countries have different ideas about this.) What functions will you find in places of different sizes?

Ideas for plenaries

1 Conduct a survey amongst your classmates to find out where people go for certain goods when out shopping. Draw a map showing the most popular settlements for shopping. This should help you see a local shopping hierarchy.

2 Where do people in your class live? Plot where they live on a map of your local area. This is the catchment area of the school – another way of showing a sphere of influence!

3 What other organisations have hierarchies? Think of three examples and draw the hierarchy with the most important person/thing at the top.

4 Use the local OS map to produce a settlement hierarchy for your area. Count settlements of different sizes. (The font size used for names is a clue to settlement size.) Then sketch a hierarchy triangle and write in the numbers of settlements of different sizes. Is there a pattern?

5 Make a graffiti wall of what students have learned today.

Further class and homework activities

Choose a number of different sized (by population) local settlements. Choose a number of different services ranging from small convience goods to larger comparison goods. Find out how many of the services the settlements offer (use *Yellow Pages*). Add up the totals and plot them on a graph against the population sizes of the settlements. Is there a pattern?

Answers to the Activities

1 A hierarchy is a ranking, usually with the biggest, the most important, or the best at the top, and the smallest, least important, or worst, at the bottom – like the football premiership, or Wimbledon seeding.

2 The pyramid is a good way of showing a settlement hierarchy because:

(i) the width of each step in the hierarchy represents the number of settlements in each category; there is usually one primate city at the top, but thousands and thousands of isolated buildings/farmsteads;

(ii) it gives a good picture of how each level feeds into the layer above.

3 UK conurbations include Greater Manchester, the West Midlands, Tyneside, and Teesside.

4 Students might think that the more urbanised a country is, the bigger a settlement has to be to be called a town, city, and so on. If the students rank the two columns, it is clear that the rankings don't fit. Japan takes 30 000 people to be a town, but has the second to last percentage urban population. Stress that population size is not the only way to rank a settlement; the number of urban functions might also be significant.

And there are oddities: Woodstock, Oxon, has a population of 2100 and is a town, but Kidlington, also Oxon, has a population of 17 500 and is a village. It doesn't want to upgrade its title because it's in the Guinness Book of Records as Britain's largest village.

5 The sphere of influence is the area surrounding a settlement where people who live within that sphere of influence depend on the settlement for education, employment, retailing, and finance. Many people in the sphere of influence work in the central place, so we can say that they 'serve' the settlement.

6 top level London;
then Bristol;
then Swindon and Reading;
then Chippenham, Newbury;
then Marshfield, Calne, Avebury, Marlborough, Hungerford, Thatcham

7 London. (It's our primate / capital city.)

8 a Bristol
b Oxford
c Swindon (probably – although it looks a smaller town, it serves a wider area)
d Salisbury (there are no other large towns near it)

The unit in brief

This unit is about urbanisation – the increase in the percentage of people living in urban areas. It looks at urbanisation in MEDCs and LEDCs. In the Activities, students draw a cumulative line graph of urban population as a percentage of the total population. They identify and explain which types of countries urbanised first and make the connection between industrialisation and urbanisation. They also identify trends in the changing locations of the world's megacities, and assess whether this trend will continue into the future.

Key ideas

◆ Urbanisation means a rise in the percentage of people living in urban areas.

◆ European and North American MEDCs urbanised during the late eighteenth and nineteenth centuries.

◆ The pattern of urbanisation in LEDCs has some similarities with what happened in MEDCs, but there are some big differences.

◆ Urbanisation is happening a lot later in LEDCs (only since the 1950s). It is also happening a lot faster.

Key vocabulary

urbanisation

Skills practised in the Activities

◆ Geography skills: explaining the term 'urbanisation'; drawing a cumulative line graph from data in a table; marking megacities on a world map and describing the changes shown

◆ Thinking skills: identifying and explaining which types of countries urbanised first; making the connection between industrialisation and urbanisation; explaining the causes of urbanisation; explaining reasons for future trends in urbanisation

Unit outcomes

By the end of this unit, most students should be able to:

◆ define the term given in 'Key vocabulary' above;

◆ explain why European and North American MEDCs urbanised in the late eighteenth and nineteenth centuries;

◆ describe the similarities and differences in urbanisation between LEDCs and MEDC;

◆ give evidence for the speed of urbanisation in LEDCs.

Ideas for a starter

1 Recap: the concept of hierarchy; urban hierarchies (related to the size of the sphere of influence).

2 Ask: What do you know about urbanisation? Who can tell me when people began moving to cities in the UK in large numbers, and why?

3 Ask: What are the factors that would make a family move from the countryside into the town? What are the factors that might prevent this movement? Draw a table on the board of factors.

4 Imagine this. It is 1805. You are 10 years old and you live in the country, but you're about to move to a mill town. Tell us why, what you think it will be like, and what you will do there.

Ideas for plenaries

1 Conduct a quick survey amongst the students in your class. Ask: Who has moved? Where have they moved from/to? What are the reasons for movement?

2 What percentage of the world's population lived in urban areas in 2000? How will that figure change by 2030?

3 What are the similarities in urbanisation between LEDCs and MEDCs? What are the differences? Tell your neighbour.

4 Investigate one of the mega-cities from the list on page 89. What are the main reasons for the growth of the city?

5 Sum up what you have learned today in 30 words.

Answers to the Activities

1 Urbanisation is a rise in the percentage of people living in urban areas (in comparison with rural areas). It's not the same as urban growth. If urban areas grow by 10%, but rural areas grow by the same percentage – or greater – then urbanisation hasn't happened.

2
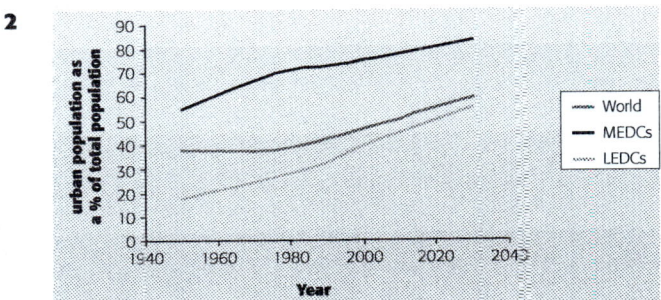

3 MEDCs urbanised first because they industrialised first.

4 Factories pulled people into jobs, these workers brought their families, and other workers moved to the city for jobs in services – like education or shops – to serve the industrial workers.

5 It's not easy to say that one of these events caused the other. If each agricultural worker produces more food, then rural workers are freed up and can move to the industrial towns. But the farmers' production is increased because of new industrial technology, like seed drills. And the surplus workers, wouldn't have any jobs to go to if there had been no industrialisation. The two factors had to come together.

6 c In 1975, there were five mega-cities: two in MEDCs (40%) and three in LEDCs. By 2000, there were nineteen mega-cities, but only four in MEDCs (21%). So the number of mega-cities has risen, but the percentage of MEDC mega-cities has fallen.

d Yes. Most MEDCs depend less on industry; more and more, goods are made in LEDCs. China, for example, is running out of workers, even though thousands of Chinese have moved to factory jobs in cities, so factory managers are recruiting workers from really distant areas, like Tibet. On top of this, services, like electricity or piped water, are poor in LEDCs, so the cities have a bigger 'pull'. You could look ahead to counter-urbanisation, on page 98 of the students' book.

The unit in brief

This unit is about land use, and how it changes across a city. In the Activities, students draw a sketch of a photo of Cardiff and label the land use zones. They have to match sentences describing activities with land use zones, and write some of their own. They write an account of a journey from a town or city centre that they know to the rural-urban fringe and suggest how twilight zones could be improved.

Key ideas

◆ Land use is what the land is used for.

◆ Many cities have a similar pattern of land use: CBD; the inner city (twilight zone); the inner suburbs; the outer suburbs; the rural-urban fringe.

◆ An urban model is a simplified diagram of the way land is used. The Burgess model is the easiest to understand and apply.

◆ Looking at transects helps us to understand how and why a city changes from the centre to the edge.

Key vocabulary

land use, Central Business District (CBD), inner city (twilight zone), inner suburbs, outer suburbs, rural-urban fringe, urban model, transect

Skills practised in the Activities

◆ Geography skills: drawing and labelling a sketch of a photo of Cardiff to show the land use zones

◆ Literacy skills: writing sentences to match land use zones; writing an account of a journey from a town or city centre to the rural-urban fringe

◆ Thinking skills: matching sentences (describing activities) with land use zones and explaining answers; suggesting how the twilight zone could be improved

Unit outcomes

By the end of this unit, most students should be able to:

◆ define the terms given in 'Key vocabulary' above;

◆ understand that land use is what the land is used for;

◆ explain the pattern of land use from the CBD to the rural-urban fringe;

◆ explain what an urban model is;

◆ describe a transect for a city or town that they know from the centre to the rural-urban fringe.

Ideas for a starter

1 Ask a few students to give a short report on their investigation into one of the mega-cities.

2 Show photographs of different parts of the urban area. Ask students to make suggestions as to where they would be found in the town/city. Ask for reasons why they would be found there.

3 Use Activity 3 as a starter.

Ideas for plenaries

1 Ask one student to draw the Burgess urban model on the board (with instructions from the rest of the class if necessary). Another student answers questions about it, posed by the rest of the class, by pointing to the correct area on the model. (For example: where are the houses likely to have the biggest gardens?)

2 Provide students with an OS map of the local area. Ask them to draw a sketch map showing land use in their settlement (or in the nearest town if their settlement is too small). Mark in the main roads, rivers, railway etc. Shade in the different land use zones.

3 If you did not use Activity 3 as a starter, it could be used as a plenary.

4 Prepare an odd-one-out for your partner on what you have learned today.

Further class and homework activities

1 Visit the UK census website and collect information about different parts of Cardiff, or another city/town of your choice. You will need to have some postcodes available. Analyse the information and try to draw some conclusions about the variation across the city/town.

2 Collect information from local estate agents about houses on the market and make a wall display showing the property and the locations. Draw concentric circles on the map and try to calculate the average price in each of the circles. What happens to the price of property as you move from the town centre? What happens to the size of the plots (maybe measured in garden size and/or number of bedrooms)?

Answers to the Activities

1 Students should be able to produce a sketch showing the CBD and identifying the inner city, inner suburbs, outer suburbs, and rural-urban fringe.

2 a A outer suburbs   B CBD   C inner city   D inner suburbs
E rural-urban fringe

b Answers should focus on: house size, street width and shape, provision of green space, and distance from CBD.

4 Refer to the inner city photo on page 90, which mentions gentrification – people 'doing up' old housing. The savings they make on commuting time and cost mean that they can afford to do up old housing. Some middle class Britons would never buy a brand new house. Then there are big redevelopment projects, like docklands in London, Liverpool, Gateshead, Glasgow, and Cardiff Bay. Urban regeneration was a key part of London's 2012 Olympic Games bid.

The unit in brief

In this unit students learn why urban zones develop, and what's happened in the CBD. In the Activities, students use a graph showing rent against distance from the CBD to explain the effect of accessibility on the cost of land. They explain why different activities locate in different parts of the city, and identify and explain the anomaly in land prices shown on the graph. They draw a sketch of a photo of the centre of Bristol to identify land issues.

Key ideas

◆ Urban zones develop due to the cost of land and accessibility.

◆ Different activities/businesses can afford to pay different amounts and therefore locate in different zones.

◆ Urban zones develop as settlements grow.

◆ The CBD is now less accessible than it used to be.

◆ Out-of-town locations may cost almost as much as those in the CBD, but they are more accessible.

Key vocabulary

urban zones, accessibility

Skills practised in the Activities

◆ Geography skills: explaining the terms 'urban zone', 'accessibility'; drawing and labelling a land use sketch from a photo

◆ Numeracy skills: analysing a line graph

◆ Thinking skills: explaining the effect of accessibility; explaining why different activities locate in different parts of a city; identifying and explaining anomalies

Unit outcomes

By the end of this unit, most students should be able to:

◆ define the terms given in 'Key vocabulary' above;

◆ understand that urban zones develop due to the cost of land and accessibility;

◆ explain why different activities/businesses locate in different zones;

◆ understand that urban zones develop as settlements grow;

◆ explain why the CBD is now less accessible than it used to be and why shops locate in out-of-town locations.

Ideas for a starter

1 Ask students to draw the Burgess model from memory and label it.

2 Ask students to think about three locations – crossroads, T junction, and between junctions. If you were a shop-owner which location would you want to develop? Which would be the most accessible? How many routes feed into the location? Now apply this to the urban area. Which will be the most sought-after location? The town centre will be the dominant crossroads in town.

3 Show photos of a range of land uses – large retail store with car park; small specialist shop; coffee shop; small office, e.g. accountants, solicitors; small factory; terraced housing; large detached house etc. Draw the Burgess model on the board and ask students which zones each type of land use would locate in (and why – you are looking to elicit accessibility and cost of land).

Ideas for plenaries

1 Provide students with information about towns which have encouraged development, such as Reading through the Oracle Centre. How have these shopping centres changed the town centre?

2 Provide students with information on out-of-town shopping centres such as Meadowhall, Sheffield (see: www.meadowhall.co.uk/). Ask them to compare these with examples such as Reading.

3 Show photos of the CBD and ask the students to judge the scenes for attractiveness. litter, signage, other management – pedestrian-only roads, street furniture, etc. Then show photos of out-of-town shopping centres, and ask them to compare them.

4 Use your own town or nearest large settlement. What shops/businesses are still in the town centre/CBD? What shops are found on out-of-town retail parks?

5 If starter 3 was not used, it could be used as a plenary.

6 Make 10-15 statements about what students have learned so far, some true, some false. Students hold up *True* or *False* cards. Where statements are false, ask students to correct them.

Further class and homework activities

Do some fieldwork in the CBD and survey the types of shops there. Are there more comparison goods shops? Measure the number of shoppers going into a variety of shops in a number of locations. Are the busiest shops at the major junctions in the CBD? This information could be presented as a wall display.

Answers to the Activities

1 a An urban zone is a distinct area of a town or city – refer students to page 90.

   b Accessibility means being able to get somewhere.

2 a Make sure that the students use the costs and distances shown on the graph. Rents fall from £130/m^2 at the CBD to £40/m^2 2000 metres from the CBD. After that, they rise, to £120/m^2 5000 metres from the CBD, but then fall dramatically to £40/m^2 6000 metres from the city centre.

   b Land is expensive in the city centre because this is where bus, tram and rail routes meet. (It's also expensive here because there isn't much of it.) Peripheral, out-of-town sites are usually located near ring roads, giving easy access for motorists.

   c HMV is happy to pay high city centre rents because this zone has large numbers of pedestrians out shopping. (They will also want to locate in the CBD because competitors locate there, too.)

3 a Coffee shops get custom from shoppers and workers in the city centre, so they can pay high rents. Solicitors don't get as many walk-in customers, and they can't afford their own car parks. Factory owners used to want to be near the centre, but these days they're more interested in cheaper land, better access and free parking. Land is cheaper in the suburbs, so richer people can buy bigger houses.

   b It's all about bid-rent; who can pay high prices, and who needs to be near the CBD. Most chain stores don't want a location more than about 250 metres from the areas of peak pedestrian flow. Marks and Spencers stores are usually located at the point of peak pedestrian flow. Others 'trade off' inner locations for cheaper land.

4 a The anomaly in the graph is the high cost of land at the periphery.

   b This is explained by the availability of land, and access to ring roads.

   c The CBD is expensive and congested. Businesses often want to provide car parks for their workers and their customers – but city-centre car parking is very expensive. Traffic congestion causes air pollution, and the crowds of people drop chewing gum and other kinds of litter. At night, CBDs are often full of young people going to pubs and clubs – refer your students to page 98 – 'My town's under attack'.

The unit in brief

This unit is about deprivation in inner cities, and about how the inner city can be improved. In the Activities, students suggest reasons why the inner city declined; list the problems of the inner city for industry and people, and suggest where new industry would locate. They also get to redevelop an inner city area of their own, using a plan of an inner city area included in the unit. Finally they prepare a presentation about their planned redevelopment.

Key ideas

◆ The inner city has high levels of deprivation and is the poorest zone of most MEDC cities.

◆ In the second half of the twentieth century, many of the original inner city factories closed and unemployment rose.

◆ Unemployment, poverty and lack of opportunities can result in high crime rates.

◆ There has been a lot of redevelopment in the UK's inner cities. Brindley Place in Birmingham is one example of large-scale redevelopment.

◆ Redevelopment is one way of bringing a higher quality of life back to inner city areas.

Key vocabulary

inner city, deprivation, redevelopment, social housing, renewal

Skills practised in the Activities

◆ Geography skills: copying a plan of an inner city area and annotating it to show how it could be redeveloped

◆ Literacy skills: preparing a presentation of the redevelopment plan – if presented using PowerPoint would become an ICT skill

◆ Thinking skills: identifying and explaining the most deprived areas of cities; giving reasons for the decline in inner cities; listing the problems of inner cities; suggesting where industry has moved to

Unit outcomes

By the end of this unit, most students should be able to:

◆ define the terms given in 'Key vocabulary' above;

◆ explain why the inner city is the poorest zone of most MEDC cities;

◆ recognise that unemployment, poverty and lack of opportunities can result in high crime rates;

◆ explain how redevelopment can bring a higher quality of life back to inner city areas, and describe one example of redevelopment.

Ideas for a starter

1 Recap: developments in the CBD and out-of-town shopping centres.

2 Ask students: What is meant by the term deprivation? Ask them to suggest variables that could be used to produce an index of deprivation in inner cities.

3 Show photographs of deprived urban areas in MEDCs. Ask students to comment on the deprivation they observe.

Ideas for plenaries

1 Work in pairs to write interviews and responses for people who might be affected by redevelopment or renewal of their area. Try to think of a range of people – old people, students, first-time buyers, etc.

2 Provide students with OS maps. Look at locations within a town and try to identify which urban zone the area belongs to. Use the patterns of buildings and streets as clues.

3 Add to the variables listed in starter 2. Classify them under these headings: Economic stress, Social stress, Housing stress, Environmental stress.

4 Think of an inner city area you know which needs redevelopment. Prepare a two-minute sound-bite for a local radio show. Describe the area, explain why it should be redeveloped, how it should be redeveloped, and who will benefit.

5 Tell me the three most important things you have learned today. Now tell me another three things which are interesting, but less important.

Further class and homework activities

1 Identify and investigate an inner city redevelopment or renewal scheme.

2 Collect information from the census at ward level for your (or any other) town to draw population pyramids. Annotate the pyramids, pointing out the main characteristics of the population. Locate them on a map of the town. This could be made into a wall display.

Answers to the Activities

1 Deprivation – not having good housing, jobs, or schooling – is high in the inner city because houses are old, and not well built.

2 a Inner cities declined because of:
- congestion;
- pollution (litter, noise, many people renting so they don't improve the houses);
- and crime (lots of houses close together, especially student areas, filled with laptops and i-pods) so better-off people move out, and low-paid or unemployed workers move in.
- many of the industries closed down due to foreign competition.

b Industry relocated from the inner city because it's dirty, run-down, and crowded.

c
- Congestion – it's difficult to manoeuvre lorries in narrow streets.
- Cost – land is still relatively expensive.
- Access – there is little space for parking.

d To the periphery.

3 Lack of living space; Adele lives in a shared house
High crime levels
Drug dealers and users
Lack of open space
Poor schools
Low-skilled, low-paid jobs

5 Presentations might include:
- providing a mix of more expensive and social housing;
- 'greening' some of the area: trees, parks, solar fuel, recycling points;
- retaining and refurbishing important old buildings;
- that redeveloping inner cities increases the number of better-off people living there, and they pay more local taxes.
- that Inner city redevelopment creates jobs for people living elsewhere in the city. It also gives people a sense of pride in their city.

The unit in brief

This unit looks at Edinburgh's transport problems, and at how the City Council is trying to solve the problems. In the Activities, students explain why Edinburgh has a congestion problem; describe what the Council has done to improve public transport and consider why the tram system may not be completed.

Key ideas

◆ Edinburgh's transport problems stem from a variety of causes, including: increasing numbers of commuters entering the city; increasing levels of car ownership and a road system which converges on the city centre.

◆ Three large park and ride areas have been created (with another two being built) which have frequent buses into the city centre.

◆ Improvements to the trains include opening new stations, building longer platforms and building park and ride areas to make the stations more accessible.

◆ A new tram system is due to come into operation in 2010.

Key vocabulary

There is no key vocabulary in this unit.

Skills practised in the Activities

◆ Thinking skills: identifying reasons for Edinburgh's traffic problems; describing improvements to public transport; suggesting reasons why the tram system may not be completed

Unit outcomes

By the end of this unit, most students should be able to:

◆ describe the causes of Edinburgh's transport problems;

◆ say how Edinburgh City Council is trying to solve the transport problems;

◆ describe the changes to the public transport system:
  – buses (park and ride schemes),
  – trains (opening new stations, building longer platforms and building larger park and ride areas to make stations more accessible),
  – trams (new system coming into operation by 2010).

Ideas for a starter

1 Record the local traffic news from the radio and play it to students. Use this as a basis for a discussion on 'Increasing traffic in the UK'.

2 Brainstorm to find out why students think traffic is such a problem – not just in Edinburgh, but everywhere.

3 Show a photo of gridlocked traffic. Create a spider diagram on the board of the problems caused by traffic.

Ideas for plenaries

1 Work in small groups. What other ways are there of reducing traffic congestion and managing traffic? Here are a few ideas: multiple occupancy lanes on roads, free public transport, pollution monitoring, restrictions on private transport entering the city, etc. Complete a table of their advantages and disadvantages.

2 Obtain information from the local planning office on the flow of traffic in the local area and provide it for students. Ask them to show the information diagrammatically, and suggest how the traffic flows could be managed.

3 Write a diary of your own and your family's personal transport over the course of the last week. Is your family dependent on the car for transport? Could this dependency be reduced? How? If your family is dependent on the car, how would family life be different if you had to rely on public transport?

4 Describe a recent scheme in your area to manage traffic. How successful has it been?

5 Create an acrostic. Write TRAFFIC down one side of the page. Make each letter the first letter of a word, phrase or sentence about traffic.

Further class and homework activities

Carry out fieldwork measuring the volume of traffic around your town. Sample the traffic (you can't count it all), and get different people to count different types of vehicle. Present the data as a series of flow diagrams. Suggest how the flow of traffic could be improved.

Answers to the Activities

1 Edinburgh has traffic congestion problems for a number of reasons: the increasing numbers of commuters travelling into the CBD every day; car ownership has risen faster (57%) than in the rest of the UK (29%); there is a radial road pattern meaning traffic converges on the CBD; there is no inner ring road, so all traffic not taking the outer city bypass is funnelled through the CBD.

2 The council has built three large park and ride areas where commuters can park for free and take a bus into the city centre. To speed up the buses they have created green bus lanes and altered traffic lights in favour of buses. The council has worked with the rail authorities to provide extra stations to allow more people to use the suburban rail routes into the city centre.

3 Some people think it will be far too expensive and funds will not be available from the Scottish Executive. With the failure of the road charging scheme where will the funds come from?

4 Two new stations were built to allow Edinburgh residents in the west and east to get into the city centre. Both new stations had large car parks. Longer platforms were built to allow longer trains to carry more passengers. Edinburgh Park, the newest station, has complimentary bus shuttles carrying rail commuters to the nearby business park.

The unit in brief

This unit is about how some of our towns and cities are expanding, and the problems this causes at the rural-urban fringe. In the Activities, students look at why people are leaving cities, and the attractions of a lifestyle at the rural-urban fringe. They name land uses that develop at the rural-urban fringe and explain why this area is attractive to them.

Key ideas

◆ Settlements are growing at the rural-urban fringe. This is called urban sprawl.

◆ In most MEDC cities, people are moving out to smaller towns and villages – the process of counter-urbanisation.

◆ Some people want to build at the rural-urban fringe, others are against it.

◆ The government has tried to protect the countryside by creating green belts.

◆ Developing brownfield sites is one alternative to building at the rural-urban fringe.

Key vocabulary

rural-urban fringe, counter-urbanisation, green belt

Skills practised in the Activities

◆ Geography skills: defining the terms 'counter-urbanisation ' and 'rural-urban fringe'

◆ Thinking skills: giving reasons why people are leaving cities; giving the attractions of the rural-urban fringe lifestyle; naming land uses at the rural-urban fringe and explaining why this area is attractive to them

Unit outcomes

By the end of this unit, most students should be able to:

◆ define the terms given in 'Key vocabulary' above;

◆ explain why people are moving out of cities;

◆ explain why some people want to build at the rural-urban fringe, and why others are against it;

◆ understand how the government has tried to protect the countryside;

◆ name one example of an alternative to building at the rural-urban fringe.

Ideas for a starter

1   Ask: Who can tell me – What does urban sprawl mean? What is the rural-urban fringe and counter-urbanisation? What are green belts?

2   Show photos of activities/developments at the rural-urban fringe, e.g. retail developments, golf courses, garden centres, housing estates, farms. Ask: What do these things have in common? Why do some of them lead to conflict?

Ideas for plenaries

1   You have been to a meeting at the planning office to discuss new developments in the rural-urban fringe. At the meeting there was: a developer wishing to build new houses; a family living in the twilight zone hoping for affordable housing in the suburbs; a farmer with land in the rural-urban fringe; a conservationist; an industrial developer wanting to build there; the local planning officer. You are a journalist and now have to write a newspaper article based on the discussions at the meeting. You have a ten-minute deadline. Off you go!

2   What plans are there for development at the rural-urban fringe in your town? What do you and other local people think about the plans?

3   Identify a new local housing estate that has been built recently, or is still being built. Show photos of the site. What was it used for before? What kind of homes are being built? What services are being provided? What effect will it have on the environment?

4   Question time! Think back over the unit and write down three questions related to what you have learned. The teacher will ask a member of the class to try to answer.

Further class and homework activities

1   Investigate the pattern of 'green belts' in the UK. Find out about the green belt around one major town.

2   The building of 'new towns' in the UK was one of the ways planners attempted to halt urban sprawl. Find out about the Abercrombie Plan and the Greater London Development Plan. Investigate the development of new towns around large urban areas such as London.

Answers to the Activities

1 a Counter-urbanisation is the fall in the percentage of the population living in towns – contrast this with urbanisation.

  b The rural-urban fringe is the area at the edge of town where the city meets the country, so it's half rural and half urban.

2   People are moving out of cities because:
  ◆ inner cities are often dirty, crowded and can feel dangerous;
  ◆ the countryside is cleaner and quieter;
  ◆ trains and fast roads make it easy to commute long distances; many people commute daily from Oxford to London, for example;
  ◆ businesses and industries are moving out-of-town.

3   Attractions of the rural-urban fringe include:
  ◆ cheaper land: for car parks and new houses;
  ◆ space for bigger retail outlets and leisure facilities;
  ◆ farmers are developing new leisure activities, like golf courses;
  ◆ business and science parks are much nicer than old industrial sites.

4 and 5   Uses include:
multiplex cinemas, sports stadiums, hypermarkets and golf courses. In all cases, stress the cheaper land, the much greater area of the periphery as opposed to the CBD, so there's more land available, good access to ring roads and motorways, and new, modern developments: business parks, science parks, affordable housing.

The unit in brief

This unit looks at the pressures on Edinburgh's green belt from developers and the growth of Edinburgh Airport. In the Activities, students describe the functions of a green belt and explain why the City Council has been forced to allow development inside the green belt. They also consider why conservationists are unhappy about the developments at Edinburgh Airport.

Key ideas

◆ A green belt is a buffer zone around a large urban area where development is limited to protect the environment and stop urban sprawl.

◆ Edinburgh's green belt has been altered several times.

◆ One of the biggest areas taken out of the green belt was for Edinburgh Park (a large shopping complex and business park), which was established in 1992.

◆ The Royal Bank of Scotland built its new headquarters on land inside the green belt, on a site previously occupied by a large mental hospital.

◆ The City Council has allowed Edinburgh Airport to expand, although it is within the green belt, because of the economic benefits it brings.

Key vocabulary

green belt

Skills practised in the Activities

◆ Geography skills: describing the functions of the green belt

◆ Thinking skills: explaining; suggesting why development has been allowed in the green belt; identifying differences; suggesting why people are unhappy about developments at the airport

Unit outcomes

By the end of this unit, most students should be able to:

◆ define the term given in 'Key vocabulary' above;

◆ explain why Edinburgh's green belt has been altered;

◆ describe the development at Edinburgh Park and explain why the City Council supported the development;

◆ explain why the planners could not stop the development of the site that the Royal Bank of Scotland chose for its new headquarters;

◆ describe the developments at Edinburgh Airport.

Ideas for a starter

1 Ask – who can remind me what a green belt is? What is an alternative to building on green belts?

2 Have there been changes in land use where you live? For example a retail park where there used to be fields, or offices or housing where there used to be a hospital? What kind of changes have there been? How does this affect a place?

3 Show before and after photos – e.g. open fields which have then been built on for housing, shops, office developments, etc.; brownfield sites which have been redeveloped and so on. What ideas can students come up with about why these sites are chosen, why these developments are allowed to go ahead, etc.?

Ideas for plenaries

1 What do you think are the main arguments for allowing development on land within the green belt?
What do you think are the main arguments against allowing development on land within the green belt?

2 Provide students with a map of Edinburgh. Ask them to annotate it to show developments in Edinburgh's green belt.

3 You're a developer looking for a site to build a new office development. What do you need to consider to make sure it's a good choice of site? Are you looking for a site in the green belt, or outside the green belt?

4 Did you find anything difficult about the work in this chapter? What? Why? What would help to make it less difficult?

5 Prepare a crossword for students including the key terms from this chapter.

Answers to the Activities

1 Dynamic means the area is changing all the time. It is coming under pressure from developers because there are a limited number of development sites within the city. Since the green belt was first planned, the boundaries and areas included have changed as some areas are added to the green belt and others are taken out. One area which used to be in the green belt was the Gyle which is now a large retail and office park.

2 The main functions of the green belt are to try and restrict the spread of urban areas out into the rural landscape, and to try and focus development on to areas within the city. Another function is to keep the separate identity of two large urban areas which are growing towards one another.

3 Edinburgh City Council has been forced to use areas of the green belt for development for a number of reasons: the creation of jobs; there are few development opportunities left within the city boundaries; land prices have been rising very rapidly; to try and

reduce the amount of traffic which has been travelling into the city centre (the traditional home of many of these office developments).

4 The new RBS headquarters is a lot more than a set of offices. The company adopted the American campus concept and has tried to create a village atmosphere by including a lot of services, shops, restaurants, gyms and other fitness facilities not normally provided by the company itself. It has also worked along with the local council to try to reduce the volume of cars being used to get to the offices.

5 Conservationists are worried about the expansion at Edinburgh Airport because even more of the green belt is being used for commercial developments. This will lead to a further reduction in the areas set aside for farming and recreation in the area around the city.

Farming

The big picture

These are the key ideas behind this chapter:

◆ Farming is a system, with inputs, processes and outputs. Systems can be applied to any farm, anywhere.

◆ Physical and human factors affect a farmer's choice about what type of farming to do. For most types of farming climate is the most important factor.

◆ Farmers face problems and challenges.

◆ Farming needs to be made more sustainable.

Note that the students' version of the big picture is given in the students' chapter opener.

Chapter outline

Use this, and the students' chapter opener, to give students a mental roadmap for the chapter.

6   **Farming** As the students' chapter opener, this unit is an important part of the chapter; see page 11 of this book for notes about using chapter openers

6.1   **Farming – what's it all about?** Introduction to farming and farming as a system

6.2   **Farming – what happens where?** Classifying farms, and different types of farming in the UK

6.3   **Farming – on the flat ... and on the edge** Case studies of an arable farm (Lynford House Farm in East Anglia), and a hill farm (Herdship Farm in Teesdale)

6.4   **Mixed farming** Case studies of two different mixed farms – Penllan Farm close to the English–Welsh border, and Ardalanish Organic Farm on the Isle of Mull

6.5   **Is farming in crisis?** A look at some of the problems facing the UK's farmers

6.6   **All change for farming** Changes to CAP, animal welfare, agribusiness and mechanisation

6.7   **Making farming sustainable** Some ways of making farming in the UK more sustainable – traditional methods, agri-environment schemes, organic farming and diversification

Objectives and outcomes for this chapter

Objectives	Unit	Outcomes
Most students will understand:		Most students will be able to:
● That farming is a system.	6.1, 6.3, 6.4	● Describe a farming system and give examples of inputs, processes and outputs.
● How farms are classified.	6.2	● Explain how farms are classified, and give examples of different types of farms.
● That climate is the most important factor affecting the distribution of farming.	6.2	● Explain the distribution of different farming types in the UK.
● The differences in farming in the UK.	6.3, 6.4	● Draw a systems diagram for a hill farm and a mixed farm; explain the differences between the different farming systems.
● That farmers in the UK face a range of problems.	6.3, 6.4, 6.5	● Describe the problems facing farmers in the UK today.
● That there are changes which affect farming in the UK.	6.3, 6.6	● Give examples of some of the changes affecting farming, and say how they affect farmers.
● How farming in the UK can become more sustainable.	6.7	● Give examples of traditional farming, agri-environment schemes and diversification; explain what organic means.

These tie in with 'Your goals for this chapter' in the students' chapter opener, and with the opening lines in each unit, which give the purpose of the unit in a student-friendly style.

Using the chapter starter

The photo on page 102 of the *geog.SG* students' book shows a dairy cow. Milk is nutritious, and kind to teeth! And from milk we get cheese, butter, ice cream, milk powders, cream, and yoghurt.

There are more than 2 million dairy cows in Britain – that's one for every thirty people. Since the mid-1980s numbers have declined, but yields have increased, so milk production has remained stable. Each cow is milked on average 300 days a year, two or three times a day. A cow can produce 20 litres of milk a day. In an average lifetime, we'll each drink over 6000 litres of milk.

The UK is largely self-sufficient in milk. We're the third-largest milk producer in Europe, the seventh-largest in the world. Our milk production is limited by EU quotas – the quota for 2005 was 14.2 billion litres.

In November 2005 about two thousand British farmers went on strike for three days in protest at the low prices paid by retailers for food and milk. Over 700 dairy farmers went out of business between 2001 and 2005. Many of them blamed the low prices they were getting from the supermarkets. As of February 2006, dairy farmers were getting 18p a litre for their milk, but they said it cost 19p a litre to produce; in the shops, milk was selling for 50p a litre.

Farming – what's it all about?

The unit in brief

This unit introduces farming, and provides a range of ideas for students to think about from 'How important is farming?' to 'How far does our food travel?'. It looks at farming as a system with inputs, processes and outputs. In the Activities, students look at the issue of battery farming, consider the role of farmers and look at the importance of physical factors in farming.

Key ideas

◆ Farmers produce food, but also look after the countryside.

◆ Farming is a system.

◆ The farming system has inputs, processes and outputs.

◆ Systems can be applied to any farm, anywhere.

Key vocabulary

agriculture, system, inputs (physical, human and economic), processes, outputs, feedback, factors

Skills practised in the Activities

◆ Thinking skills: thinking about the importance of battery farming; considering whether people are influenced by others' opinions; thinking about the role of farmers; explaining the importance of physical factors in farming

Unit outcomes

By the end of this unit, most students should be able to:

◆ define the terms given in 'Key vocabulary' above;

◆ understand that farming is a system;

◆ know that the farming system has inputs, processes and outputs;

◆ explain why physical factors are important in farming.

Ideas for a starter

1 Brainstorm to find out what students know about farming, and what farmers do. Write all the ideas on the board, then try to group them together.

2 Use the photos of Rick Stein and battery hens on page 104 to promote a discussion of different methods of farming and issues to do with farming in the UK.

3 Use the spider diagram on page 104 to introduce the topic. Draw it on the board and ask students for answers to each of the questions.

Ideas for plenaries

1 Provide students with a blank systems diagram. Students can work in pairs to complete a diagram for different types of farming.

2 If you did not use starter 3, divide the class into groups. Give each group one 'bubble' question for the spider diagram on page 104 and ask for responses from each member of the group.

3 Are physical or human factors likely to be most important in farming?

4 Which of the physical factors is most important in farming?

5 Can students think of a type of farming where climate isn't important?

6 Ask your neighbour to name three physical inputs and three human and economic inputs into the farming system. Ask for examples of two processes and two outputs.

Answers to the Activities

1 Increased pressure due to the demand for cheap food. Increased demand overall. This type of farming is the most cost effective, enabling farmers to produce high levels of output.

2 This will depend in part on students' opinions. It is not likely that Rick Stein will have a great influence over farmers. The same could be argued for the general public. There is the counter argument that, as a celebrity chef with forthright views, he will be able to influence a significant minority of consumers.

3 This is only partially true. It is correct to state that, obviously, much of our food is grown in the UK. It is important to note, however, that a large minority of food in the UK is imported. There is also the point that, increasingly, the farmland of this country is used for purposes other than the cultivation of food.

4 The answer will depend upon the factors chosen. Any answer should relate to the importance that physical factors have upon farming.

Farming – what happens where?

The unit in brief

This unit looks at how farms are classified, and the pattern of farming types in the UK. In the Activities, students look at farm classification; identify the factors needed for wheat growing; think about why climate is often the most important physical factor influencing farming; and also consider the role of technology for today's farmers.

Key ideas

◆ Farms can be classified in three ways – by processes, input and output.

◆ There's a pattern to the distribution of farming types in the UK.

◆ For most types of farming, physical factors are more important than human factors.

◆ Climate is the most important physical factor.

◆ There are five main types of farming in the UK.

Key vocabulary

processes, arable farms, pastoral farms, mixed farms, market gardening, inputs, intensive, extensive, outputs, commercial, subsistence, pattern, factors, climate, hill sheep farming, dairy farming

Skills practised in the Activities

◆ Thinking skills: explaining; identifying physical and human factors; thinking

Unit outcomes

By the end of this unit, most students should be able to:

◆ define or explain the terms given in 'Key vocabulary' above;

◆ understand that farms can be classified by processes, inputs and outputs;

◆ recognise the pattern of distribution of farming types in the UK;

◆ understand that for most types of farming physical factors are more important than human factors;

◆ explain why climate is the most important physical factor;

◆ identify the five main types of farming in the UK.

Ideas for a starter

1 Show, or give, students maps of annual rainfall in the UK, January and July temperatures, and relief of the UK. Ask for suggestions about where sheep farming, cereal farming, dairying and beef-rearing might be found.

2 Ask students where market gardening might be found. What is the difference between this and other types of farming in terms of physical and human factors? (It is not dependent on climate.)

3 Show photos of rural areas and ask students to identify types of farming. Are there any obvious physical factors influencing farming?

Ideas for plenaries

1 Draw the distribution of farming types on a blank map of the UK. Add annotations to explain the distribution.

2 What do the following terms mean:
- ◆ arable
- ◆ pastoral
- ◆ mixed
- ◆ market gardening
- ◆ intensive
- ◆ extensive
- ◆ commercial
- ◆ subsistence

3 How can the words in plenary 2 be used to classify different types of farming systems?

4 What is pastoral farming? Why are most pastoral farms in the UK found in the north and west?

5 Where do you find arable farms in the UK? And why?

6 Make a graffiti wall of what students have learned today.

Answers to the Activities

1 Farming in the UK is usually classified as commercial because the outputs – crops and animals – are mostly sold to make a profit; farming in the UK is geared to selling produce for profit. This contrasts with subsistence farming, where farmers produce food for themselves and their family, with nothing left to sell for a profit.

2 a Physical factors to grow wheat: warm sunny summers, a dry harvest time, flat land suitable for large machines, fertile soils. Human factors to grow wheat: money to buy large machines, good transport links.

b Wheat isn't grown in the Lake District because the climate is too cold and wet for the wheat to grow well, the soils tend to be poor and thin, and the land is too steep and hilly for large machines.

3 Climate is often the most important physical factor influencing farming because it determines what plants and crops will / won't grow – climate largely determines growth conditions. Other physical factors, such as soils, are more easily modified.

4 Technology helps the modern farmer in the UK to overcome or moderate many physical factors, even climate (think about greenhouses, heaters, irrigation, fertilisers, new strains of crops, fast transport links to markets) – but overall, climate continues to play an important determining role: until a strain of wheat is developed that will grow in cold, wet, windy conditions wheat will never be grown in areas with this type of climate.

The unit in brief

This unit consists of two case studies – Lynford House Farm in East Anglia (an example of a commercial arable farm) and Herdship Farm in Teesdale (an example of an extensive hill sheep farm). In the Activities, students look at the issue of subsidies; think about how farmers could make more money, draw a systems diagram for Herdship Farm and compare it with that for Lynford House Farm.

Key ideas

◆ All farms can be described as systems.

◆ The inputs, processes and outputs differ for different types of farming.

◆ Farms and farmers have to change if they are to survive.

◆ Farmers consider themselves as land managers – not just farmers.

Key vocabulary

arable farming, hill farming

Skills practised in the Activities

◆ Geography skills: drawing a farming systems diagram and comparing it with another

◆ Thinking skills: explaining and thinking about subsidies; suggesting ways for farmers to make money

Unit outcomes

By the end of this unit, most students should be able to:

◆ explain the terms given in 'Key vocabulary' above;

◆ understand that inputs, processes and outputs differ for different types of farming;

◆ realise that farmers have to change if they are to survive;

◆ understand why farmers consider themselves as land managers, not just farmers.

Ideas for a starter

1 Recap: the classification of different types of farm.

2 Show headlines from newspaper articles relating to arable farming and hill farming to generate discussion that farming is not an easy life – and the farms and farmers are changing.

3 Draw two blank systems diagrams on the board. Ask for suggestions for inputs, processes, and outputs for an arable farm and a hill farm.

Ideas for plenaries

1 Activity 4 could be used as a plenary.

2 Using the systems diagram for Lynford House Farm, show what decisions the farmer might have to make if the price of the main crop was reduced, or the price of fertiliser increased. Give a number of different scenarios to the students so that they can use their diagrams to think about the different decisions farmers may be forced to make.

3 Should farmers be considered to be 'managers' of the countryside?

4 Make 10-15 statements about farming in the UK based on what students have learned so far, some true, some false. Students hold up True and False cards. Where statements are false, ask students to correct them.

Further class and homework activities

1 Investigate other types of farms and draw systems diagrams like those used in the students' book. Compare your models with those given in the students' book. You could use the following website to help you: www.face-online.org.uk/

2 Investigate the current system of farm support in the UK. Look at the DEFRA website.

Answers to the Activities

1 A payment made to support farmers, usually to make something profitable.

2 Many products will otherwise result in a loss to farmers. Subsidies therefore help to keep farmers in business, and farming the land.

3 The farm could diversify into other food products. It could diversify into an increasingly wide range of non-food activities, from activity trails to corporate hospitality events.

4 Inputs would include:
   ◆ 229 hectares of land, which comprises 22 hectares of flowering hay meadow and the rest is mostly rough pasture
   ◆ 580 mm of rain a year
   ◆ 12 cows
   ◆ sheep: 268 Swaledales and 180 North Country Cheviots

Processes would include:
   ◆ sheep-dipping
   ◆ sheep-shearing
   ◆ lambing

Outputs would include
   ◆ silage
   ◆ hay
   ◆ lambs

Herdship farm will have fewer inputs, processes and outputs.

The unit in brief

This unit consists of two case studies of two very different mixed farms – Penllan Farm in Herefordshire on the English-Welsh border, and Ardalanish Organic Farm on the Isle of Mull. In the Activities, students draw a systems diagram for each of the two farms to compare them; they consider why organic farms should provide more jobs than non-organic farms (and therefore help to boost rural economies), and why most organic farms are mixed farms.

Key ideas

◆ All farms can be described as systems.
◆ The inputs, processes and outputs differ for different types of farming.
◆ Penllan Farm is a large mixed farm – over 400 hectares, and 2700 sheep and cows.
◆ The Morgans – who run Penllan Farm – have seen their income fall. 25% of their income now comes from grants and subsidies.
◆ Most organic farms are mixed farms.
◆ At Ardalanish Organic Farm the way the animals graze helps to conserve the natural environment; crops are grown on the 'inbye' land.

Key vocabulary

mixed farming, organic farming

Skills practised in the Activities

◆ Geography skills: drawing farming systems diagrams and comparing them
◆ Thinking skills: thinking about why organic farms provide more jobs than non-organic farms, and why most organic farms are mixed

Unit outcomes

By the end of this unit, most students should be able to:
◆ explain the terms given in 'Key vocabulary' above;
◆ understand that inputs, processes and outputs differ for different types of farming;
◆ draw systems diagrams for two different mixed farms;
◆ identify the similarities and differences between the different mixed farms;
◆ explain why most organic farms are mixed farms.

Ideas for a starter

1 Ask: who can remind me what an arable farm is, what a pastoral farm is, and what a mixed farm is?

2 If you did not use starter **2** for Unit 6.3 you could do something similar here – show headlines from newspaper articles relating to farming to generate discussion that farming isn't an easy life – farms and farmers face a variety of problems and are having to change.

3 Mental map time! Ask: Where are East Anglia, Teesdale, Herefordshire, and the Isle of Mull (the locations of the four case study farms in Units 6.3 and 6.4)? With all books closed, and no clues on the walls, ask students to mark them on a blank outline map of the UK and on the board.

Ideas for plenaries

1 Activity **1** could be used as a plenary.

2 If starter **2** is not used, ask students: Why are farm incomes falling? Why do farmers need grants and subsidies? What other problems do farmers face?

3 Create an acrostic. Write FARMING down one side of the page. Make each letter the first letter of a word, phrase or sentence about farming.

4 Prepare an odd-one-out for your partner on what you have learned today.

Answers to the Activities

1 **Penllan Farm**.
Inputs include: fertiliser; pesticide; animal feed; fuel; labour; machinery; farm buildings; animals.

Processes include: calving/lambing; sowing crops; spraying crops; shearing, drenching, footbathing; harvesting; ploughing; feeding.

Outputs include: wheat, barley, oats; potatoes; turnips, straw, hay; cattle, lambs, ewes; wool.

Ardalanish Organic Farm.
Inputs include: Kyloe cattle and Hebridean sheep; animal feed; labour.

Processes include: feeding; calving/lambing; shearing; ploughing; planting; harvesting.

Outputs include: lambs, calves; wool and woollen products; oats, hay; turnips for the animals; tatties, carrots and winter vegetables for sale.

Ardalanish Farm has fewer inputs and processes than Penllan Farm. It uses no artificial fertiliser or pesticide. Vegetables are grown to sell locally. The Mackays use their own wool on the farm and are likely to do most of the work on the farm themselves.

2 It is the actual methods used on organic farms that create most of the additional jobs when comparing organic and non-organic farms. Organic farms require more people and skills to manage crops, the soil and farm animals.

3 Organic farms do not use artificial fertiliser, etc. but use farmyard manure as fertiliser instead on arable land. Mixed crop and livestock farming, crop rotation and traditional practices are an inherent part of organic farming.

The unit in brief

This unit is about some of the problems facing the UK's farmers – from those caused by the supermarkets, to things like soil erosion and the use of chemicals. In the Activities, students carry out a small survey to discover where their food comes from; consider why farmers have difficulty competing with those abroad; evaluate the factors causing problems for farmers; and look at two past (but fairly recent) problems – BSE and foot and mouth disease.

Key ideas

◆ People want to buy different food, and buy food more cheaply, than in the past.

◆ Supermarkets are very powerful. They choose their suppliers and decide how much to pay farmers.

◆ The food market is global – farmers in the UK compete with those abroad.

◆ Soil erosion, use of chemicals and the number of farmers leaving the industry are further problems.

Key vocabulary

global, soil erosion, overproduction, pesticides, fertiliser

Skills practised in the Activities

◆ Geography skills: carrying out a survey and describing the results; researching past farming problems

◆ Thinking skills: thinking of reasons; assessing how far various factors have caused problems for farmers

Unit outcomes

By the end of this unit, most students should be able to:

◆ define the terms given in 'Key vocabulary' above;

◆ understand the connection between supermarkets and farmers;

◆ understand why the food market has become global;

◆ explain other farming problems such as soil erosion, use of chemicals and numbers of farmers leaving the industry.

Ideas for a starter

1 Go to:
news.bbc.co.uk/1/hi/special_report/1999/09/99/farming_in_crisis/442787.stm
Summarise the main points for students: Do they agree that farming is in crisis?

2 Use a variation on Activity **1** as a starter. Ask students what they ate at home last night for dinner. List the fresh items with their (likely) source. What does that tell us? How much of the food produced abroad is also grown in the UK? How much of it was not in season?

3 Brainstorm to find out what students know about problems facing farmers in the UK. Some may have already been touched on in previous units, but see how many others they can come up with. Some will be dealt with in Unit 6.6.

Ideas for plenaries

1 If you used starter **2**, students can continue to work on the idea of food miles. For each item work out how far their food has travelled to reach their shop or supermarket. They can use an atlas to measure the distance the food has travelled.

2 Ask students to produce a world map showing where the food listed in starter **2** came from.

3 Would we be prepared to pay more for our food to support local farmers? The countryside is managed by the farmers. If we value the countryside should we be prepared to pay more for our food?

4 Write an email from Robin Spence to his local supermarket. He wants to know why he is paid so little for his milk, compared to the price it is sold for. Now write a reply from the supermarket.

5 Take two minutes with a partner to think up one interesting question about some of the problems facing the UK's farmers that we have not covered today.

Answers to the Activities

1 The answer to this will depend upon the results of students' research.

2 Much production abroad has significantly lower input costs, for example cheap labour. Conditions may be more favourable for certain products, and there could be significant government support.

3 This will depend upon students' opinions. Look for some evaluation of the different factors.

4 Answers will vary according to the example chosen. BSE led to the mass slaughter of cows, the banning of certain cuts of meat, an export ban on British beef; foot and mouth resulted as well in destruction of livestock, in the 'closure' of much of the countryside to the public during the outbreak.

The unit in brief

This unit is about changes in farming in the UK. It looks at changes to the Common Agricultural Policy (CAP) and other changes such as concern about animal welfare, agribusiness and increased mechanisation (with the subsequent loss of jobs). In the Activities, students look at the differences between the CAP subsidies and the new Single Payment System (SPS) and consider whether farmers will benefit from the SPS; think about the set-aside policy; consider the advantages and disadvantages of agribusiness, and look at the benefits of increased mechanisation for larger farms.

Key ideas

◆ CAP reform means that farmers no longer get several different subsidies – they now get one single payment a year.

◆ There is increased concern about animal welfare in farming.

◆ Some farms are getting bigger as they become agribusinesses.

◆ People working on farms continue to be replaced by machinery, and the numbers employed full-time continue to fall.

Key vocabulary

Common Agricultural Policy, subsidies, Single Payment System, quota, set-aside, animal welfare, agribusiness, mechanisation

Skills practised in the Activities

◆ Thinking skills: explaining food mountains and lakes; explaining differences; thinking of reasons; stating whether they agree with a policy; considering benefits and disadvantages of agribusiness; explaining benefits of increased mechanisation

Unit outcomes

By the end of this unit, most students should be able to:

◆ define or explain the terms given in 'Key vocabulary' above;

◆ understand the differences between CAP subsidies and the SPS;

◆ recognise that there is concern about animal welfare in farming;

◆ describe the advantages and disadvantages of agribusiness;

◆ understand why full-time jobs continue to be lost in farming.

Ideas for a starter

1 How could the following change farming:
- ◆ a guaranteed price for the produce
- ◆ greater consideration for animal welfare
- ◆ amalgamation of farms to create bigger farms
- ◆ increased mechanisation
- ◆ greater diversification – country walks, golf courses and camp-sites instead of farmland.

Invite discussion of each scenario, and decide how they would change farming.

2 Show a photo of set-aside land (like the one at the bottom of page 114). Ask: Is this farming?

Ideas for plenaries

1 Draw a spider diagram showing changes in farming in the UK. What are the advantages and disadvantages of the changes?

2 Consider this: 'Farming needs to change to stay profitable. Are there any changes which are not acceptable?' Discuss this with a neighbour and be ready to participate in a whole-class discussion on this issue.

3 Show a photograph of traditional hedge and field farmland. Ask students: How would this area change with greater mechanisation and increased farm sizes? Draw a sketch showing how the area could be changed, and annotate it to show the changes.

4 What is the CAP? What are subsidies? What is the SPS?

5 What are quotas and set-aside?

6 Did you find anything difficult about the work in this unit? What? Why? What would help to make it less difficult?

Answers to the Activities

1 As farmers were guaranteed a minimum price for their produce – no matter how much of it there was – there was no market incentive to cap their production.

2 The SPS is a single annual payment that doesn't rely on huge production to ensure significant payment. It breaks the link between production and payment, and is linked more closely to market needs.

3 This will depend in part upon students' opinions. It is likely that larger farms will benefit more.

4 To encourage them not to over produce on their land. To maintain some of their land in a semi-natural state.

5 Agribusiness benefits from economies of scale, and so can meet the ever increasing demand of the British public for cheap food. It can, however, strangle competition and stifle choice. Agribusiness landscapes can be monotonous and unattractive, and lead to a lack of bio-diversity.

6 Economies of scale mean that large farms are likely to benefit most. Benefits are likely to be least in hill farming regions in the north of Britain.

The unit in brief

This unit looks at some of the ways in which farming in the UK can be made more sustainable. It covers a variety of approaches ranging from using traditional methods and agri-environment schemes through to diversification and producing potato chips. In the Activities, students consider the importance of making farming sustainable; think about the attractions and disadvantages of agri-environment schemes; find out how farmers diversify; and think about whether farmers should be paid to look after the countryside.

Key ideas

◆ Farming needs to become more sustainable to overcome some of the problems and challenges facing farmers today.

◆ There are a number of ways in which farming can become more sustainable:
  - using traditional methods
  - agri-environment schemes
  - changing to organic farming
  - diversification.

Key vocabulary

sustainable, agri-environment, Tir Gofal, Environmental Stewardship, Rural Stewardship Scheme, organic, diversify

Skills practised in the Activities

◆ Thinking skills: considering the importance of making farming sustainable; explaining; thinking of disadvantages; finding out; justifying opinions

Unit outcomes

By the end of this unit, most students should be able to:

◆ define or explain the terms given in 'Key vocabulary' above;

◆ understand what 'making farming sustainable' means;

◆ understand that there are a number of ways in which farming can become more sustainable;

◆ give some examples of how farmers can diversify.

Ideas for a starter

1 Recap: What challenges has farming faced in the UK in recent years?

2 Ask: What does sustainable mean in a geographical context? Can you think of some examples you have already studied which have included sustainable options?

3 Draw a farming system diagram (inputs/processes/outputs) on the board. Ask for suggestions on how a farm could become more sustainable.

4 Show photos of a variety of ways farming can become more sustainable, e.g. using traditional farming methods, traditional meadows, and examples of diversification. Ask: What do these have in common?

Ideas for plenaries

1 Making farming more sustainable might bring some unwanted problems with it. Food may become more expensive and local farmers may lose if we can buy food produced more cheaply abroad. We might only be able to buy food when it's in season rather than all year round. Can you think of any other problems? Do you think farming should become more sustainable?

2 Draw a poster advertising locally grown organic vegetables in preference to the vegetables from the supermarket.

3 Write a letter to the local newspaper advising local people to demand local products on sale in the supermarkets in the interest of sustainability.

4 Do you value the countryside? Is it worth spending our money collected in taxes to maintain it or should more be spent in the towns? Work out your argument with your neighbour before putting it to the rest of the class. Have a vote and see what the result is.

5 Write 'Making farming sustainable' in the middle of your page. Create a mind map around the phrase. How many ideas can you come up with in two minutes?

Further class and homework activities

1 Visit this website to find information on the relationship between farming and wildlife www.rspb.org.uk/countryside/farming/hopefarm/index.asp

Write up a small case study of Hope Farm remembering to include location, enterprises, and the plans for the future.

2 Look up sustainability on the websites of Friends of the Earth, Greenpeace, and the UK government department DEFRA. Collect examples of sustainable systems and compare the various definitions.

Answers to the Activities

1 To preserve the countryside for future generations. To maintain biodiversity.

2 a To help maintain the environment and preserve biodiversity. There may be economic benefits through grants and subsidies.

b Inability to compete against large producers. Labour intensive work. May not be sustainable in the long term.

3 This will depend upon students' research. Examples may include adventure trails, childrens' playgrounds, corporate hospitality, etc.

4 This will depend upon students' opinions. Ensure that opinions are justified.

7 Industry and economic change

chapter overview

The big picture

These are the key ideas behind this chapter:

◆ Industry and employment is classified as primary, secondary, tertiary and quaternary.

◆ Industry, or a factory, is a system, with inputs, processes and outputs.

◆ As industry changes different location factors become more important, so industry locates in different places, and employment structures change.

◆ Some areas of traditional heavy industry, such as South Wales, the Ruhr and West Lothian, have gone through a cycle of decline, followed by a cycle of growth. They have attracted new industries and jobs.

◆ Economic change has social and environmental consequences.

Note that the students' version of the big picture is given in the students' chapter opener.

Chapter outline

Use this, and the students' chapter opener, to give students a mental roadmap for the chapter.

7 **Industry and economic change** As the students' chapter opener, this unit is an important part of the chapter; see page 11 of this book for notes about using chapter openers

7.1 **Industry, jobs and systems** Classifying industry, and industry as a system

7.2 **Industry – deciding where to put it** Industrial location factors – using the example of the Toyota factory in Burnaston, Derbyshire, and how as industry changes location factors change

7.3 **Industry – traditional and heavy** Case studies of traditional heavy industry in South Wales and the Ruhr

7.4 **Industry – footloose (and fancy free)** High-tech and footloose industries, and where they locate

7.5 **All change for industry** Deindustrialisation, changing employment structures and government help

7.6 **Industry – about turn** Another look at the case studies of South Wales and the Ruhr to see how these two declining areas have changed

7.7 **Industrial change – West Lothian** How West Lothian embraced change and became one of Scotland's fastest growing local economies

Objectives and outcomes for this chapter

Objectives	Unit	Outcomes
Most students will understand:		Most students will be able to:
● That industry and employment are classified as primary, secondary, tertiary and quaternary.	7.1	● Give examples of primary, secondary, tertiary and quaternary industry.
● That industry is a system.	7.1	● Describe a system and give examples of inputs, outputs and processes.
● That there are a range of factors which affect where an industry locates, and that as industry changes location factors change.	7.2, 7.3, 7.4, 7.7	● List the factors affecting industrial location; give an example of one industry and say what factors were important in its location; explain why the location of industry has changed in the UK.
● Why areas of traditional heavy industry have declined, and the social and environmental consequences of this.	7.3, 7.7	● Give an example of an area of traditional heavy industry and explain why the industry declined; describe the effect that the decline of industry had on people and the environment.
● That high-tech and footloose industries can choose where to locate.	7.4	● List the factors important for the location of footloose industries; explain why these factors are important; give examples of the types of places footloose industries can locate; give an example of a science park and describe and explain its location.
● How industry has changed in the UK, and how declining areas have turned around.	7.2, 7.4, 7.5, 7.6, 7.7	● Define deindustrialisation; draw pie charts to show changing employment structures in the UK from 1800 onwards; explain how and why governments help areas in decline; describe how a declining area has changed.

These tie in with 'Your goals for this chapter' in the students' chapter opener, and with the opening lines in each unit, which give the purpose of the unit in a student-friendly style.

Using the chapter starter

The photo on page 118 of the *geog.SG* students' book shows a factory worker in Malaysia inspecting latex surgical gloves that have been manufactured. The gloves will be exported to countries around the world.

Latex is a type of rubber made from the sap of rubber trees found in Malaysia and other countries including Indonesia, Thailand, and Brazil

A number of companies in Malaysia manufacture and export latex gloves for use in the medical profession – and Malaysia is currently the world's leading exporter of surgical gloves.

The Malaysian government is encouraging the growth of the medical device industry. It is currently concentrated on rubber-based products, but more companies are starting to manufacture products made from plastics, silicone, and metal alloys.

The unit in brief

This unit introduces industry. It is about how industry is classified, and how it works as a system. In the Activities, students come up with a list of jobs and classify them as primary, secondary etc. They compare the number in each sector with national UK figures. They also look at the links between different industries.

Key ideas

◆ Industry provides employment.

◆ Industry can be classified into primary, secondary, tertiary and quaternary industry.

◆ Industries aren't separate – they are linked together.

◆ An industry, or a factory, is a system with inputs, processes and outputs. There may be some feedback, e.g. where profits are put back into the system.

◆ Inputs (factors) affect where an industry, or factory, will locate.

Key vocabulary

employment, classifying, primary, secondary, tertiary, quaternary, raw materials, manufacture, service, system, inputs (physical, human and economic), processes, outputs, feedback, factors, waste

Skills practised in the Activities

◆ Numeracy skills: comparing numbers of jobs with UK figures

◆ Thinking skills: listing and classifying jobs; explaining links between industries

Unit outcomes

By the end of this unit, most students should be able to:

◆ define or explain the terms given in 'Key vocabulary' above;

◆ understand that industry provides employment (jobs);

◆ classify jobs into primary, secondary, tertiary and quaternary sectors;

◆ use examples to explain how industries are linked together;

◆ understand that industry, or a factory, is a system with inputs, processes and outputs;

◆ understand that inputs (factors) affect where an industry, or factory, will locate.

Ideas for a starter

1 Provide students with job advertisements from the local newspaper. Ask them to classify jobs under the headings Primary, Secondary, Tertiary and Quaternary. Compare the local pattern with the national picture (2% in primary, 28% in secondary, and 70% in tertiary in 2000).

2 Quiz students about parents/family employers and classify jobs using the same headings as in **1** above.

3 Discuss job aspirations with the class and classify these under the headings given in **1** above. Consider what a similar group of young people would have aspired to 150 years ago.

4 Give a systems model (inputs/processes/outputs) to the class and ask them to work in pairs and add as much information as possible about a firm or industry they have studied before. Ask one group to work on the interactive whiteboard to present to the rest of the class.

Ideas for plenaries

1 Ask students to classify the following jobs into primary, secondary, tertiary, and quaternary:

IT consultant	Deep sea fisherman	Nurse
Financial consultant	Printer	Journalist
Football professional	Garage Mechanic	Baker
Stock market broker	Headhunter	Personal Trainer

Make sure the students understand what each of the jobs entails.

2 Give students the percentages of people working in the various sectors in the UK. How will this compare with employment structures in LEDCs?

3 How has employment structure in the UK changed in the past? How is it likely to change in the future?

4 Use Activity **2** as a plenary. Use the board to draw/write the answer and ask students to come and add links to other factories/businesses.

5 Provide students with a photo of a factory. Ask students to annotate the photo showing how it affects the environment.

6 Ask your neighbour to name three physical inputs and three human and economic inputs into the industry system. Ask for examples of two processes and two outputs.

7 Tell me the three key things you have learned today. Now tell me another three things you have learned.

Further class and homework activities

Research using the internet and other resources a local factory or firm to develop the input/process/output model.

Answers to the Activities

1 The answers provided here will depend upon students' choice of jobs.

2 The car factory is an assembly plant, building from many already manufactured components. If the factory closes, the suppliers will also lose out. There are many examples, including the manufacture of engines, windscreens, and tyres.

The unit in brief

This unit is about industrial location factors, and uses the Toyota plant at Burnaston in Derbyshire as a real-life example of how decisions are made about where to locate an industry or factory. It also looks at how industry has changed and how this has been matched by changes in locations. In the Activities, students describe the distribution of industry and explain why location has changed. They give reasons for the decline in coal mining and textiles; explain the change in importance of different location factors; and think about whether Toyota would have located in the UK without government help.

Key ideas

◆ There are a range of physical and human and economic factors which people have to think about when deciding where to locate an industry or factory.

◆ As industry changes, different location factors become more important.
   - Many of the old traditional UK industries needed raw materials and energy provided by coal, so they located near coalfields or ports.
   - New industries have developed with different locational requirements.

◆ Changes to the type or location of industry can mean factories or businesses close and people lose their jobs.

Key vocabulary

locate, factors, raw materials, site and land, energy supply, labour (workers), transport, capital (money), markets, government policy, environment

Skills practised in the Activities

◆ Geography skills: describing the distribution of industry

◆ Thinking skills: explaining the change in the location of industry and the changing importance of location factors; giving reasons for the closure of industries; thinking and giving reasons for answers

Unit outcomes

By the end of this unit, most students should be able to:

◆ define or explain the terms given in 'Key vocabulary' above;

◆ name the physical and human and economic factors which affect the location of industry;

◆ explain why the importance of various location factors has changed;

◆ understand that changes to the type or location of industry can mean factories or businesses closing and people losing their jobs.

Ideas for a starter

1 Brainstorm to identify as many factors as possible which could affect the location of a factory. Write them in two columns on the board and ask students what the two columns are (physical inputs/factors and human and economic inputs/factors).

2 Use the photograph of the Toyota car plant at Burnaston on page 122 and the text boxes to recap the previous lesson.

3 Use an industrial simulation/game to demonstrate how changing factors may lead to changes in location.

Ideas for plenaries

1 Provide students with a resource such as that found at http://news.bbc.co.uk/1/hi/programmes/working_lunch/education/1804227.stm and get them to draw a spider diagram summarising the information about the siting of the car plant.

2 Why might companies decide not to move even when their factors of location have changed? Discuss in pairs and then open up to the whole class.

3 A quick test. Call out a student's name and term: primary industry; secondary industry; tertiary industry; quaternary industry; classification; inputs; outputs; processes; feedback; location etc. The student has 10 seconds to give you a definition.

4 Look at the two maps on page 123. Close your book. Describe the patterns before 1970 and today to your neighbour. Your neighbour can tell you why industrial location has changed.

5 Question time! Think back over the lesson and write down three questions related to what you have learned. The teacher will ask a member of the class to try to answer.

Further class and homework activities

1 Research the traditional industry of South Wales and/or the Ruhr in preparation for the next lesson.

2 Go to the Toyota website www.toyotauk.com and find out more about the company. This will be useful for the work on transnational companies in Units 10.4 and 10.5.

Answers to the Activities

1 Most industry was concentrated in the northern and western parts of the UK; London was the principal exception to this.

2 There has been a decline in some traditional industrial areas, such as mining and shipbuilding in North-east England. There has also been a growth in high-tech industry in areas such as the M4 corridor, due to good transport links and a trained workforce.

3 The decline has been due to factors such as the exhaustion of raw materials, cheaper foreign production, and the substitution in some cases of alternative products (such as gas for coal).

4 Many industries in the past were based around raw materials – heavy industries. As such, physical factors were important. More modern industries are more footloose, and depend much less upon the natural environment – more upon human factors such as the location of markets and suitably qualified workers.

5 This depends upon students' opinions, but it is important to stress the role played by government incentives that attract foreign investment.

The unit in brief

This unit includes two case studies of traditional heavy industry – South Wales and the Ruhr in Germany. In the Activities, students think about why the steelworks at Port Talbot and Llanwern remained open while others closed, and the reasons for Llanwern's eventual closure. They draw a consequences map to show how people are affected by the closure of industries; and look at the similarities and differences between the examples of South Wales and the Ruhr.

Key ideas

◆ In the nineteenth century South Wales had supplies of all the raw materials for making iron and steel.

◆ By the 1990s all the steelworks had closed except for Port Talbot and Llanwern (both integrated steelworks). All local raw materials had run out (only Tower Colliery remained working).

◆ Llanwern closed in 2001 with the loss of 3000 jobs.

◆ The Ruhr iron and steel industries developed as there were huge amounts of coal and local supplies of iron ore.

◆ Other industries were attracted to the Ruhr.

◆ Since the 1970s, coalmines, steelworks and other heavy industries have closed in the Ruhr with massive job losses.

◆ Unemployment is far higher in the Ruhr than in the rest of Germany, and the environment has been ruined.

Key vocabulary

integrated steelworks

Skills practised in the Activities

◆ Thinking skills: identifying disadvantages; suggesting reasons; thinking about reasons for Llanwern's closure; drawing a consequences map; identifying similarities and differences between South Wales and the Ruhr

Unit outcomes

By the end of this unit, most students should be able to:

◆ explain the term given in 'Key vocabulary' above;

◆ explain why iron and steel industries developed in South Wales and the Ruhr;

◆ explain why other industries were attracted to the Ruhr;

◆ explain why coalmines, steelworks and other heavy industries closed in South Wales and the Ruhr;

◆ describe the effect of the decline of traditional heavy industry on people in South Wales and the Ruhr.

Ideas for a starter

1   Ask students to brainstorm in groups – either about South Wales or the Ruhr – to find out what they know about the area's industrial past, and how things are changing (in preparation for Unit 7.6). When students have finished, produce a concept map for each area which shows the links between different ideas.

2   Show students a map extract of an area with a traditional industrial past. Ask them to draw a sketch map of the area and label the main features of the industry and associated features such as spoil heaps and housing. This could be done on an interactive whiteboard.

Ideas for plenaries

1   Ask students to draw a cycle of decline showing how an area can become run-down following the closure of the main industry. Ask them to add facts and figures to their cycle. Share the best ones with the class.

2   Students work in twos and threes to produce either a storyboard for a video or a script for a TV/radio programme developing the theme of industrial decline. These should include an interview with someone who has lived in the area for a long time.

3   Use Activity 6 as a plenary.

4   Close your books. Draw a sketch map to show either South Wales or the Ruhr. Annotate your map to show why industry developed there and what has happened since the decline of heavy industry.

5   Make 10-15 statements about industry based on what students have learned so far, some true, some false. Students hold up *True* or *False* cards. Where statements are false, ask students to correct them.

Answers to the Activities

1   The area is inaccessible – it is very hilly with steep valleys.

2   Combination of factors, relating to accessibility – via motorway, large area of flat land, proximity of workforce.

3   Imports of steel could be cheaper from countries where the costs of extracting raw materials is cheaper, or where the labour costs are lower.

4   Students' consequence maps should include loss of income, less money to spend, purchasing of fewer 'luxury' goods, decline in other local industries, particularly services, decline in social facilities.

5   Cheaper costs of production in other countries. Costs of industrial pollution. People moving away from the area to better environments.

6   Similarities include reasons for growth and development of the areas – such as presence of raw materials, and the decline in the traditional industries. Differences include the scale, as the Ruhr is much larger, and the range of industries being greater in the Ruhr e.g. the chemical industries.

The unit in brief

This unit is about the location of high-tech and footloose industries, and includes Heriot-Watt Research Park as an example of a science park. In the Activities, students find out where high-tech products are made; choose important location factors for a high-tech company; explain why some industries are footloose and others are tied to particular locations; identify what problems footloose industries could cause the UK economy.

Key ideas

◆ The new industries that have developed and begun to replace the traditional heavy industries are often high-tech.

◆ Industries which can choose where to locate are called footloose.

◆ Footloose industries need: cheap land with large sites to expand; good transport links; workers with special skills nearby; other industries nearby to swap ideas and information.

◆ Footloose industries are often found on science parks, business parks and industrial estates.

◆ Areas where footloose industries have located include the M4 corridor, Silicon Glen in central Scotland, Silicon Valley in California.

Key vocabulary

high-tech, footloose, science parks, business parks, industrial estates

Skills practised in the Activities

◆ Geography skills: finding out where products are made

◆ Thinking skills: choosing factors and justifying choices; explaining why some industries are footloose and others aren't; identifying the problems of footloose industries; identifying similarities and differences in location choices

Unit outcomes

By the end of this unit, most students should be able to:

◆ define or explain the terms given in 'Key vocabulary' above;

◆ understand that the new industries that have developed and begun to replace traditional heavy industries are often high-tech;

◆ know that industries which can choose where to locate are called footloose;

◆ describe what footloose industries look for when deciding where to locate;

◆ explain the differences between science parks, business parks and industrial estates;

◆ name three areas where footloose industries have located.

Ideas for a starter

1 Recap: where industry traditionally located and where industry is located now.

2 Students work in pairs to decide what they would want from an area if they worked for a high-tech company. Try to draw out behavioural factors. 'What facilities would you want to be close to?'

3 Provide students with catalogues of high-tech products. Ask them to define high-tech products, and decide what would be the most important factors in the location of factories producing them.

Ideas for plenaries

1 Give students an odd-one-out exercise. Give them four words, three are obviously linked and the fourth one is not. They have to discuss and give reasons for rejecting the fourth word, e.g. Labour, Raw Materials, Energy, Footloose (Answer = Footloose because all the others are factors of production).

2 Provide students with photos of a traditional industrial area and a Science Park. Ask them to produce a table comparing the environments shown in the photos.

3 Write a letter to an unemployed miner living in a traditional industrial area describing the benefits of retraining and working in an area of footloose industries.

4 Consider some of the reasons why the miner might not want, or be able, to move.

5 Provide students with advertisements from newspapers and magazines advertising new industrial location opportunities. Ask students what type of industry would locate in these places, and why.

6 Quick-fire test. Close your books. Ask your neighbour how many location factors they can come up with for footloose industry in 15 seconds.

7 What is the difference between science parks, business parks, and industrial estates?

8 Make a graffiti wall of what students have learned today.

Further class and homework activities

1 Investigate other science parks such as the Cambridge Science Park http://www.cambridge-science-park.com/home.htm and produce an annotated map of the park including location features.

2 Research a local business park or industrial estate and produce a wall display.

Answers to the Activities

1 This will depend upon students' own choices, but will probably include products related to computing and electronics, for example the latest in game products, telephones and other communication devices.

2 This will depend upon students' choices. Ensure that choices are justified.

3 There are many industries that have a freedom of choice of location. Examples include many types of communication industries, telephone and internet sales, etc. Industries tied to a particular location will include those dependent upon raw materials, and to a certain extent those dependent upon a certain type or size of market.

4 An increase in competition from companies making use of the advantages of locating abroad, such as cheaper labour, and importing their products cheaply into the UK - thus costing UK jobs.

5 This will depend upon students' choices.

The unit in brief

This unit is about how industry has changed in the UK. It looks at deindustrialisation, what the government has done to help, and where the new jobs are. In the Activities, students look back at the list of jobs they wrote in Unit 7.1 to find evidence of deindustrialiastion, and explain why it has happened in the UK. They compare the maps showing the changing location of industry in Unit 7.2 with the pie charts of employment structure to see the relationship between them, and suggest why the government should help industries that are in decline or struggling.

Key ideas

◆ Deindustrialisation is the decline in manufacturing (secondary) industry, and the growth in tertiary and quaternary industry.

◆ The Industrial Revolution in the 19th century changed the jobs people did. In the 20th century there was a shift to service industries (and then quaternary).

◆ 'New' jobs are in high-tech and footloose industries, often located on the edges of towns and cities.

◆ The government has tried to help areas in decline by providing grants and subsidies to create new jobs and to protect existing ones.

◆ The Environment Agency is responsible for maintaining and improving water quality in England and Wales. In Scotland the Scottish Environment Protection Agency (SEPA) is responsible for controlling pollution.

Key vocabulary

deindustrialisation, productivity, rural-urban fringe

Skills practised in the Activities

◆ Geography skills: comparing maps and pie charts

◆ Thinking skills: identifying evidence of deindustrialisation and explaining why it has happened; identifying positive impacts; thinking of reasons

Unit outcomes

By the end of this unit, most students should be able to:

◆ define the terms given in 'Key vocabulary' above;

◆ explain how industry and employment in the UK has changed since 1800;

◆ describe the consequences of industry locating at the rural-urban fringe;

◆ describe how the government has tried to help areas in decline;

◆ identify how industry affects the environment and who is responsible for controlling pollution.

Ideas for a starter

1 Recap: Industrial changes in the UK over the last 200 years:
   ◆ Heavy industry on coalfield locations
   ◆ Footloose industries – the development of Science Parks etc.

2 How has Government influenced the location of industry? Refer back to the work on South Wales.

3 What changes are happening locally to industry and employment structure? Are factories closing down and new businesses starting up? Are these in different locations in and around the town? What types of industries are closing? What types of businesses are starting up?

Ideas for plenaries

1 Do you think there are problems in changing jobs from one industry to another? Would the skills in steel making transfer to the IT industry?

2 How might the government help the transition from one industry to another?

3 Activity **5** could be used as a plenary.

4 Close your book. Describe how employment structure has changed in the UK since 1800. Draw pie charts to show the employment structure in 1800, 1900, and 2000.

5 Close your book. Why does industry locate at the rural-urban fringe? What are the consequences of this?

6 A quick-fire test. Call out a student's name and the definition of a term related to changes in industry in the UK. The student has 5 seconds to give you the term.

Further class and homework activities

Find out about industrial change in South Wales. How did the Welsh Development Agency help? What new industries have been attracted to South Wales?

Answers to the Activities

1 This will depend upon students' choices from 7.1, but it is likely that most of the chosen jobs will have come from the tertiary sector, supporting the argument that deindustrialisation is taking place.

2 Partly due to increased competition from industries abroad, which are often able to offer products at a cheaper price. Also due to increased efficiency in the UK, which means that fewer jobs are needed in industry. Also an increased demand for tertiary and quaternary industries.

3 The growth areas on the map are principally those areas where there has been the greatest growth in service industries.

4 Increased wealth to the local area, creation of new jobs. Less congestion than if located further into urban areas.

5 To support the local economy by sustaining jobs. Saving jobs has the knock-on effect of preserving other local industries. Products may still be needed, they may be too expensive – government help may make them economical.

The unit in brief

This unit looks again at the two case study areas included in Unit 7.3 – South Wales and the Ruhr – to see how these two declining areas have changed. In the Activities, students explain whether new industries would have located in South Wales without government help, and why most of them are from outside the UK. They consider the advantages and disadvantages of foreign companies locating in South Wales, and compare the changes in South Wales and the Ruhr.

Key ideas

◆ New industry was attracted to Wales through the setting up of the Welsh Development Agency and Urban Development Corporations. The region also became a Development Area and had help from the EU.

◆ New industries (particularly tertiary) were attracted to South Wales, including a lot from overseas.

◆ South Wales and the Ruhr have been through a cycle of decline and are now in a cycle of growth.

◆ In the 1960s the state and central government wanted to encourage new types of employment and improve the environment in the Ruhr. Now 65% of the Ruhr's workers are in the tertiary sector.

◆ New universities and colleges were set up, business parks developed and the environment cleaned up.

Key vocabulary

There is no key vocabulary in this unit.

Skills practised in the Activities

◆ Geography skills: comparing South Wales and the Ruhr; finding out about another area in the UK where new industries have located

◆ Thinking skills: thinking of reasons; explaining answer; identifying advantages and disadvantages; identifying similarities and differences

Unit outcomes

By the end of this unit, most students should be able to:

◆ compare the changes that have taken place in South Wales and the Ruhr, and identify the similarities and differences;

◆ draw a diagram to show a cycle of decline and a cycle of growth for either South Wales or the Ruhr;

◆ describe how one place in South Wales has changed;

◆ give one example of how the environment has been improved in the Ruhr.

Ideas for a starter

1 Ask: Who can remind me about traditional heavy industry in South Wales? And who can remind me about the Ruhr?

2 Draw a map of South Wales showing iron and steel production.

3 Show 'before' and 'after' photos of either South Wales or the Ruhr (photos of traditional heavy industry and modern industry). Ask: Is this the same place? How has it changed?

Ideas for plenaries

1 Write an email to a friend describing the changes in South Wales or the Ruhr over the last 40 years.

2 Ask students to draw a cycle of growth showing how an area can be regenerated.

3 Give two examples of new industry attracted to South Wales. Where did the companies come from?

4 Activity **4** could be used as a plenary.

5 Take two minutes with a partner and think up one interesting question about how South Wales and the Ruhr have changed that we have not covered today.

Further class and homework activities

1 Produce a wall display of all information collected on South Wales (and/or the Ruhr if possible).

2 Investigate the TNCs which have moved to Wales. Where have they located and what do they manufacture? Have they all stayed or have some of them moved away?

Answers to the Activities

1 This will depend upon students' opinions – ensure that they are justified. Considerable government help has been given to regenerate industry in the area.

2 Government incentives and grants. Wanting to get a foothold in the UK / European markets. Larger scale multi-nationals. There may be some reluctance by UK firms to invest in an area of economic and industry decline.

3 Advantages would include more jobs, positive impact on other industries in the local area. Disadvantages would include that many of the best-paid jobs will go to foreign workers, many are not well paid or with good career development prospects.

4 Similarities are many, including references to the role of the government. The Ruhr is a larger-scale system, and one difference is the creation of the Emscher Landscape Park, the largest in Europe.

5 Answers will depend upon the example chosen by students.

The unit in brief

This unit looks at the changing fortunes of West Lothian. Some parts of Scotland's Central Belt have struggled to cope with the decline of traditional heavy industries, but West Lothian has embraced change and is one of Scotland's fastest growing local economies in terms of both population and job creation. In the Activities, students consider the reasons for the closure of British Leyland's plant in 1986 and the success of the authorities in replacing the lost jobs; draw a timeline of job losses and gains in West Lothian; explain why distribution companies are being encouraged to locate in the region, and describe how the Heartland Project will help to regenerate the local economy.

Key ideas

◆ In the 1960s British Leyland located its new truck and tractor plant in Bathgate with the creation of 7000 jobs. Closure of the plant in 1986 along with Plessey's electronics factory left the area with high unemployment.

◆ Bathgate became a growth point in 'Silicon Glen'. The Motorola and NEC plants closed in 2002 – at their peak they had employed nearly 5000 people.

◆ West Lothian is now being marketed as an ideal location for distribution centres. Over 100 have located there.

◆ The British Leyland site is now being used for housing – creating jobs in the construction industry.

◆ The Heartlands Regeneration Project is one of the largest industrial regeneration schemes in the UK.

◆ West Lothian is now one of Scotland's fastest growing local economies in terms of both population and job creation.

Key vocabulary

There is no key vocabulary in this unit.

Skills practised in the Activities

◆ Thinking skills: suggesting reasons for the closure of the British Leyland plant; considering the success of the authorities in replacing jobs; drawing a timeline of job losses and gains; explaining why distribution companies are being encouraged to locate in West Lothian; suggesting how the Heartlands Project will help to regenerate the local economy

Unit outcomes

By the end of this unit, most students should be able to:

◆ understand the causes of industrial change in West Lothian;

◆ explain how new industry has been attracted to the area since the 1960s;

◆ draw a timeline to show job losses and gains in West Lothian;

◆ explain why West Lothian is now one of Scotland's fastest growing local economies.

Ideas for a starter

1 Have atlases ready. Give students five minutes to find Bathgate, and write down five geographical facts about it. Record facts on a spider diagram on the board.

2 Ask: What kind of gadgets do you own that did not exist a hundred years ago? Prompt answers such as mobile phones, computers, calculators, MP3 players. What do they all have in common? (The computer chip.) Discuss how these new products mean that new factories get opened – somewhere!

3 Is your local area suffering from decline? If so, ask: How is this affecting the area? What is being done about it?

Ideas for plenaries

1 The manufacturing sector is continuing to shrink in the UK. Thousands of people lose jobs every year as factories close. Do you think this is a good thing, or a bad thing? Why is it happening? What problems might there be if we had no factories left?

2 Draw a cycle of decline and a cycle of growth for West Lothian.

3 Activity **3** could be used as a plenary.

4 Close your books. Draw a sketch map to show West Lothian. Annotate your map to show how industry has changed there.

5 Create an acrostic. Write INDUSTRY AND ECONOMIC CHANGE down the side of the page. Make each letter the first letter of a word, phrase or sentence about industrial change in the UK or Europe.

Answers to the Activities

1 The British Leyland plant closed for a number of reasons: the huge costs in transporting raw materials and components to the plant; the cost of transporting vehicles to the main markets; production costs rose because of damaging labour disputes; in some cases it was cheaper to import similar vehicles.

2 The local council was able to attract more than 5000 jobs in the electronics industry, but never reached the numbers employed in the British Leyland plant.

3 The timeline should start in 1986 with the loss of 7000 jobs at the British Leyland plant. Students should mark all the major job losses and gains, e.g. those created by the location of distribution centres, even if they cannot give specific job numbers.

4 Distribution companies are being encouraged to locate in West Lothian because of its excellent transport links; it is close to the container terminal at Grangemouth and the ferry terminal at Rosyth (meaning European destinations are accessible); it is also accessible to the major population centres in Central Scotland (it's within an hour's drive of 60% of Scotland's population).

5 The building of the new motorway junction will help to make the area more accessible to businesses. Building the golf courses and hotel should create jobs in an area with high unemployment.

8 Population

The big picture

These are the key ideas behind this chapter:

◆ The world's population hit 6.5 billion in the spring of 2006, and continues to rise.

◆ The world's population is unevenly distributed – some places are densely populated, others are sparsely populated.

◆ Population is rising fastest in LEDCs.

◆ Migration is the movement of people from one place to another. It can be voluntary, forced (creating refugees), permanent or temporary. It affects population sizes and structures, and of course, individuals.

◆ Countries carry out population censuses to enable them to plan for the future.

◆ Changes in population can cause problems for countries – rapid population growth and too many young people in LEDCs slows down development, an increasingly elderly population in MEDCs could mean the economy runs short of workers.

◆ Immigration may help to solve the shortage of workers in the UK; population control and family planning - and economic development - can help in LEDCs.

Note that the students' version of the big picture is given in the students' chapter opener.

Chapter outline

Use this, and the students' chapter opener, to give students a mental roadmap for the chapter.

8   **Population** As the students' chapter opener, this unit is an important part of the chapter; see page 11 of this book for notes about using chapter openers

8.1   **Where in the world is everyone?** Why the world's population is unevenly distributed

8.2   **World population increase** How quickly the world's population is rising, but it's not rising at the same rate everywhere

8.3   **The demographic transition model** The model shows the pattern of population increase over time

8.4   **Population contrasts** Population increase varies between MEDCs and LEDCs; how to use population pyramids

8.5   **Migration** What it is; voluntary, forced, permanent and temporary migration; international and internal migration; push and pull factors

8.6   **Migration and the UK** International and internal migration, and how it affects the UK

8.7   **Darfur refugee emergency** A case study of international refugees

8.8   **Population census – count me in!** Why countries carry out population censuses, and the problems faced by LEDCs

8.9   **The problems of population change** Rapid population growth in LEDCs, and ageing populations in MEDCS

8.10   **How the UK is coping** An ageing population means immigration is needed to fill labour and skills shortages

8.11   **How India is coping** India has tried to control population growth. Population control is now seen in the context of improving social and economic conditions

Objectives and outcomes for this chapter

Objectives	Unit	Outcomes
Most students will understand:		Most students will be able to:
• That a range of physical and human factors affect population density, and that some areas are densely populated and others are sparsely populated.	8.1	Give physical and human factors which affect population density; give four examples of densely populated areas and four examples of sparsely populated areas.
• How quickly the world's population is rising; population increase is not the same everywhere.	8.2, 8.4	Draw the graph to show world population growth; define birth rate, death rate and natural increase; name the six countries which account for the largest increase in population; name countries where population is declining.
• That many countries have had similar patterns of population increase, as shown by the demographic transition model.	8.3	Draw the demographic transition model and explain what happens at Stages 1–5; give examples of countries at each stage of the model.
• That population pyramids show the age and structure of a country's population, and that they can be used for planning, and comparing countries.	8.4	Say what population pyramids show and what they can be used for.
• What migration is; why people migrate; the differences between international and internal migration.	8.5	Define migration; describe four types of migration; explain the differences between international and internal migration; give four push and four pull factors for migration between countries.
• How migration affects the UK.	8.6	Describe the pattern of international migration to and from the UK; describe where migrants to the UK come from; explain the effect migration has on the UK's population; describe the pattern of internal migration in the UK.
• That forced migration creates refugees.	8.7	Give an example of international refugees; explain why they have become refugees; describe what is being done to help them.
• What a census is; why countries carry out population censuses; the problems LEDCs face carrying out censuses.	8.8	Explain what a census is and why countries carry them out; describe the problems that Sudan has carrying out a census.
• That rapid population growth in LEDCs is slowing down their development, and that countries such as India have tried to control their population growth.	8.9, 8.11	Explain why death rates have fallen, but birth rates remain high in LEDCs; describe the problems LEDCs with growing populations face; describe how India tried to control population growth in the past, and the policies and targets in place today.
• That ageing populations in MEDCs could lead to labour shortages, and that in countries such as the UK immigrants could help to fill the shortages.	8.9, 8.10	Explain why populations in MEDCs are ageing and why this is a problem; explain how immigration can help to fill the labour and skills shortage in the UK; describe the problems migrants face.

These tie in with 'Your goals for this chapter' in the students' chapter opener, and with the opening lines in each unit, which give the purpose of the unit in a student-friendly style.

Using the chapter starter

The photo on page 134 of the geog.SG students' book shows newborn babies in a hospital nursery. It's not in the UK, where nurseries like this are no longer used; they're tightly-swaddled, which suggests it could be Eastern Europe. These babies are likely to live more comfortable and more varied lives than previous generations. They're likely to produce fewer children than previously. They're also likely to live longer – life expectancy is going up in most countries (the exceptions being countries devastated by HIV/AIDS).

The unit in brief

In this unit students learn why the world's population is unevenly distributed. In the Activities, students use the map of population density to describe the distribution of the most densely populated parts of the world. They explain why population density is low in certain parts of the world and why it varies so much within the UK. They also consider a statement about population density and physical and human factors.

Key ideas

- The population density of an area or place is the average number of people per square kilometre.
- The world's population isn't evenly distributed. Some places are densely populated, some are sparsely populated.
- Population density is affected by a range of physical and human factors.
- There can be variations in population density within a country.

Key vocabulary

population density, densely populated, sparsely populated, physical factors, human factors

Skills practised in the Activities

- Geography skills: describing the distribution of densely populated areas
- Thinking skills: explaining low population density; explaining variations in population density in the UK; identifying and explaining sparsely populated environments likely to remain so; stating how far they agree with a statement

Unit outcomes

By the end of this unit, most students should be able to:

- define or explain the terms given in 'Key vocabulary' above;
- understand that the population density of an area or place is the average number of people per square kilometre;
- understand that the world's population isn't evenly distributed;
- describe the distribution of the most and least densely populated parts of the world;
- explain how population density is affected by a range of physical and human factors;
- explain why there can be variations in population density within a country.

Ideas for a starter

1 Ask students to stand up, and move around the room, to a spot where they feel comfortable. Explain that they are showing population distribution. Which parts of the room are densely populated? Which are sparsely populated? Ask why they chose those places – perhaps near a window or radiator (light, sunshine, heat, fuel) etc. Ask whether people might settle in different parts of the world for similar reasons.

2 What factors will encourage people to live there? Which factors will discourage people from living there? Use the responses to develop a model of population distribution.

3 Why is Western Europe densely populated?

4 Why do we study population distribution in Geography?

Ideas for plenaries

1   Apply the population distribution model developed in starter **2** to the UK. Use your atlas to draw a sketch map of the UK showing the distribution of population. Annotate it giving likely reasons for the distribution.

2   What's the difference between **population distribution** and **population density**?

3   Question time! Think back over the lesson and write down 3 questions related to what you have learned today. The teacher will ask a member of the class to try to answer.

Further class and homework activities

Go to this news page:

news.bbc.co.uk/1/hi/special_report/1999/06/99/world_population/top

print off the report. It's part of a special report on world population. Read the report carefully and find answers to the following questions:

◆   Why should the birth of babies be a cause for worry?

◆   Why have births continued to rise?

◆   What is the projected stabilized population for 2050?

◆   Why is it important to meet this figure?

◆   Why will the economy suffer?

◆   In what ways is Kenya a success story?

◆   Why has family planning not been as successfu as hoped?

◆   What other key areas have got to be addressed?

◆   Why is a high population increase damaging the environment?

◆   Why is population a political issue?

Answers to the Activities

1 a Much of Asia, notably India, SE Asia, NE China and Japan. Western Europe and NE USA.

b This will depend upon the locations chosen, but will generally refer to the nature of the physical environment, e.g. extreme heat in the Sahara, extreme cold in Greenland.

2 Australia and Canada – size of the country and inhospitable nature of some environments. Others relate to the nature of the environment.

3 Although a temperate climate overall, the north and west of the British Isles have a harsher climate. These areas also have the most mountainous conditions, and least fertile soils. In combination, this leads to a relatively low population density. Conversely, in the milder, undulating and fertile south and east, population densities are much higher.

4 This depends upon students' opinions and is difficult to predict, but it is probable that Antarctica will remain an area unpopulated for the foreseeable future.

5 The physical environment is the basis for variations in population density, although the situation is made increasingly more complicated by human factors such as accessibility and relative wealth.

The unit in brief

This unit is all about the increase in the world's population. In the Activities, students use the table showing world population increase to see when the world's population was increasing fastest both in percentage terms and in actual numbers. They identify when the world's population explosion began and think about its causes, particularly in relation to LEDCs. They also have to think about the situation in various named countries.

Key ideas

- The world's population is growing by just over 70 million people a year.
- You can measure a country's birth rate, death rate and natural increase.
- The world's population explosion began around the beginning of the nineteenth century.
- There are big differences in population increase from country to country.
- A few countries have a falling population.
- Overall the rate of world population growth is slowing down slightly.

Key vocabulary

birth rate, death rate, natural increase

Skills practised in the Activities

- Numeracy skills: analysing tables of population data; identifying the beginning of the population explosion from a graph and tables
- Literacy skills: completing a writing frame about world population increase
- Thinking skills: thinking about the causes of the population explosion; stating whether a statement is correct

Unit outcomes

By the end of this unit, most students should be able to:
- define the terms given in 'Key vocabulary' above;
- describe the growth in the world's population from 1800;
- describe some of the differences in population increase between countries;
- be aware that a few countries have a falling population;
- understand that overall the rate of world population growth is slowing down slightly.

Ideas for a starter

1 If the population of the world was like the water in a bath, then the taps letting water in would be like the birth rate, and water emptying out of the plug hole would be like the death rate.

   ◆ If the birth rate increased and the death rate stayed the same what would happen to the size of the population?

   ◆ If the birth rate stayed the same and the death rate decreased what would happen to the size of the population?

2 Ask: About how many people are on the earth right now? How many will there be in a minute from now? About how many more by this time tomorrow?

3 Who can define the terms **birth rate** and **death rate**?

4 What does the natural increase of the population depend on?

Ideas for plenaries

1 World population obviously has to be estimated for the future. What changes may take place to speed up or slow down population increase?

2 What will happen to the natural increase in these situations:

   ◆ The age of first marriage goes up

   ◆ Contraception is readily available

   ◆ People eat a much better diet

   ◆ Better health care is provided

   ◆ Better water supplies are installed in most houses.

3 What problems might there be if the population keeps on rising? What will we need more of? What will the effect be on the world's ecosystems? Students could work in pairs and draw a spider map.

4 Do you think the world's population could fall? What might cause that?

Further class and homework activities

Look up data about population change on the internet and try to find the population change figures by continent. Draw a suitable graph to show the relative sizes of the population in the continents and try to show how they have been changing over time.

Answers to the Activities

1 You should expect something like this:

   Birth rate is the number of births in a country in a year, per 1000 people. Death rate is the number of deaths in a country in a year, per 1000 people. Natural increase is the birth rate minus the death rate, often given as a percentage. The world's population is currently increasing by over 70 million people a year. Overall, the rate of increase is slowing down a bit.

2 a The twentieth century.

   b The second half of the twentieth century.

3 About 1950.

4 a Both Afghanistan and Angola have a high rate of natural increase because they have very high birth rates / their birth rates are much higher that than their death rates.

   b Both Germany and Bulgaria have a negative rate of natural increase – so the population of both countries is falling.

   c Both Brazil and China have more young people in their population, so they have fewer deaths per 1000 / both the UK and Germany have an ageing population.

5 The rapid growth was more due to declining death rates, particularly in many LEDCs.

6 It is true to say that population growth has been most rapid in LEDCs – the USA being a notable exception of an MEDC undergoing continuous population growth.

The unit in brief

This unit is about why the world's population is increasing, and looks at the demographic transition model. In the Activities, students analyse the five stages, predict whether there might be a stage 6 of the model in the future and suggest what it might look like. They also explain why some LEDCs might not follow the stages of the model.

Key ideas

◆ Many countries have had a similar pattern of population increase over time. The demographic transition model shows and explains this pattern.

◆ The demographic transition model is divided into five stages:
  – Stage 1 High stationary = high death rate, high birth rate, low natural increase in population.
  – Stage 2 Early expanding = death rate starting to fall, birth rate still high, high natural increase.
  – Stage 3 Late expanding = death rate still falling, birth rate starting to fall, some natural increase but lower than before.
  – Stage 4 Low stationary = death rate remains low, birth rate is low, little or no natural increase.
  – Stage 5 Declining = death rate could go up (greater proportion of population is elderly), birth rate remains low and could get lower, if more people die than are being born population falls.

◆ The demographic transition model fits what happened in the UK, the rest of Europe and other MEDCs, but LEDCs might not follow the same pattern.

Key vocabulary

demographic transition model, high stationary, early expanding, late expanding, low stationary, declining

Skills practised in the Activities

◆ Thinking skills: analysing the demographic transition model; predicting a stage 6 for the demographic transition model; explaining why some LEDCs might not follow the model

Unit outcomes

By the end of this unit, most students should be able to:

◆ define or explain the terms given in 'Key vocabulary' above;

◆ explain what happens in the five stages of the demographic transition model;

◆ explain why some LEDCS might not follow the stages of the demographic transition model.

Ideas for a starter

1 Ask: Who can remind me what factors affect natural increase?

2 How has the population of the UK changed over time? Has it increased? Are families smaller or larger than families of 80 years ago? Do people live longer nowadays? What else can you tell me about the population of the UK?

3 Have other countries' populations changed in the same way as the UK? Who can tell me two countries that have changed in the same way, and two that are different?

Ideas for plenaries

1 Read the following statements and try to fit them in the various stages of the DTM:

   ◆ The church no longer wants a team of full time grave diggers.

   ◆ A family planning clinic has just opened up in the area.

   ◆ Jill has just set herself up as a silver wedding arranger.

   ◆ Families are now very big and children have to share rooms.

   ◆ The family are very sad as they are burying their fifth out of 12 children.

   ◆ The area is growing very quickly as more houses are being built.

   ◆ The town can be very proud of its new sewage system.

   ◆ The planners were very surprised as many big houses only had a couple living in them.

   ◆ Very few children know their grandparents.

   ◆ The pension age is probably going to go up and people will have to work longer.

2 In some countries population is falling. Is this a good thing? What benefits or problems could it cause?

3 Quick fire test: call out a student's name and ask them to tell you what is happening in Stage 1 of the demographic transition model. Then move on to another student for Stage 2 and so on.

Further class and homework activities

Look up the birth rates and death rates of a range of countries – say 10 in all – and try to position them on the demographic transition model. Now look up their rate of population increase. Has the model accurately predicted the growth rate? If not try to explain why this country does not fit the model.

Answers to the Activities

1 a Stage 1 has a high birth rate and a high death rate.

  b MEDCs are in Stages 4 and 5.

  c LEDCs are in Stages 2 and 3.

  d Population will increase in stages 2 and 3 – because the birth rate is higher than the death rate.

2 In theory there could be many more stages - a stage 6 could see the gap between death rate and birth rate grow, leading to a continued decline in world population. There are, of course, numerous other possibilities.

3 Some countries, such as Brazil, have rapidly industrialised – it is possible that such countries could miss out one or more of the stages.

The unit in brief

In this unit students learn how population increase varies between MEDCs and LEDCs, and about population pyramids. In the Activities, students explain the causes of population increase in LEDCs and think about the reasons for high birth rates. They explain why many MEDCs have an ageing population; compare population pyramids for India and the UK; make the connection between population pyramids and the demographic transition model, and draw an outline population pyramid.

Key ideas

◆ Population increase isn't happening everywhere at the same rate. Over 90% of the world's increase is in the LEDCs, in many MEDCs population increase is slowing down (in some European countries population is declining).

◆ In the last 50 years the world's population has increased from 2.5 to 6.5 billion.

◆ Population pyramids show the age structure of a country's population.

◆ Population pyramids tell us how the population might develop which can be used for planning for the future. They also enable us to compare populations of countries.

◆ The stages of the demographic transition model produce different shaped pyramids.

Key vocabulary

population pyramid

Skills practised in the Activities

◆ Geography skills: comparing and contrasting population pyramids; drawing an outline population pyramid

◆ Thinking skills: explaining the causes of population increase; thinking about reasons for high birth rates; explaining ageing populations in MEDCs; making connections between population pyramids and the demographic transition model

Unit outcomes

By the end of this unit, most students should be able to:

◆ define the term given in 'Key vocabulary' above;

◆ explain where population is increasing the most and what is happening to population in many MEDCs;

◆ describe what population pyramids show and what they can be used for;

◆ understand that the stages of the demographic transition model produce different shaped pyramids.

Ideas for a starter

1 Remind the class how the demographic transition model works. Ask them to identify countries at different stages of the model.

2 Ask students to work in pairs. How does population increase vary between MEDCs and LEDCs? Allow pairs to discuss before opening up to the class.

3 Ask: Does anybody know what a population pyramid is? What does it show?

Ideas for plenaries

1 Work in pairs. Take 30 seconds to look at the world map of population change on page 142. Close your book and describe the patterns on the map to your neighbour.

2 Look at the population pyramids for Stages 1-4 of the demographic transition model on page 143. Close your book and draw the pyramids from memory.

3 You are going to look at migration in the next unit. How many types of migration can you think of? Can you classify them into different groups?

4 Tell your neighbour the three key things you have learned today. Then tell them another three things which are less important.

Further class and homework activities

1 Go to www.census.gov/ipc/www/idbpyr.html and find the following countries which represent different stages in the DTM – Burkina Faso (Stage1), India (Stage 2), Brazil (Stage 3), UK (Stage 4) and Sweden (Stage 5). Draw the pyramids for 2005 and copy them to a Word document. Project them on to 2050 and compare the pyramids. Describe the changes in each pyramid and make suggestions about the planning and resource issues in these countries over time e.g provision of hospitals and schools.

2 What do you think are the problems faced by countries in the 5th stage of the DTM? Look up a number of web references, for example:

 ◆ news.bbc.co.uk/1/hi/uk/2287650.stm

 ◆ news.bbc.co.uk/1/hi/uk/2288275.stm

 ◆ news.bbc.co.uk/1/hi/world/asia-pacific/1083097.stm

 Read the articles and write a short commentary on the issues raised.

Answers to the Activities

1 Continued high birth rates, for reasons such as lower use of contraception, desire to have larger families, etc. together with decline in death rates, due to factors such as improved medical facilities and better diet.

2 Tradition to have large families linked to culture and / or religion. Slow take up / realisation of effects of declining death rates. Continued dominance of subsistence agriculture in some countries leading to the wish to have children to work in the family.

3 Low birth rates, improving medical facilities, diet, care, etc, leading to people living for longer.

4 UK = stage 4, with low birth and death rates, high life expectancy. India = Stage 2, with high birth rate, falling death rate, increasing life expectancy but still lower than UK.

5 The DTM links the changes in birth and death rates in each country to the shape of the pyramids. In the UK, the 'stable' nature of the pyramid is due to both birth and death rates being low. In India,

the shape of the pyramid is explained by reference to the changing nature of the birth and death rates, with the former being high and the latter declining.

6

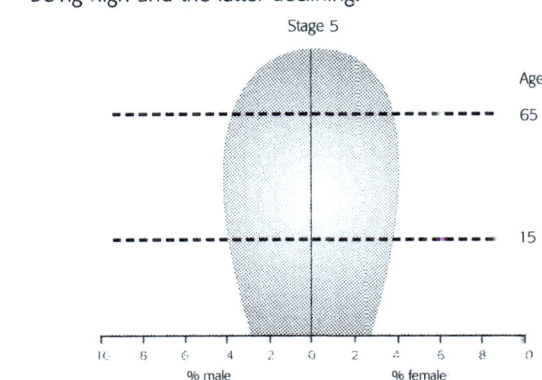

Stage 5

The unit in brief

In this unit students learn what migration is, what causes it and about population change due to migration. In the Activities, students look at the effects which economic migration is likely to have on the migrants' home countries and identify different push and pull factors for different types of migration. They also consider a statement about push and pull factors and migration.

Key ideas

◆ Migration is the movement of people from one place to another.

◆ There are four types of migration: voluntary, forced, permanent and temporary.

◆ Migration between countries is international migration. It can affect the population size and structure of the countries involved.

◆ Migration inside a country is internal migration. It affects the distribution of the country's population.

◆ International and internal migration take place because of push and pull factors.

   - Push factors encourage people to leave their own country or area.

   - Pull factors attract people to a new country or area.

Key vocabulary

migration (voluntary, forced, permanent, temporary), economic migrants, refugees, environmental refugee, international migration, push and pull factors, net migration, population growth rate, internal migration, rural-urban migration

Skills practised in the Activities

◆ Thinking skills: suggesting the effects of economic migration on migrants' home countries; identifying different push and pull factors; stating how far they agree with a statement

Unit outcomes

By the end of this unit, most students should be able to:

◆ define or explain the terms given in 'Key vocabulary' above;

◆ give examples of voluntary, forced, permanent and temporary migration;

◆ explain how international migration can affect the population size and structure of the countries involved;

◆ describe the push and pull factors causing internal migration in LEDCs.

Ideas for a starter

1. Ask: Why do people migrate? What might encourage them to leave their own country (push factors) and attract them to a new country (pull factors)? What obstacles are there to migration?

2. Brainstorm to identify as many types of migration as possible. Ask students to devise a way of classifying the types of migration.

3. Show a video/news clip which shows people on the move fleeing war, environmental disasters etc.

Ideas for plenaries

1. Draw a spider diagram showing the impact migrants have on the country they migrate to. Use different colours to show the positive and negative effects on the host country.

2. Test each other. Work with your neighbour (close your books) and ask them to identify as many push factors as possible. Your neighbour then asks you to identify as many pull factors as possible.

3. Quick fire test: call out a student's name and a definition (of one type of migration). The student has 5 seconds to tell you which type of migration it is. Then move on to another student and another definition.

4. What's the difference between international and internal migration? What is net migration? What is the population growth rate?

Further class and homework activities

1. Migration is highly political. Try to collect as many newspaper cuttings as possible that feature aspects of migration. Make a wall display of your findings.

2. Many families may have recently migrated. Conduct a survey of people in your year group to find the percentages of people who have moved within the last two or three generations. Find out where people have come from and if possible the reason for moving. Remember you may find out about people who have migrated short distances as well as international migration.

Answers to the Activities

1. It will have a negative effect, because the migrants leaving the country are economically active and, therefore, could contribute most to their own country.

2. Much internal migration in LEDCs is determined by rural push – people leaving poor conditions in the countryside, often knowing that conditions will be far from ideal in the cities. In MEDCs and with international migration, push and pull factors will be important and there will often be more choice over locations.

3. This depends upon students' opinions, and should follow on from the previous question. Ensure that students justify their answers. It is possible to argue in line with the statement, or against. Migrants from the new EU countries, for example, may be 'pulled' by the attractions of higher wages and better living conditions in western Europe. On the other hand, internal migrants in LEDCs move to urban areas generally with an understanding of what life will be like in the cities – they are pushed from conditions in the countryside, rather than pulled to the city.

The unit in brief

This unit is all about how migration is affecting the UK. It looks at international and internal migration. In the Activities, students describe changes in international migration patterns and the pattern of internal migration within the UK. They look at the effect of international migration on the total UK population and the expansion of the EU in terms of migration to the UK. They consider the consequences of large-scale migration from other countries, and why London is a special case in migration terms.

Key ideas

◆ Immigration now accounts for about 80% of the UK's population growth.

◆ 50% of immigrants settle in South-East England (mostly London).

◆ People coming to the UK are mainly economic migrants.

◆ A new trend has been the arrival of workers from the EU's new eastern countries.

◆ Internal migration in the UK has two clear trends:
  - A movement of people from the north to the south.
  - A movement of people out of cities.

◆ We need to know about internal migration so that we can see how the distribution of the population is changing.

Key vocabulary

economic migrants, counter-urbanisation

Skills practised in the Activities

◆ Geography skills: describing changes in international migration patterns; describing and explaining patterns of internal migration

◆ Thinking skills: identifying the consequences of migration and the expansion of the EU on migration; thinking; suggesting reasons; justifying answers

Unit outcomes

By the end of this unit, most students should be able to:

◆ define the terms given in 'Key vocabulary' above;

◆ explain the effect immigration has on the UK's population;

◆ describe how international migration patterns to and from the UK have changed since the early 1970s;

◆ explain the effect that expansion of the EU has had on migration to the UK;

◆ describe and explain the pattern of internal migration in the UK;

◆ explain why it is important to know about internal migration.

Ideas for a starter

1 Recap: the factors that bring about migration (push and pull factors, and other factors).

2 Brainstorm to find out what students know about migration in the UK. Where do international migrants come from? How many come? What about internal migration? Where do people move from and to?

Ideas for plenaries

1 What benefits do migrants bring to an area? Make a list and include a wide range of socio-economic-cultural factors.

2 Provide census figures for the class so students can see where the migrants come from, and how many there are. Ask them to map where they live. Is there an obvious pattern?

3 Ask students to investigate the refugee situation in Darfur, Sudan. Two of them will be asked to present a short talk (maximum 3 minutes) at the beginning of the next lesson.

4 What are economic migrants? What is the new trend in migration to the UK?

5 Make a graffiti wall of what students have learned today.

Further class and homework activities

Read the article "Greying west 'needs immigrants'" found at
news.bbc.co.uk/1/hi/world/685332.stm
What is the article's main message?

Answers to the Activities

1 From 1973 to 1982 430 000 net migration out of the UK. From 1983 to 1992, a smaller number net inward migration – 240 000. Since then, much higher inward flow – 1 million.

2 Over the last decade, half of the UK's population growth has been due to migration – it now accounts for 80% of the country's growth in population.

3 It has increased the number of people migrating to the UK, as well as the range of countries from which people migrate.

4 Increased population, greater diversity of population, increased need for services, e.g. languages and schooling, to assist migrants, possible tensions over new migrants, loss to the donor countries of often the economically active people.

5 Illegal immigrants by definition have arrived secretly, so records are not kept of their arrival and estimates have to be based on the numbers being caught.

6 Growth continues to be in the south of the UK, greatest in the south west. Population continues to decline in the north. The exception to this in Northern Ireland, where population is growing.

7 The largest city in Europe, and where a disproportionate number of people head.

8 This will depend upon students' opinions, which need to be justified. The case for building more homes refers to the increase in demand in certain areas of the country where there are shortages, notably the south east (particularly affordable housing). There are also more single occupancy dwellings, and people are living longer.

The unit in brief

This unit provides an example of international refugees – those from Darfur fleeing the fighting in Sudan. In the Activities, students describe the background to the conflict in Darfur; explain why the UNHCR moved the refugees who fled to Chad, and why the government of Chad had to rely on the UNHCR. They think about the effects on Chad of having such a large number of refugees, and find about about another example of international refugees to compare with the situation in Darfur and Chad.

Key ideas

◆ By early 2006 there had been three years of civil war in Sudan's western region of Darfur.

◆ Black African Darfuris say that Sudan's government favours the ruling Arab elite in the north of the country. There has been tension in Darfur for many years over land and grazing rights.

◆ Refugees fled to Chad. In 2004 the UNHCR moved most of them to camps a safe distance from the border.

◆ Aid workers have warned that many thousands are at risk of starvation and disease in the camps, with 1 million children threatened by malnutrition.

◆ No one knows if, or when, the refugees will be able to go home.

Key vocabulary

There is no key vocabulary in this unit.

Skills practised in the Activities

◆ Geography skills: describing the background to a conflict; finding out about another example of international refugees and comparing and contrasting it with the situation in Darfur and Chad

◆ Thinking skills: thinking; explaining; identifying the effects of large numbers of refugees on Chad

Unit outcomes

By the end of this unit, most students should be able to:

◆ describe the background to the conflict in Darfur;

◆ explain why the UNHCR moved most of refugees to camps a safe distance from the border;

◆ understand why the government of Chad had to rely on the UNHCR for emergency relief and aid;

◆ describe the problems the refugees face in the camps.

Ideas for a starter

1 Ask two students to give their talks on the refugee situation in Darfur.

2 Refer back to the classification of migration (starter **2**, Unit 8.5). Did you include refugees? What are the push/pull factors involved in refugee migration? What is the difference between political refugees and economic migrants?

3 Ask students to find Sudan and Chad in an atlas. Ask them to identify the Darfur region of Sudan.

Ideas for plenaries

1 You are a news reporter travelling in Darfur reporting the refugee emergency to the world. Prepare a 2 minute report on the refugee crisis.

2 Close your books. Tell your neighbour why there is fighting in Darfur; how many people have been killed; where the refugees are going and what the situation is like for the refugees.

3 Write down as many words as you can relating to today's work.

Answers to the Activities

1 People living in Darfur say that the government favours the ruling classes in the north of Sudan. Tension over land and grazing rights escalated as Darfur rebels started attacking government targets in 2003. The government struck back, directly and through the support of other groups against Darfur.

2 Entrenched attitudes and beliefs of both sides. War crimes committed cause reprisals and make the conflict worse. The difficulties faced by external mediators trying to intervene in the situation.

3 They were originally within easy reach of the militia groups, so they were still not safe. They were also in an area where there was little water.

4 Better resources – Chad was not well placed to deal with the refugee situation using its own limited resources.

5 Putting enormous strain on the resources of the country.

6 This will depend upon the example chosen.

The unit in brief

This unit looks at why countries carry out population censuses, and at some of the problems they face. It focuses on the UK and Sudan – an LEDC which last held a census in 1993. In the Activities students describe some of the problems LEDCs face when trying to collect census information, and consider why different organisations as diverse as supermarkets and governments might need the information which a census provides.

Key ideas

◆ Governments need to know a range of information about their populations including:
  – how many people there are,
  – where they live,
  – how many are of school age,
  – what jobs they do.
◆ Governments need census information to help plan services such as health, education and housing.
◆ Most countries hold a census every ten years.
◆ LEDCs face particular problems collecting census data caused by:
  – the size of the population,
  – cost,
  – difficulties of landscape,
  – different languages.
◆ Problems faced by LEDCs can make it hard to guarantee the accuracy and quality of the census results.

Key vocabulary

census

Skills practised in the Activities

◆ Thinking skills: suggesting reasons why different organisations need census data; describing problems; explaining

Unit outcomes

By the end of this unit, most students should be able to:

◆ define the term given in 'Key vocabulary' above;
◆ describe the type of information collected in a census;
◆ suggest why governments and other organisations need information about the population;
◆ know that most countries hold a census every ten years;
◆ describe the problems LEDCs face trying to collect census data in general, and those faced by Sudan in particular.

Ideas for a starter

1   Ask: Who can tell me what a census is? What does it count? What is the information used for?

2   Who counts people? Why? How? Can they be sure of the exact number of people they've counted? What problems might they have?

3   Conduct a mini census in the classroom. Ask:
    – How many people are there in your household?
    – How many rooms are there in your home?
    – Is your home owned or rented?
    – How many cars or vans are owned by your household?

Ideas for plenaries

1   Activity 1 could be used as a plenary.

2   Draw a sketch map of Sudan. Close your book. Annotate your map to show why Sudan has problems carrying out a census.

3   Governments carry out censuses because … Go around the class getting students to add to the information on censuses without repetition or hesitation.

4   Tell me the three main things you have learned today. Now tell me another three things that are interesting but less important.

Answers to the Activities

1   A company like Tesco will be looking for an area with a growing population as the best site for a new store. They may also look for particular age profiles of potential customers.

2   Collecting census information in LEDCs can be difficult for various reasons: the physical environment – some areas are difficult to reach either due to the nature of the terrain, lack of roads or wet season making roads impassable; some countries do not have the money to carry out a census for a country which could be half the size of Europe (eg Sudan), or where a number of different languages are spoken (adds to the cost); countries with large total populations or large nomadic populations have difficulty counting people.

3   The Scottish Executive needs to keep accurate records of population growth and decline so that they can target resources for schools, health services, social services etc when they are allocating money to local government.

4   The Scottish parliament is devolved from Westminster. The Scottish Executive is responsible for the planning and delivery of services in Scotland and therefore holds its own census to provide the information it needs.

This unit is about the different problems caused by population change in LEDCs and MEDCs. In the Activities, students explain why birth rates have remained high in LEDCs and suggest solutions to reduce them. They draw scattergraphs and describe and explain the patterns shown. They look at the problems of ageing populations and suggest how some of these could be solved, and the consequences if they're not.

Key ideas

- Over 90% of the world's population increase is in LEDCs.
- LEDCs have falling death rates and high birth rates.
- Rapid population growth is slowing down the development of LEDCs and putting pressure on resources.
- Population growth in most MEDCs is slow – in some MEDCs population is falling.
- MEDCs have low birth rates and low death rates, with an increasing life expectancy.
- MEDCs have ageing populations with increasing numbers of elderly dependents, and a possible shortage of labour in future.

Key vocabulary

family planning, fertility rate, infant mortality, life expectancy, ageing population, elderly dependents

Skills practised in the Activities

- Geography skills: describing and explaining patterns shown on scattergraphs
- Numeracy skills: drawing scattergraphs; comparing graphs
- Thinking skills: explaining; suggesting solutions; explaining answer

Unit outcomes

By the end of this unit, most students should be able to:
- define or explain the terms given in 'Key vocabulary' above;
- understand that over 90% of the world's population increase is in LEDCs;
- explain why birth rates are still high in LEDCs;
- describe how rapid population growth is slowing down the development of LEDCs;
- explain why population growth in most MEDCs is slow;
- explain the problems that an ageing population can cause, and suggest ways in which some of these problems could be solved.

1 Recap: the demographic transition model – the stages and countries at different levels of development at each stage.

2 Ask different students to draw sketches of stylised population pyramids for stages 1-4 of the demographic transition model on the board. Ask the class for comments on the likely birth rate (the base), the death rate (the slope) and the life expectancy (the height) of pyramids at different stages of the model.

3 Show images to convey the idea of large numbers of children/young people in LEDCs and few children but large numbers of elderly people in MEDCS.

Ideas for plenaries

1 Work in a small group to prepare a short report on the problems of either population growth in LEDCs or population change in MEDCs. Divide up the work between you. Use the format:

- ◆ the basic problem
- ◆ a series of solutions
- ◆ an analysis of each of the solutions – maybe use a table for this section
- ◆ a conclusion and recommendation.

2 Why are people living longer in MEDCs? Why is population still growing rapidly in LEDCs?

3 Choose a student to be in the hot seat. Another student asks him or her a question about population change in LEDCs and MEDCs. Then nominate two different students (4-6 pairs in total). There's one golden rule – questions cannot be repeated.

4 Write down as many words as you can relating to today's work.

Answers to the Activities

1 Family, cultural and religious traditions. Lag time between fall in death rates and acceptance that change is taking place. Subsistence farmers still wanting large families to work on the farm.

2 Improvement in education, particularly for women. Increased economic prosperity and stability.

3 a
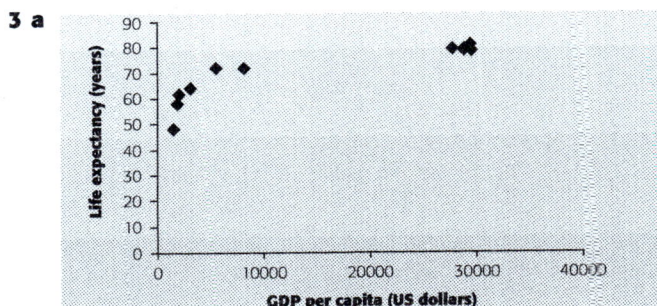

b Per person.

c There is a positive correlation between life expectancy and GDP.

d It is a good indicator in that it gives an overall measure of wealth in a country. It is, however, a crude measure as it does not take in to account actual standards of living, or other social indictors that will influence quality of life (cf. the HDI).

4 a
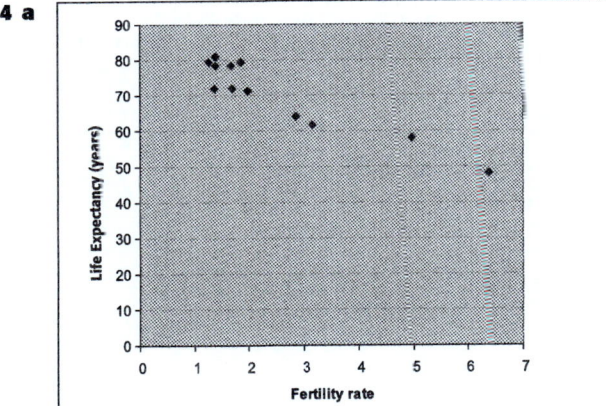

b There is a negative correlation – generally, as fertility rates increase, life expectancy decreases.

5 Both show that there is a connection between the two indicators, one positive and one negative – high GDP is an indicator of relatively high life expectancy, and high fertility rates of relatively low life expectancy.

6 Social and other facilities, such as pensions, are insufficient to cope with the growing number of dependent people. As the number of older people grows, there is a declining number of people of working age to support them, and the economy of the country.

7 By providing better services for older people. By overhauling pensions systems. By encouraging migration of working age people from other countries. Without this, a growing pensions and economic crisis.

The unit in brief

This unit looks at how the UK is coping with population change and how individuals are coping. In the Activities, students explain why life expectancy in the UK is increasing and identify the problems this might cause. They look at why the UK is relying more on foreign workers and consider comments made (some by immigrants) about living and working in the UK. They draw graphs of population data and explain the trends shown.

Key ideas

- The UK's population is getting older. There will be an increasing number of elderly dependents and a possible shortage of workers.
- We may have to work longer, save more towards our pensions, and pay higher taxes.
- The labour and skills shortages are being filled by immigrants.
- New migrants to the UK can face a whole host of problems.

Key vocabulary

elderly dependents

Skills practised in the Activities

- Numeracy skills: drawing graphs and explaining the trends shown
- Thinking skills: explaining increasing life expectancy; identifying problems; explaining why the UK is relying on foreign workers; justifying opinions

Unit outcomes

By the end of this unit, most students should be able to:

- define the term given in 'Key vocabulary' above;
- describe the problems that the UK's ageing population may cause;
- explain why the UK will have to rely on immigrants to fill our jobs;
- identify the problems migrants to the UK might face.

Ideas for a starter

1 Draw a mind map to show what is happening to population in the UK.

2 Brainstorm the issues created by an ageing population.

3 Visit the census website to check on the current state of the UK population. Put the statistics you find on the board and ask students: What do these figures tell us?

Ideas for plenaries

1 If you used starter **1**, ask students to return to their mind map at the end of the unit and make any additions/deletions.

2 One of the UK's advantages according to many people is its multicultural nature. What is your view?

3 What problems do immigrants face in the UK?

4 Sum up what you have learned today in 35 words.

5 Investigate population growth in India. Prepare a short presentation focussing on how the government has tried to control population.

Further class and homework activities

Download the following article on international migration and summarise the major points. Discuss these with your neighbour and then put them to the rest of the class.

news.bbc.co.uk/1/hi/world/europe/1003324.stm

Answers to the Activities

1 a Better health care, education, diet, life styles.

b Fewer people of working age to support the economy, increased demand for services from older people, problems of managing pensions for a growing elderly population.

2 To help fill the gap of a declining number of people of working age.

3 This will depend upon students' opinions – ensure that they are justified.

4 One graph type students could draw would be a cumulative graph, as shown here.

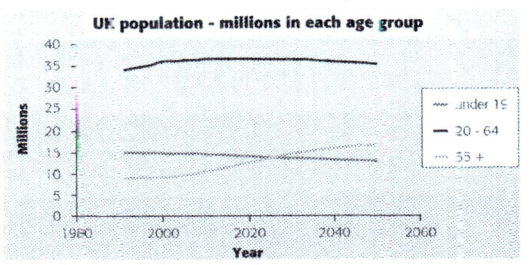

UK population - millions in each age group

The principal changes are – a decline in the under 19 population, together with an increase in the over 65 population.

The unit in brief

This unit looks at how India has tried to use family planning to tackle population growth. In the Activities, students think about and explain whether India's new policies are likely to be more successful than the old ones in controlling population growth; consider the importance of population policies in India and China, and explain what is happening in Kerala – a state with a fertility rate as low as some European countries.

Key ideas

◆ Some LEDCs are relying on economic development to bring higher living standards, many see population control as more important.

◆ In the early 1950s, India's population was growing very quickly. In 1952 the government started offering contraception advice and sterilisation.

◆ By the 1970s, the population was still growing too quickly – forced sterilisations began (first on men, then changed to women), a policy to encourage only two children per family led to abortions and the killing of children.

◆ The population still trebled between 1952 and 2006.

◆ Better health and welfare (including education) is now seen as a key part of controlling population growth.

◆ There are big variations in fertility rates and rates of population increase across India.

Key vocabulary

population control, family planning

Skills practised in the Activities

◆ Thinking skills: explaining the partial success of early attempts at population control, and whether new policies will be more successful; thinking about the effect of population policies in China and India; explaining the low fertility rate in Kerala

Unit outcomes

By the end of this unit, most students should be able to:

◆ explain the terms given in 'Key vocabulary' above;

◆ explain why India has tried to control population growth since the 1950s;

◆ describe the policies India has used to try to control population growth;

◆ suggest why the new policies are likely to be more successful than the old ones;

◆ list some of the targets set for 2010 (at least 4).

Ideas for a starter

1 Quick fire key terms check – call out a student's name and a definition of any key terms covered in this chapter. The student has 5 seconds to give you the term.

2 If you have not recapped the demographic transition model, then do so here. Recap the stages and countries at different levels of development at each stage.

3 Ask a few students to show their presentation on population growth in India. What have they discovered about how the government has tried to control population?

Ideas for plenaries

1 What type of population policies are likely to be most successful in terms of slowing population growth?

2 What targets has India set for 2010?

3 Question time! Think back over the lesson and write down 3 questions related to what you have learned. The teacher will ask a member of the class to try to answer.

4 Think about all the work you have done in this chapter. *Why do we study population?*

5 Create an acrostic. Write POPULATION down the side of a page. Make each letter the first letter of a word, phrase or sentence about Population.

Further class and homework activities

Other countries of South East Asia have tried to implement population policies to slow down the growth of population. China is a well known example but there are others. Download and read the following articles and see if there are any lessons for India:

◆ www.cpirc.org.cn/en/eindex.htm

◆ news.bbc.co.uk/1/hi/world/asia-pacific/906114.stm

◆ news.bbc.co.uk/1/hi/world/asia-pacific/1011799.stm

Answers to the Activities

1 Rigidly set targets, that were attempted to be enforced through sterilisation and contraception. Did not have sufficient parallel economic growth and development.

2 As these countries are the largest in the world, they will increasingly dominate the world economy and the global environment.

3 Population is stable, birth rates have fallen to European levels, with relative economic prosperity.

4 This will depend upon students' opinions, which need to be justified. More recent policies are likely to be more successful as they are based on a broader based approach – and include the importance of factors such as education, the role of women, and building economic prosperity.

9 Development and international alliances

The big picture

These are the key ideas behind this chapter:

◆ Countries around the world are at different stages of development. There is a huge gap between the most and least developed countries.

◆ Development can be measured using a variety of indicators. The UN Human Development Index measures quality of life.

◆ Resource use is rising at an increasingly faster rate due to population growth and economic development.

◆ Huge numbers of people in LEDCs don't have access to the basics – sufficient food, or clean water.

◆ International influence is the ability of a country, or group of countries, to control or affect other countries. It can be measured in a variety of ways.

◆ Countries join forces to form an alliance in order to increase their international influence.

Note that the students' version of the big picture is given in the students' chapter opener.

Chapter outline

Use this, and the students' chapter opener, to give students a mental roadmap for the chapter.

9   **Development and international alliances** As the students' chapter opener, this unit is an important part of the chapter; see page 11 of this book for notes about using chapter openers

9.1   **Development – a world of two halves** What development means, and how it can be measured

9.2   **Mind the gap** Why some countries are more developed than others, and measuring development using the Human Development Index

9.3   **The rate of resource use** How population growth and economic development leads to increased resource use

9.4   **Food** Millions of people don't have enough food to eat

9.5   **Water – a matter of life and death** The difference in the availability and use of clean water in MEDCs and LEDCs

9.6   **Water – increasing supply** A large scale scheme – the High Aswan Dam in Egypt, and a small scale scheme – a well and pump in Ghana

9.7   **International influence** What international influence is, and how it can be measured

9.8   **International alliances** Examples of trading (the EU), social (UNICEF) and defence alliances (NATO)

Using the chapter starter

The photo on page 158 of the *geog.SG* students' book shows a technician from a MAG bomb clearance team, revealing an unexploded mine in the ground, in Cambodia. MAG (Mines Advisory Group) is a humanitarian organisation, operating as a charity, and funded by a number of governments. Its members aim to help people who are affected by the remains of war, in the form of unexploded mines. MAG claims to be the only organisation that employs women as well as men to clear mines, with around 30% of their workers being female.

Objectives and outcomes for this chapter

Objectives	Unit	Outcomes
Most students will understand:		Most students will be able to:
● That different aspects of development can be measured.	9.1	● Give at least six examples of development indicators including GDP per capita (PPP).
● That the UN's Human Development Index combines three aspects of human development to measure quality of life.	9.2	● State which aspects of human development the HDI measures.
● Why some countries are more developed than others.	9.2	● Give at least four reasons why there is a gap in development between the LEDCs and MEDCs; name at least three MEDCs and three LEDCs.
● How population growth and economic development increases resource use.	9.3	● Describe how population growth puts pressure on resources; describe three ways in which economic development leads to increased resource use.
● Where people are suffering from malnutrition; why, and what can be done about it.	9.4	● Describe where in the world people are suffering from malnutrition; explain why people are malnourished and what can be done.
● That there is a difference in access to clean water in MEDCs and LEDCs.	9.5	● Explain: why water supplies and access to clean water vary globally; why countries need water to develop; what can be done to improve water supply and access to safe water.
● That water supply can be increased through large scale schemes (but these can have benefits and disadvantages), and small scale schemes.	9.6	● Give an example of a large scale scheme to increase water supply in an LEDC and list the benefits and disadvantages; give an example of a small scale scheme in an LEDC and say why it is an example of appropriate technology and sustainable development.
● What international influence means and how it can be measured.	9.7	● Explain what international influence is: give five ways of measuring international influence.
● Why countries join forces in international alliances; that there are different types of international alliances.	9.8	● Explain why countries form international alliances; name three types of international alliances and explain how they work.

These tie in with 'Your goals for this chapter' in the students' chapter opener, and with the opening lines in each unit, which give the purpose of the unit in a student-friendly style.

Using the chapter starter continued

Cambodia is one of the world's most heavily mine-contaminated nations, as a result of conflicts over the last 35 years. There is often no record of where the mines are laid, so the conflict continues to have tragic consequences for years afterwards, with most of the victims being civilians, and many of those, children. Unexploded mines make useless large areas of land that could have supported families, and deprive people of basic amenities such as water. As such, these minefields stand in the way of the development of some of the world's poorest countries.

International organisations funded by wealthier countries such as Japan, the UK, and Australia are able to use their resources (in the form of money and people with specialist training) to assist in clearing the land of mines, and making it suitable for living on again.

The people are then able to re-establish their lives, growing sufficient crops to feed their own families. They are poor, but able to look after themselves once again. Organisations such as MAG are also able to train and employ local people, who are then able to sustain their family on their wages alone.

Development – a world of two halves

The unit in brief

This unit introduces students to development. It explains what development means and shows students how it can be measured using a variety of indicators. In the Activities, students use the map of GDP per capita to describe the distribution of wealth and analyse a table of development indicators.

Key ideas

◆ Development is about improving people's lives.

◆ Development can be measured.

◆ GDP per capita (PPP) is one of the easiest ways of measuring development.

◆ Social indicators (population and health) can be used to measure development, as well as other things.

Key vocabulary

development, more economically developed countries (MEDCs), less economically developed countries (LEDCs), GDP (Gross Domestic Product), GDP per capita, PPP (purchasing power parity), social indicators

Skills practised in the Activities

◆ Geography skills: describing the distribution of wealth shown on a GDP map; describing differences between countries

◆ Numeracy skills: analysing a table of development data

◆ Thinking skills: explaining anomalies; suggesting reasons; identifying indicators; explaining choices, explaining differences

Unit outcomes

By the end of this unit, most students should be able to:

◆ define or explain the terms given in 'Key vocabulary' above;

◆ understand that development is about improving people's lives;

◆ understand how development can be measured;

◆ describe the world distribution of GDP per capita (PPP);

◆ explain which indicators are good measures of development.

Ideas for a starter

1 What are the aspects of your home environment that you value - clean water, comfortable housing, employment, leisure, good diet, safety etc? Brainstorm to find out what students value, and use these to come up with a definition of what development is about.

2 Ask students to look at a list of development indicators in an atlas. Which indicators do they think will give the best definition of development?

3 Ask students: do you consider that you live in a developed part of the world? What is it that makes the UK developed?

4 Use the two photos and speech bubbles at the top of page 160 as a starter. What do they tell students about development?

Ideas for plenaries

1 Why do we need to know how developed or less developed certain countries are?

2 The Brandt report initially divided the world into 'north' and 'south'. Is this a useful way of describing the levels of development in the world?

3 Continue the conversation between the two teenagers at the top of page 160. Remember, the boy is from a MEDC and the girl is from an LEDC.

4 What is sustainable development? Try to come up with your own definition. Look at different resources to try to refine your definition. You could try these websites:

www.worldbank.org/depweb/ (click on What is sustainable development?)

www.sustainable-development.gov.uk/

www.johannesburgsummit.org/

news.bbc.co.uk/hi/english/static/in_depth/world/2002/disposable_planet/

5 Investigate two LEDCs. Why are these countries so poorly developed?

6 Make a graffiti wall of what students have learned today.

Further class and homework activities

Use data from the following sources to put together indicators of development. You will have to sample counties – use a maximum of 20.

www.cyberschoolbus.un.org/

http://web.worldbank.org/WBSITE/EXTERNAL/DATASTATISTICS/0,,menuPK:232599~page
PK:64133170~piPK:64133498~theSitePK:239419,00.html

Answers to the Activities

1 Richer countries are mostly the MEDCs, which are in the north. Obvious exceptions to this simple geographical pattern include Australia and South Africa. The division becomes increasingly indistinct when considering countries such as India and China, one above and one below the 'Brandt' line. There are also some countries that are developing very quickly, such as Brazil.

2 This is the currency that is most widely used for global trading purposes, and so is the most easy to use for comparisons.

3 This will depend upon students' choices – ensure that reasons are given.

4 a Students describe the figures for Kenya and the UK – the reasons for the differences are those that distinguish LEDCs and MEDCs, such as better health care and education, diet, etc.

b Students describe the figures for Japan and China. Although many differences are like those between the UK and Kenya, they are generally not as great.

5 Although it is possible to broadly divide the countries, it is quite difficult to give exact definitions because there are several measures used, and not all countries fit neatly into the 'rich' or 'poor' category for all of the indicators.

The unit in brief

In this unit students find out why some countries are more developed than others. The unit also introduces the HDI – the UN measure of quality of life. In the Activities, students describe the pattern of development on the map showing HDI and compare it with the GDP map in Unit 9.1; think about whether the HDI is a better measure of development than other indicators; and consider how the HDI rankings of countries will change in the future.

Key ideas

- There is a gap in levels of development between LEDCs and MEDCs.
- The gap in development has been caused by a variety of factors, including history, politics, industry, debt, environment and hazards.
- The UN produces the HDI which measures quality of life, not just wealth.
- The HDI for countries around the world is improving except for the new countries of central Asia and sub-Saharan Africa.

Key vocabulary

Human Development Index (HDI), quality of life

Skills practised in the Activities

- Geography skills: describing the pattern of development shown on the HDI map; comparing patterns on maps and explaining differences
- Thinking skills: explaining reasons; thinking; suggesting reasons

Unit outcomes

By the end of this unit, most students should be able to:

- define the terms given in 'Key vocabulary' above;
- recognise that there is a gap in levels of development between LEDCs and MEDCs;
- understand what has caused the gap in development;
- describe the pattern of development shown by the HDI;
- understand why the HDI for certain countries is getting worse.

Ideas for a starter

1 Ask students to report back from their investigation into why some LEDCs are so poorly developed (plenary **5** unit 9.1).

2 Brainstorm to find out what students know about why there's a gap in levels of development between MEDCs and LEDCs. Produce a spider diagram of students' responses.

3 Recap: the definition of development from unit 9.1. Does it contain an element of sustainability in it?

4 Ask students: How do you think the world's development is viewed from one of the poorer LEDCs? Put yourself in their shoes. Tell me why they are so poorly developed.

Ideas for plenaries

1 Work in groups of 3. Role play an interview between one person representing an LEDC and one person representing a MEDC. The interview is about the reasons for differences in levels of development. (The third person is the interviewer).

2 Go back to the spider diagram you produced in starter **2**. Ask students to modify it if necessary.

3 Choose one of these: History, Industry, Environment and Hazards, Debt and Politics. Draw a consequence map to show how it has held back development in the LEDCs.

4 Did you find anything difficult about the work in this Unit? What? Why? What would help to make it less difficult?

Further class and homework activities

1 Go to the United Nations Human Development Report website: hdr.undp.org/ Click on 'What is HD' and read the description of human development. Compare this with the definition you came up with in Unit 9.1 starter **1**.

2 Use the statistics part of the website to look at different countries and different aspects of development. Use this information to produce a wall display showing the level of development in developed and developing countries. Add pictures and flags to make your display look more interesting.

Answers to the Activities

1 In many ways this is similar to the rich / poor divide usually seen on LEDC / MEDC maps – Western Europe, Australia / New Zealand, and North America have the highest figures. Typically, it is the LEDCs that have the lower figures. An exception to this is the southern part of South America, where Argentina and Chile have high HDI figures.

2 Africa is the continent that comes out poorest in both maps. North America and Western Europe are high on both, as is Australia. There are some differences, for example parts of South America – where the HDI is relatively high, and variations in Asia.

3 Before industrialisation and globalisation. MEDCs grew rapidly and became wealthy very quickly. Part of this growth was at the expense of colonies of these countries – many of today's LEDCs – which became poorer as a result of poor trade relations and increasing dependence on a small number of products for export.

4 The HDI gives a composite measure, and so is likely to be more accurate than just one measure such as GDP. It is meant to give a broader picture of quality of life. GDP does not really do this, particularly as it does not give an indication of real wealth within a country.

5 One possibility is that it will not change much – actual figures may change as countries develop, but the ranking may not. Alternatively, some of the LEDCs may move up the rankings as their quality of life improves – perhaps, for example, Brazil.

The unit in brief

In this unit students learn why resource use is growing at an increasingly faster rate. It looks at population increase and economic development as the drivers for increased resource use. In the Activities, students complete a flow diagram to show the effect which population increase can have on the environment. They draw a bar chart of China's earnings (per head) and explain the connection between what the bar chart shows and China's increased fuel use. They also describe how economic development leads to increased resource use.

Key ideas

◆ A resource is anything – natural or man-made – which people can use.

◆ Population growth puts pressure on soil and water resources.

◆ In places like Burkina Faso, trying to feed growing populations has led to overcultivation and overgrazing. Declining rainfall and droughts exacerbate the problems.

◆ As countries become richer, resource use increases.

◆ LEDCs want to improve standards of living through economic development. As they develop, energy use increases.

Key vocabulary

overcultivation, overgrazing, evapotranspiration, drought, materialism, economic development

Skills practised in the Activities

◆ Geography skills: explaining the term 'overcultivation'; describing how economic development leads to increased resource use

◆ Numeracy skills: drawing a bar chart of earnings

◆ Thinking skills: completing a flow diagram to show how population increase can damage the environment; explaining a bar chart and making a connection between earnings and fuel use

Unit outcomes

By the end of this unit, most students should be able to:

◆ define the terms given in 'Key vocabulary' above;

◆ understand that a resource is anything which people can use;

◆ explain how population growth puts pressure on soil and water resources;

◆ describe how, as countries develop and become richer, resource use increases.

Ideas for a starter

1 Ask students: What do you understand by the term resources? Make a list of all the resources you can think of.

2 What resources do you use as a family? Make a spider diagram of the resources your family uses.

3 Write the dictionary definition of resources on the board. Ask students to develop it for Geography.

4 Show photos of a range of resources – food, water, fuel, soil, etc. Ask students what these things have in common.

Ideas for plenaries

1 What is the relationship between economic development and resource use?

2 Are the amounts of resources finite? Can you think of situations and factors that may lead to more of a resource being available? Think of a resource such as oil. Make a list and discuss it with the rest of the class.

3 Investigate energy resources in the UK. How is our energy produced? What issues do we face in terms of energy use and production? Prepare a short presentation of your findings.

4 What do you think are the five resources which would help a country to develop?

5 Take two minutes with a partner and think up one interesting question about resource use that we have not covered today.

Further class and homework activities

Find out about *ecological footprints*. Some websites such as www.bestfootforward.com/footprintlife.htm will help you calculate your ecological footprint.

Answers to the Activities

1 Overcultivation means growing too much; usually this means not letting the soil revive between sowings, so that there is no time for the soil to recover the nutrients lost.

2 upper box, middle: overgrazing; lower box, middle: overcultivation; end box: soil erosion

3 a Bar graph of China's earnings per head showing earnings generally rising, but a dip for 1986. Ensure that scales are appropriate, and check that both axes are labelled and graph has a title!

  b This is a good exercise in stressing that numbers should be extracted from the table. The answer would identify rises and falls; for example:

  In 1981, China's earnings per head were $285. This fell by $10 in 5 years, but improved to $351 by 1991. After that earnings per head rose steadily to $921, and were forecast to reach $1553 in 2006.

  c Total energy demand, million tonnes coal equivalent is:
  ◆ 1981 1100
  ◆ 1986 1700
  ◆ 1991 1750
  ◆ 1996 1770
  ◆ 2001 1800
  ◆ 2006 2500

  But answers will vary, according to the accuracy of graph-reading.

To look at possible correlations, students could **either** draw a scattergraph – this makes a nice plot, with only 1986 as an anomaly – **or** rank the data:

year	energy demand million tonnes coal equivalent	rank energy demand, highest ranked 1	earnings per head, $US	rank earnings per head
1981	1100	6	285	5
1986	1700	5	275	6
1991	1750	4	351	4
1996	1770	3	667	3
2001	1800	2	921	2
2006	2500	1	1553	1

. . . this shows that in four out of the six years energy demand and earnings have the same rank.

4 Economic development leads to increased resource use because:

  ◆ as we get richer we buy more things – like computers, plasma televisions, or play stations; all these use up resources

  ◆ many of the things we buy need energy, like electric toothbrushes or four-wheel drive vehicles

  ◆ as our living standards rise, we live longer, so use more resources.

The unit in brief

This unit is about food – how much and what type we need, what happens if you don't have enough, who's suffering from malnutrition and why. In the Activities, students identify the causes of the 2005 food crisis in Niger and the possible long-term problems that giving food aid to countries such as Niger might cause. They also consider whether poverty and hunger are problems for Africa or all LEDCs.

Key ideas

◆ The dietary energy supply is the number of calories each person needs.

◆ A balanced diet is necessary to maintain health.

◆ People with insufficient food suffer from malnutrition.

◆ The percentage of people suffering from malnutrition has fallen everywhere since 1970, except for sub-Saharan Africa, but the number of people affected has risen.

◆ Helping countries to develop and making trade fairer are two ways of increasing food for LEDCs.

Key vocabulary

malnutrition, dietary energy supply (DES), marasmus, kwashiorkor

Skills practised in the Activities

◆ Thinking skills: identifying causes; identifying problems; thinking; explaining reasons for agreeing or disagreeing with a statement

Unit outcomes

By the end of this unit, most students should be able to:

◆ define or explain the terms given in 'Key vocabulary' above;

◆ understand that a balanced diet is necessary to maintain health;

◆ explain what happens if you don't have enough food;

◆ explain what has happened to the percentage and number of people suffering from malnutrition since 1970;

◆ say what can be done to increase food supply in LEDCs.

Ideas for a starter

1 Ask students: What do you eat in an 'average' day? How much food do we need? (In MEDCs at least 2600 calories. Our average intake in the UK is 3317 calories.)

2 What do we mean by a 'balanced diet'? Is your diet balanced? What problems might people in MEDCs have by eating an unbalanced diet? What problems might people in LECDs have by eating an unbalanced diet? (Things like rickets, blindness.)

3 Use photographs and video clips from the news to highlight the problem of starvation and poverty. Contrast this with photographs from the affluent west to highlight the problem of distribution rather than production.

Ideas for plenaries

1 Draw a consequence map to show how providing huge amounts of food aid can result in low prices for local farmers and an eventual reduction in the number of farmers and locally produced food.

2 Should rich farmland in LEDCs be used to grow cash crops for export or food crops for the local market? Cash crops bring in money from abroad. Food crops will be consumed locally and produce a better nourished population. What is more important? Prepare a 3 minute speech to the rest of the class on this issue.

3 Draw a basic cycle of poverty and give to students. Ask them to annotate the cycle. Where would they break the cycle? Whose responsibility is it to break the cycle? Who can help?

4 Look at the map on page 167 showing the percentage of the population which is malnurished. Close your book and describe the pattern shown on the map to your neighbour.

5 Make 10-15 statements based on what students have learned so far, some true, some false. Students hold up True or False cards. Where statements are false, ask students to correct them.

Answers to the Activities

1 Repeated drought made worse by destruction of crops by plagues of locusts.

2 Dependency upon foreign aid, lack of sustainability of own agricultural systems.

3 This will depend upon students' opinions, which should be justified.

4 This will depend upon students' opinions, but it is true that the situation in Africa is worse than other developing parts of the world – there is little change in many African nations, while many other LEDCs are developing rapidly.

The unit in brief

This unit is about the availability of clean water in MEDCs and LEDCs. It considers what can be done to increase the supply of, or access to, safe, clean water. In the Activities, students study a map showing access to safe water to describe the pattern shown and compare this with a map of population distribution; contrast the way water is used in individual countries; and look at some of the solutions to water shortages.

Key ideas

◆ The supply of water is finite.

◆ The amount of water available varies spatially and temporally.

◆ One billion people don't have access to safe clean water.

◆ As countries develop and population increases, so does the demand for water.

◆ Solutions to the water supply problem range from UN initiatives to countries doing deals for water.

Key vocabulary

There is no key vocabulary in this unit.

Skills practised in the Activities

◆ Geography skills: describing the pattern of access to safe water shown by a map; comparing patterns on maps; contrasting water use

◆ Thinking skills: suggesting reasons and explaining them

Unit outcomes

By the end of this unit, most students should be able to:

◆ recognise that the supply of water is finite;

◆ understand that the amount of water available varies spatially and temporally;

◆ describe the global pattern of access to safe clean water;

◆ understand why the demand for water is increasing;

◆ suggest whether the solutions to the water supply problem are more suitable for LEDCs or MEDCs.

Ideas for a starter

1 Brainstorm to find out what students think are the basic necessities of life. Try to come up with an agreed list.

2 Hold up a bottle of muddyish (river) water. Ask: How would you like to drink this? Explain that over 1 billion people don't have access to clean piped water.

3 Recap: the water cycle. Draw it on the board and label it. Use it to explain why there is a crisis when water is in a closed system.

4 Use this article to discuss the water problem – Crisis? What crisis?

news.bbc.co.uk/hi/english/static/in_depth/world/2000/world_water_crisis/default.st

Ideas for plenaries

1 Compare the map on page 168 showing the percentage of population with access to safe water with the map on page 160. What does it tell you?

2 Write a letter from the United Nations to member states urging them to provide more support for water aid projects. Explain the importance of water supply to them.

3 It is 2015. Has the UN been successful in halving the number of people without access to safe drinking water and basic sanitation?

4 How will water use change around the world in the future? How can we make water supplies go further?

5 Find out about the different types of irrigation in Egypt. Be ready to give a short report at the beginning of the next lesson.

6 Tell me the 3 main things you have learned today. Now tell me another 3 things which are interesting but less important.

Further class and homework activities

1 Contact your local water supply company to find out if there are any water supply problems, or measures in place such as hose pipe bans to reduce water use.

2 Use the information in this unit and from the websites below to find out more information on the problem of access to water:

news.bbc.co.uk/1/hi/world/1887451.stm

www.wateraid.org/uk/

Write a short essay on the problem.

3 Do a survey to find out how much water is used by your family over the course of a week. What suggestions could you make to reduce the amount of water used?

4 Contact your local water supply company. Do they have suggestions on how water supply could be better used?

Answers to the Activities

1 Generally best in MEDCs, and also those countries where the HDI is relatively high, for example Argentina and Chile.

2 Some of the areas of greatest population density are those with lowest access to safe water, for example Bangladesh. Problems of increased diseases due to unclean water, particularly where population is increasing. Possibility of disputes and conflicts over future water supplies.

3 In MEDCs – UK and USA – greatest amount of water usage is for industry. In LEDCs it is for agriculture. The exception is Brazil, where there has been rapid industrial growth.

4 'Use less' is often cited as a means by which MEDCs may solve the problems of water shortages, although arguably this applies increasingly to LEDCs as well. 'Get technical' is usually more appropriate for MEDCs, and the others for MEDCs.

The unit in brief

In this unit students find out about two different approaches to increasing water supply. The examples included are the High Aswan Dam in Egypt (a large-scale project) and the provision of wells and pumps in Ghana (a small-scale example of appropriate technology). In the Activities, students complete a cost-benefit table for the High Aswan Dam and compare the advantages of large-scale projects like this with those of the small-scale scheme in Ghana.

Key ideas

◆ Water supply can be increased using both large and small-scale projects.

◆ The High Aswan dam brought both benefits and disadvantages.

◆ The provision of wells and pumps for rural communities in LEDCs are examples of appropriate technology and sustainable development.

Key vocabulary

non-governmental organisation (NGO), appropriate technology, sustainable development

Skills practised in the Activities

◆ Geography skills: completing a cost-benefit table for the High Aswan Dam and evaluating its impact; comparing schemes

◆ Thinking skills: explaining; identifying benefits and disadvantages; thinking

Unit outcomes

By the end of this unit, most students should be able to:

◆ define the terms given in 'Key vocabulary' above;

◆ understand that water supply can be increased using both large and small-scale projects;

◆ give examples of the benefits and disadvantages of the High Aswan Dam;

◆ explain why the provision of wells and pumps for rural communities in LEDCs are examples of appropriate technology and sustainable development.

Ideas for a starter

1 Ask students to report back on irrigation in Egypt. How efficient are the different types of irrigation? Draw up a table showing the advantages and problems associated with each type of irrigation. Use photographs to help the debate.

2 Read out the first paragraph under the heading 'Or is small beautiful' on page 171. then read the first two paragraphs of Apoyanga's story. Ask students to complete the story. What happened next? How did Apoyanga and her family's life improve? Who helped and how?

3 The expression 'small is beautiful' was first used by an economist called E.F. Schumacher. Ask students what they think it means. (He believed the best results in LEDCs came from small projects involving local people, as opposed to large projects controlled by international organisations which may exploit people and be unsustainable.)

Ideas for plenaries

1 Try to produce a model of linked boxes showing the positive effects of the High Aswan Dam. Be ready to draw yours on the board and discuss it with the class.

2 Why have people in Apoyanga's village not built a well themselves, ages ago? (Lack of knowledge/experience, lack of tools and materials, no spare money.)

3 Large scale development schemes can bring a great deal of prestige to a country and can bring rapid economic development, while small scale schemes may bring lasting sustainable development. Discuss this amongst the class, and come to a conclusion.

4 Activity 4 could be used as a plenary.

5 Create an acrostic. Write WATER SUPPLY down the side of a page. Make each letter the first letter of a word, phrase or sentence about water supply.

Further class and homework activities

Investigate an NGO which is working in developing countries to improve water supplies. Find out about some of the projects they are involved with.

Answers to the Activities

1 The majority of Egypt is a desert, and so therefore lacking in a reliable supply of water. The Nile provided that only source.

2 Benefits include those to agriculture, HEP from the dam and economic growth. Disadvantages include lack of silt downstream affecting soil fertility, increased pollution through the use of fertilisers for agriculture, and the re-settlement of people due to the creation of Lake Nasser.

3 Suitable for the people who use it – it is not expensive, complicated, or reliant upon input from other people.

4 This will depend upon students' opinions – ensure that they are justified.

The unit in brief

This unit is about international influence – how a country or group of countries can control or affect other countries. It looks at how international influence is measured, and how Scotland compares with the UK, EU, USA and Japan. In the Activities, students explain the meaning of the term 'international influence' and complete a spider diagram showing how it is measured. They also draw a bar chart showing the wealth of Scotland, the UK, EU, USA and Japan and rank them according to their wealth.

Key ideas

◆ International influence is the ability of a country, or group of countries, to control or affect other countries.

◆ International influence can be measured by:
  – wealth, measured as GDP per capita (PPP),
  – population, countries with big populations are considered to have more influence,
  – land area, if the land or sea surrounding the country is rich in natural resources,
  – military strength, meaning either the number of people in the armed forces, or the power of its weapons, or both,
  – natural resources – oil, gas, coal and iron ore being the most important.

◆ The USA is the world's wealthiest country; China has the greatest population and military strength; Russia wins in terms of natural resources and land area.

Key vocabulary

international influence

Skills practised in the Activities

◆ Geography skills: explaining the term international influence

◆ Numeracy skills: drawing a bar chart and ranking countries according to wealth

◆ Thinking skills: completing a spider diagram to show how international influence can be measured

Unit outcomes

By the end of this unit, most students should be able to:

◆ explain the term given in 'Key vocabulary' above;

◆ describe and explain how international influence is measured;

◆ draw a bar chart to show countries' wealth;

◆ rank countries according to their wealth.

Ideas for a starter

1 Ask the class: How can you influence other people? Write the ideas generated on the board as a spider diagram. Now ask: How can one country influence another? You should be looking for some commonality in the ideas generated, such as being the richest (countries with a high GDP per capita (PPP)); strongest (a country having a big military capability); biggest (a country with a large land area might contain valuable natural resources).

2 Show photos similar to those on page 172 of the student book, e.g. large scale open cast mining (to illustrate wealth of natural resources), and military parade (to illustrate military strength). Ask: How do these affect a country's wealth?

Ideas for plenaries

1 What are the five ways in which we can measure international influence?

2 Who is the world's number one in terms of wealth, population, resources, military strength and land area?

3 If you used starter 1 students can go back to the ideas generated for the question 'How can one country influence another?' and modify, or add to, them.

4 Choose a student to be in the hot seat. Another student asks him or her a question about international influence. Then nominate two different students (four to six pairs in total). There's one golden rule – questions cannot be repeated.

5 Write down as many words as you can relating to today's work.

Answers to the Activities

1 The definition given in the students' book is 'International influence is the ability of a country, or group of countries, to control or affect other countries'. Any explanation which gets across this idea is acceptable.

2 The spider diagram should include the five measures of international influence given in the students' book:
– wealth – usually measured as GDP per capita (PPP). Wealthy countries can spend money on things which improve people's lives, e.g. education, health care, new industry, etc. and so they have more influence.
– countries with large populations means the people can work to improve the economy and increase the country's wealth, etc. – but only if they are healthy.
– land area – this is important as the land or surrounding sea could contain valuable natural resources, such as coal, iron ore

or oil. If the land doesn't contain natural resources or the climate is unsuitable for crop growth, then a big country won't necessarily have more influence.
– military strength – a country with a strong military can protect itself and help other countries. Some countries spend money on the military instead of health care and education.
– natural resources, e.g. oil, gas, coal and iron ore. A country can increase its wealth by producing electricity and making manufactured goods.

3 a Students use the figures in the table in the students' book to draw the bar chart.
b Rankings are: 1 USA, 2 UK, 3 Japan, 4 USA, 5 Scotland.

The unit in brief

This unit introduces students to different types of international alliances. It looks at three different types of alliance and includes examples of each: a trading alliance – the EU; a social alliance – UNICEF; and a defence alliance – NATO. In the Activities, students explain how these three types of alliances work; find out which countries belong to NATO, mark them on a blank world map, and describe the pattern shown.

Key ideas

◆ Countries join with others in an alliance to increase their international influence.

◆ There are many different types of international alliances, including trading, social and defence alliances.

◆ In a trade alliance, countries work together to make it easy to buy and sell goods to, and from, one another, and make trading harder for non-member countries by using quotas and tariffs.

◆ In a social alliance, countries work together to develop policies on a social interest that they have in common, e.g. education, religion, etc.

◆ In a defence alliance, countries help each other if one is attacked by another outside the alliance.

Key vocabulary

international alliances, quotas, tariffs

Skills practised in the Activities

◆ Geography skills: marking and naming NATO members on a world map and describing the pattern shown

◆ Thinking skills: explaining how international alliances work

Unit outcomes

By the end of this unit, most pupils should be able to:

◆ define the terms given in 'Key vocabulary' above;

◆ name three different types of alliance and explain what they do;

◆ give an example of a trading alliance, a social alliance and a defence alliance;

◆ describe the pattern of NATO countries marked on a world map.

Ideas for a starter

1 Ask: What does the word 'alliance' mean? Give me examples of different types of alliance. Give me examples of those involving different countries.

2 Using the examples of alliances involving different countries in starter **1** ask students to group them into different categories including trading, social and defence alliances.

3 Show photos of European leaders at an EU meeting. Ask: Who are these people? What are they doing? Why? When did this organisation (the EU) start? What is it for?

Ideas for plenaries

1 What are the EU, UNICEF and NATO examples of? What do they do?

2 Make 10-15 statements about international influence and international alliances based on what students have learned so far, some true, some false. Students hold up True or False cards. Where statements are false, ask students to correct them.

3 Take two minutes with a partner and think up one interesting question about international alliances that we have not covered today.

4 Do an alphabet run from A-Z, with a word to do with development for each letter

Answers to the Activities

1 a Trading alliances, social alliances and defence alliances – these are the three covered in the students' book.

b In a trade alliance, the countries work together to make it easy to buy and sell goods to, and from, one another, and make trading harder for non-member countries by using quotas and tariffs. In a social alliance, countries work together to develop policies on a social interest that they have in common. In a defence alliance, countries promise to help each other if one country is attacked by another outside the alliance.

2 a Students can easily find out the NATO member states using the internet.

b Students may need to use an atlas to help them

c NATO members cluster either side of the North Atlantic, but also include other European countries, e.g. Poland and Hungary, and those in the eastern Mediterranean – Greece and Turkey.

 # Development, trade and aid

The big picture

These are the key ideas behind this chapter:

◆ There's a global imbalance in trade with the MEDCs taking the lion's share. So the rich North gets richer, and the poor South gets poorer.

◆ LEDCs have a number of trading problems, and trade isn't fair. But, fairer trade and more effective aid can help countries develop.

◆ Industry is global – TNCs often have their headquarters and main factory in an MEDC with smaller factories and offices in LEDCs.

◆ LEDCs need aid for a variety of reasons, but different types of aid have advantages and disadvantages.

◆ Many LEDCS work with aid agencies to introduce self-help schemes. These enable people to help themselves.

Note that the students' version of the big picture is given in the students' chapter opener.

Chapter outline

Use this, and the students' chapter opener, to give students a mental roadmap for the chapter.

10 Development, trade and aid As the students' chapter opener, this unit is an important part of the chapter; see page 11 of this book for notes about using chapter openers

10.1 Development and trade What trade is, and patterns of trade

10.2 Trade – problems and partners LEDCs trade problems, and trading partners

10.3 Trade – it's not fair Free trade, tariffs and quotas; trading blocs; the WTO; and Fair Trade

10.4 Global industry Transnational corporations – their advantages and disadvantages

10.5 Global industry – Ford in India A case study of one TNC in India

10.6 Aid – closing the gap What aid is, different types of aid, and why it is needed

10.7 Aid – is it all good news? Advantages and disadvantages of aid, and other ways to help LEDCs develop

10.8 Self-help schemes What self-help schemes are, examples, and how they work

Objectives and outcomes for this chapter

Objectives	Unit	Outcomes
Most students will understand:		Most students will be able to:
● What trade is, and patterns of trade.	10.1	● Define imports, exports, trade balance, trade surplus, trade deficit; describe the pattern of trade between MEDCs and LEDCs.
● That LEDCs face a number of problems when it comes to trade.	10.1, 10.2 10.3	● Describe and explain the problems facing LEDCs, e.g. the export of primary products, reliance on one or two export products, free trade, tariffs, quotas, trading blocs, etc.
● That MEDCs and LEDCs have different trading patterns and partners.	10.1, 10.2	● Describe the exports, imports and trading partners of Kenya and Japan.
● That it is possible to improve people's lives with Fair Trade.	10.3	● Explain how Fair Trade helps to improve people's lives in LEDCs.
● That TNCs can bring advantages and disadvantages.	10.4, 10.5	● Explain what TNCs are and where they operate; give examples of the advantages and disadvantages they can bring to LEDCs; give one example of a TNC in an LEDC; explain why it located there and whether it benefits the economy of the LEDC.
● What aid is, why it is needed, and that there are different types of aid.	10.6	● Define aid; explain why LEDCs need aid; describe different types of aid (government – bilateral, international organisations – multilateral, and voluntary); explain that aid can be short-term/emergency or long-term/sustainable.
● That different types of aid have advantages and disadvantages.	10.7	● List the advantages and disadvantages of different types of aid.
● What else can be done to help people in LEDCs to help themselves, develop their economy and reduce poverty.	10.7, 10.8	● Give at least two examples of self-help schemes and four examples of different things which can help LEDCs develop their economies and reduce poverty.

These tie in with 'Your goals for this chapter' in the students' chapter opener, and with the opening lines in each unit, which give the purpose of the unit in a student-friendly style.

Using the chapter starter

The photo on page 176 of geog.SG students' book shows a container ship leaving the port of Seattle, USA. It could be sailing for anywhere in the world.

Container ships carry most of the world's 'dry' cargo – this means manufactured goods. (Cargoes like metal ores, coal, and wheat are carried in bulk carriers.) They can carry up to 10 000 containers. The biggest are nearly 400 metres long – only oil tankers are bigger. Larger container ships are being built – they will be able to service a line of lorries over 60 miles long.

Containerisation made the shipping of products around the world cheaper. Consequently it has been said that without the container, there would be no globalisation.

Development and trade

The unit in brief

This unit introduces trade. Using Japan and Kenya as examples, it investigates what trade is, and patterns and imbalances of trade. In the Activities, students look at the differences in Japan and Kenya's exports; identify the problems caused by trade deficits; look at the effect of price fluctuations on countries exporting primary products; and think about the trade imbalance.

Key ideas

◆ Trade consists of imports and exports.

◆ The difference between imports and exports is the trade balance.

◆ Where the value of exports is higher than the value of imports then a country has a trade surplus.

◆ Where the value of imports is higher than the value of exports then a country has a trade deficit.

◆ Generally, MEDCs export manufactured goods and import primary products, while LEDCs export primary products and import manufactured goods.

◆ There is an imbalance in trade.

Key vocabulary

trade, imports, exports, trade balance, trade surplus, trade deficit, interdependent

Skills practised in the Activities

◆ Thinking skills: explaining differences; identifying problems; thinking

Unit outcomes

By the end of this unit, most students should be able to:

◆ define or explain the terms given in 'Key vocabulary' above;

◆ understand that trade consists of imports and exports and that the difference between imports and exports is the trade balance;

◆ give examples of countries with a trade surplus and a trade deficit, and identify the problems caused by a trade deficit;

◆ identify the problems caused for people in LEDCs which export primary products;

◆ say why the imbalance in trade doesn't appear to be improving.

Ideas for a starter

1 Ask students: where does your food come from? Mark the origin of the food on a blank map of the world. What pattern does this show?

2 Go to the World Trade Organisation website (www.wto.org/). Download the 10 advantages they give for world trading arrangements. Put these up on the board for the class. Do students agree with them?

3 Read the paragraph 'Trouble brewing' to the class, omitting the last sentence. Ask students: How are your lives linked to Elizabeth and Ibrahim's? This can lead to a discussion of 'what is trade', 'how does it work', etc.

Ideas for plenaries

1 Produce a flow map using the figures at the top of page 179 to show Kenya and Japan's imports and exports.

2 Work with a partner. Try to come up with a set of principles which could give the LEDCs a better and fairer trade deal. Try to build in the principles of sustainability as well.

3 Write 'development and trade' in the middle of your page. Create a concept map around the phrase. How many ideas can you come up with?

4 Test your partner! Ask your partner to close their book. Test them to see if they know the definitions of the key words in this unit (the ones in bold).

Further class and homework activities

1 Investigate the trade patterns of other LEDCs and see if they are similar to those shown for Kenya.

2 Investigate the prices of a selection of primary products over the last few years and compare these with the prices for manufactured items.

Answers to the Activities

1 Most of Japan's exports are manufactured goods – many hi-tech – and those from Kenya are primary.

2 Many countries, particularly LEDCs, rely heavily upon manufactured imports to balance their export of primary products – and so are likely to have a trade deficit.

3 More money leaving than entering the country, reliance on imports.

4 Variable demand, as some products may go in and out of fashion. In times of high production, prices may fall – as they may rise when production is limited. Prices may also change as alternative / substitute products become available.

5 It is in the interests of MEDCs to preserve the imbalance. Many LEDCs owe so much in interest charges on debt that it will be a long time before they are able to make progress. Exports of primary products, typically from LEDCs, are of lesser value than manufactured goods and services, predominantly from MEDCs.

help at a glance

The unit in brief

This unit looks at some of the other trading problems LEDCs have, and at trade partners – again using Japan and Kenya as examples. In the Activities, students look at the problems of LEDCs which rely on a small number of export products; suggest why the price of primary products fluctuates; classify statements about trade and describe and explain the pattern of trading partners for Japan and Kenya.

Key ideas

◆ Many LEDCs rely on the export of one or two goods.

◆ Trade creates winners and losers.

◆ All countries have trading partners.

◆ Japan imports most from China, and exports most to the USA.

◆ Kenya imports most from UAE and Saudi Arabia, and exports most to Uganda.

Key vocabulary

trading partners

Skills practised in the Activities

◆ Geography skills: describing and explaining the pattern of import and export partners

◆ Thinking skills: identifying and explaining problems; suggesting reasons; classifying statements

Unit outcomes

By the end of this unit, most students should be able to:

◆ explain the term given in 'Key vocabulary' above;

◆ explain the problems for LEDCs which rely on a small number of export products;

◆ understand that trade creates winners and losers;

◆ describe and explain the pattern of trading partners for Japan and Kenya.

Ideas for a starter

1 Recap: the problems of trade for LEDCs covered in Unit 10.1.

2 Show photos of container ships, freight planes, lorries, freight trains, etc. Ask: What are these vehicles doing? Where have they come from? Where are they going to? Introduce the idea of trading partners.

3 Your name is Ibrahim Shikanda (boys)/Ellizabeth Mitreso (girls). The cost of fertiliser is rising, and the price of sugar/tea is falling. What will happen to you? What does it mean for Kenya? Who will benefit at your expense?

Ideas for plenaries

1 Activity 3 could be used as a plenary.

2 Japan imports most from China – why?
   Japan exports most to the USA – why?

3 Which have a higher percentage of their workforce in farming – MEDCs or LEDCs? So, which will be harmed more by falls in the prices of primary products like sugar, bananas, cocoa, tea and coffee? Which will benefit most?

4 Draw a mind map about world trade. Include all the ideas and issues you have met so far.

5 Investigate the World Trade Organisation and trading blocs. Be ready to report back at the beginning of the next lesson.

Answers to the Activities

1 There could be a fall in prices and a reduction in earnings for the economy if there is a reduced demand for a product, a substitute comes on to the market, or there is surplus production.

2 Due to much greater fluctuation in demand, possible loss of crops due to natural hazards or diseases.

3 Political factors are dominant, with environmental factors being of greater weight in LEDCs.

4 Japan is closely linked to Asian neighbours, and the USA. Kenya has some links with Africa, then some European nations and some links with Asia.

Trade – it's not fair

The unit in brief

This unit explains some of the reasons why trade isn't fair. It looks at free trade, tariffs and quotas, trading blocs and the WTO. It also looks at Fair trade as a way of making trade fairer. In the Activities, students explain why countries introduce tariffs and quotas; why they form trading blocs; consider free trade and think about Fair trade products.

Key ideas

◆ Countries control trade with tariffs and quotas.

◆ Countries group together to form trading blocs to improve their trade balance.

◆ The WTO polices and promotes free trade, settles trade disputes and organises trade negotiations.

◆ Fair trade is where producers in developing countries get a guaranteed price for their product.

Key vocabulary

tariffs, quotas, free trade, trading blocs, World Trade Organisation (WTO), Fair trade

Skills practised in the Activities

◆ Thinking skills: explaining; identifying products; expressing opinions; giving reasons

Unit outcomes

By the end of this unit, most students should be able to:

◆ define or explain the terms given in 'Key vocabulary' above;

◆ understand that countries control trade with tariffs and quotas;

◆ understand why countries group together to form trading blocs;

◆ explain the role of the WTO;

◆ explain how Fair trade works.

Ideas for a starter

1 Ask students what they have found out about the World Trade Organisation and trading blocs.

2 Bring in some Fair Trade products and try them with the class. Compare them with other products – the chocolate is particularly nice!

3 Trade benefits MEDCs at the expense of LEDCs. Ask: How could the situation be changed to make it fairer to LEDCs? Would students be prepared to pay more for goods from LEDCs to make trade fairer for them?

Ideas for plenaries

1 What is to stop LEDCs grouping together to form their own trading group? Discuss this possibility with the class.

2 Should we in the UK try to eliminate trading blocs over time? How could this be done? Would it benefit the LEDCs?

3 Write to your local supermarket. Ask them what their policies are on stocking more produce from a fair trade source. Will they increase their fair trade produce?

4 Remember Elizabeth and Ibrahim on page 178? How could fair trade help them?

5 Write a short essay entitled *How can we make International Trade fairer for all?*

6 Choose a student to be in the hot seat. Another student asks him or her a question about trade. Then nominate two different students (4-6 pairs in total). There's one golden rule – questions cannot be repeated.

Further class and homework activities

1 Investigate the Fair Trade organisation. Survey your local supermarket and identify products that are traded under the Fair Trade brand.

2 Go to some of the NGO websites and try to find information about schemes which have been set up with farmers selling their produce through fair trading organisations.

Answers to the Activities

1 To protect their own products by making those from other countries more expensive, or limiting the number allowed to enter the country.

2 Any country may set tariffs and quotas. Typically, an MEDC will set a tariff or quota to protect its industries from cheaper goods from abroad – although often from other MEDCs, these increasingly may come from other parts of the world such as the NICs.

3 To gain greater strength in protection than is possible for individual countries.

4 This will depend upon students' opinions, which need to be justified.

5 Russia has not met the agreed rules of the WTO. This matters due to the size, power and influence of Russia.

6 This depends upon students' own opinions.

The unit in brief

This unit looks at transnational corporations and considers their advantages and disadvantages. In the Activities, students look at why TNCs have their headquarters in MEDCs and their factories in LEDCs. They identify the advantages and disadvantages for a country of having a branch of a TNC located there; and think about whether the arguments for and against TNCs are the same for the UK having a Toyota factory in Derbyshire, as for an LEDC. They also give their opinions on the increasing trend towards globalisation.

Key ideas

◆ Britain is part of a global market.

◆ Transnational corporations (TNCs) have offices and factories around the world.

◆ The revenue of some TNCs is larger than the total GDP of some countries.

◆ Some of the first businesses to become TNCs were car manufacturers.

◆ TNCs can have advantages and disadvantages for the countries they locate in.

Key vocabulary

global market, transnational corporation (TNC)

Skills practised in the Activities

◆ Thinking skills: explaining the location of TNC headquarters and factories; identifying the advantages and disadvantages of TNCs; giving opinions

Unit outcomes

By the end of this unit, most students should be able to:

◆ define or explain the terms given in 'Key vocabulary' above;

◆ understand that Britain is part of a global market;

◆ explain what a TNC is and why they manufacture products in LEDCs;

◆ describe how the revenue of some TNCs is larger than the total GDP of some countries;

◆ give six examples of TNCs;

◆ identify the advantages and disadvantages for an LEDC of having a TNC factory located there.

Ideas for a starter

1 Ask students: what are transnational corporations? Then brainstorm to find out how many transnational corporations they can come up with.

2 Brainstorm the range of products made by the transnational corporations listed in **1**.

3 Ask students to guess the names of the largest companies in the world and then look at the figures on page 185. You could also look up the current rich list on the internet.

4 Ask students which items at home are likely to have been manufactured by TNCs? Which country are they likely to have been made in? You could plot the countries on a blank world map.

Ideas for plenaries

1 Activity **3** could be used as a plenary.

2 Prepare a speech given by the head of a TNC outlining the reasons for opening a new factory in an LEDC.

3 What is globalisation? What are the arguments for and against globalisation?

4 Read Chen Ernu's story. Put yourself in her shoes. Write a letter from her to the boss of the company she works for. She's worried that she can't feed her family on the amount she earns and that her job isn't safe. Now write a reply from the boss.

5 Write an email from the head of the TNC which Chen Ernu's garment factory supplies. Tell the boss of the company that the factory must produce more goods, more cheaply, or else they will get the garments elsewhere.

6 Sum up what you have learned today in 35 words.

Further class and homework activities

Investigate some TNCs. Put together a fact file showing where they have their head office, where Research and Development is located, and what they manufacture.

Answers to the Activities

1 The capital needed to invest in and develop these industries is in MEDCs. The initial market for the products is in MEDCs. The infrastructure, knowledge and management expertise required to run a TNC is generally in MEDCs.

2 Cheaper labour costs, often cheaper other costs of production. Regulations may be less strict regarding labour rights, pollution, etc.

3 a Costs will include foreign ownership, most benefits going abroad, most management jobs being occupied by people from the HQ country. Benefits will include the creation of jobs, probably including hi-tech, possible improvements to the country's infrastructure to support the development of industry, economic growth.

b The evaluation will be a matter of judgement.

4 a This will depend upon students' opinions – ensure that they are justified. Costs and benefits will be similar to those outlined for question 3.

b There will be differences related to matters such as pollution and labour controls, which will be stricter in MEDCs. Arguably, MEDCs will gain more from having a TNC than will LEDCs.

5 This will depend upon students' opinions, which need to be justified.

The unit in brief

This unit is a case study of one TNC in an LEDC – the Ford Motor Company in India. In the Activities, students explain why Ford decided to locate in India; give their opinions about statements to do with Ford in India; and consider whether Ford has benefited the Indian economy. They also look at how far Ford has tried to make its car manufacturing in India sustainable.

Key ideas

◆ The Ford Motor Company is the world's second largest car manufacturer.

◆ In 1995 Ford went into partnership with Mahindra and Mahindra Limited in India and opened a high-tech manufacturing plant at Maraimalai Nagar.

◆ Ford chose to locate in India because of its huge population, high-earning middle class and cheap labour costs.

◆ Ford India pays its workers more than others in similar industries in India, and all are offered training.

◆ Ford India is working towards the sustainable use of resources in its production processes.

Key vocabulary

There is no key vocabulary in this unit.

Skills practised in the Activities

◆ Thinking skills: explaining why Ford located in India; giving opinions; thinking about benefits to the Indian economy and sustainable manufacturing; explaining answer

Unit outcomes

By the end of this unit, most students should be able to:

◆ explain why Ford chose to locate in India;

◆ describe how far Ford has benefited the Indian economy and identify any disadvantages of Ford locating in India;

◆ explain how far Ford India has tried to make its car manufacturing sustainable.

Ideas for a starter

1 Recap: the advantages and disadvantages that a TNC brings to an LEDC.

2 Ask: What do you know about the Ford Motor Company? Write the answers on the board as a spider diagram. Then ask: What has Ford got to do with geography? You can return to this for plenary **4**.

3 Show students a photo of a very old Ford car; a Jaguar, Land Rover or Aston Martin; and a map of India. Ask: What's the connection between these three images?

Ideas for plenaries

1 Draw up a table which compares the benefits that Ford locating in India brings to India, and the benefits that it brings to Ford. How much does India benefit?

2 How could the development of the car plant boost the local area? Draw a diagram to show this.

3 Use Activity **3** as a plenary. Discuss the notion of sustainability and see how it could be applied to a car plant.

4 If you used starter **2**, ask students to return to the spider diagram and add to, or modify, it.

5 Create an acrostic. Write 'transnational corporations' down one side of a page. Make each letter the first letter of a word, phrase, or sentence about TNCs.

Answers to the Activities

1 Large market – over 1 billion people. Despite being an LEDC, India has a large middle class and is therefore a relatively prosperous car buying country. Labour is cheaper than in MEDCs.

2 Answers will depend upon students' opinions.

3 Most components come from the local area. It is trying to bring in a more sustainable use of resources. Reuse of waste products. Cars are being designed so that the components may be re-cycled.

The unit in brief

This unit is about aid, with Niger and Zimbabwe used as examples. It looks at different types of aid, why aid is needed, what exactly aid is and how much aid we give. In the Activities, students identify when it is best to give short-term and long-term aid; look at the similarities and differences between the aid given to Niger and Zimbabwe and consider 'tied' aid and how much MEDCs should give in aid.

Key ideas

◆ Aid is when one country or organisation gives resources to another country.

◆ Aid can be provided by governments (bilateral aid), international organisations (multilateral aid) or voluntary organisations (NGOs like Oxfam).

◆ Aid can be short-term/emergency or long-term/sustainable.

◆ LEDCs need aid because of imbalances in trade, differences in levels of development and to recover from hazards.

◆ The UN's target is that MEDCs should give 0.7% of their income in aid to LEDCs.

Key vocabulary

aid, government (bilateral), international organisations (multilateral), voluntary, short-term/emergency, long-term/sustainable, donor, recipient

Skills practised in the Activities

◆ Thinking skills: thinking and giving examples; identifying similarities and differences; justifying opinions

Unit outcomes

By the end of this unit, most students should be able to:

◆ define or explain the terms given in 'Key vocabulary' above;

◆ understand that aid is when one country or organisation gives resources to another country;

◆ explain different types of aid;

◆ know that aid can be short-term/emergency or long-term/sustainable;

◆ understand why aid is needed;

◆ say whether MEDCs should give more or less than the UN's target of 0.7% of their income as aid.

Ideas for a starter

1. Brainstorm to find out if students know what aid is, what types of aid there are and why aid is needed. Who needs aid? Do MEDCs ever need aid?

2. Do students or their families give money to help people in LEDCs? E.g. do they contribute to Children in Need or did their family purchase one of the *Feed the World* singles? Should aid be something that individuals give or should it be up to governments to provide it?

3. Recap: the cycle of poverty used in plenary **3** Unit 9.4. How can aid help to break the cycle?

Ideas for plenaries

1. Draw up a table showing the advantages and disadvantages of different types of development aid – bilateral aid, multilateral aid and voluntary aid from NGOs.

2. Think of situations when aid from the different sources would be most appropriate.

3. Consider the following schemes for LEDCs and prioritise them. Bear in mind that any developments you make need to be sustainable.

   ◆ A gift of tractors and other farm machinery
   ◆ Electrical pumps for providing irrigation water
   ◆ Toilets built by a volunteer organisation such as World Challenge
   ◆ Two school teachers and school text books from UK
   ◆ Two places for teacher training at a UK university for two local people
   ◆ Irrigation equipment
   ◆ Supplies of grain
   ◆ A communal meeting house
   ◆ A medical centre and medical equipment
   ◆ A doctor funded for 12 months.

   Present your findings to the rest of the class and be ready to discuss the outcomes.

4. Draft an email to the American and UK governments. Tell them why you think they should give more money in aid than they did in 2003 (be polite!). Say how much they should give and what it could be spent on.

5. Sum up what you have learned today in 35 words.

Answers to the Activities

1 a In the case of an emergency situation to provide immediate help.

b For the longer-term benefit and development of a country.

2 Both are LEDCs receiving aid. Niger was in danger of becoming reliant on short-term food aid. Zimbabwe was being provided with longer-term techniques and tools to promote sustainable development.

3 a Aid that has to be spent on a particular purpose.

b Lack of opportunity to use the money for other purposes, or to respond to changing needs.

4 This will depend upon students' opinions, which should be justified.

The unit in brief

This unit looks at the advantages and disadvantages of different types of aid and asks students to think about whether aid is the best way to help LEDCs develop. In the Activities, students pull together what they have learnt about aid in Units 10.6 and 10.7; think about debt cancellation and consider whether trade and not aid is what LEDCs need to develop.

Key ideas

- ◆ Different types of aid have advantages and disadvantages.
- ◆ Aid has a role in helping LEDCs develop.
- ◆ There are many other things which can be done to help LEDCs develop.
- ◆ Many people think that trade, not aid, is the key to economic development.

Key vocabulary

government (bilateral), international organisations (multilateral), voluntary, short-term/emergency, long-term/sustainable,

Skills practised in the Activities

- ◆ Literacy skills: summarising information on aid
- ◆ Thinking skills: identifying benefits of aid; thinking; justifying opinions

Unit outcomes

By the end of this unit, most students should be able to:

- ◆ define or explain the terms given in 'Key vocabulary' above;
- ◆ identify the advantages and disadvantages of different types of aid;
- ◆ understand that while aid has a role in helping LEDCs to develop, there are many other things which can be done to help them;
- ◆ say whether they think that trade, not aid, is what is needed to help LEDCs to develop.

Ideas for a starter

1 Produce a word search for students. Include the following terms – trade, imports, exports, interdependence, trading bloc, WTO, tariffs, quotas, aid, donor, recipient. Once students have found the words ask them to write definitions for them.

2 Who can remind me what aid is? And who can remind me what different types of aid there are?

3 Brainstorm to find out how many ideas students can come up with of different ways of helping LEDCs develop their economies and reduce poverty. Record students' ideas on a spider diagram.

Ideas for plenaries

1 LEDCs have often borrowed lots of money to try to develop. The money has to be repaid – with interest. Should MEDCs cancel all the debts of the LEDCs? Do you think this is the best way to help LEDCs develop?

2 If students completed plenary 1 Unit 10.6, ask them to look back at the table of advantages and disadvantages of different types of aid they drew up. Ask them to modify their table if necessary.

3 Activity 4 could be used as a plenary.

4 If students completed plenary 3 Unit 10.1, ask them to look back at the concept map they created. Now ask them to write 'Development, Trade and Aid' in the middle of their page and create a concept map around the phrase. How many other ideas can they add?

5 Do an alphabet run from A-Z, with a word or phrase to do with development, trade or aid for each letter.

Further class and homework activities

Find out about the G8 summit and Live8 concerts in 2005. Do you think these initiatives were successful? What has changed since then?

Answers to the Activities

1 This will depend upon students' own learning. Government aid is often tied, so it has to be spent on particular projects. International aid is not usually tied, so it can be more flexible. Voluntary aid is useful, but is dependent upon how much people give. Short-term aid can provide help for emergencies, while long-term aid aims to be sustainable.

2 Poorer regions of the EU receive help from the EU, so strictly speaking they are receiving aid. It is a matter of opinion as to whether all aid to should go to the poorest nations, or some should be spent in MEDCs.

3 This will depend upon students' opinions, which should be justified.

4 Students should refer to the imbalance of trade between LEDCs and MEDCs, and it is clear that a more equitable system of trading would be in the long-term interest of LEDCs.

Self-help schemes

The unit in brief

This unit introduces students to the idea of self-help schemes. Many LEDCs work with aid agencies to introduce self-help schemes. These are where local people take part in activities to help themselves and their communities. In the Activities, students give examples of self-help schemes and explain how they could improve people's lives, and create a mind map about self-help schemes.

Key ideas

◆ Many LEDCs work with aid agencies such as Oxfam to introduce self-help schemes.

◆ Self-help schemes are where local people take part in activities to help themselves and their communities.

◆ Self-help schemes tend to be small-scale and use simple technology.

◆ Diguettes (magic stones), afforestation, education and training, and providing families with goats (or anything which multiplies), are examples of successful self-help schemes.

◆ Sustainable development means improving people's lives without wasting resources or harming the environment.

Key vocabulary

self-help schemes, diguettes, sustainable development, barefoot doctors

Skills practised in the Activities

◆ Geography skills: defining terms

◆ Thinking skills: suggesting other examples of self-help schemes and explaining how they would improve people's lives; creating a mind map about self-help schemes

Unit outcomes

By the end of this unit, most students should be able to:

◆ define or explain the terms given in 'Key vocabulary' above;

◆ explain why self-help schemes such as building diguettes and afforestation are appropriate for people in LEDCs;

◆ give at least two examples of successful self-help schemes;

◆ explain why the examples of self-help schemes chosen are successful.

Ideas for a starter

1 Ask: Who can tell me what self-help schemes are? Do they involve big organisations or local people? Are they big schemes or small ones?

2 An economist called E F Schumacher used the expression 'Small is beautiful' about development projects in LEDCs. What do you think it means?

   (E F Schumacher promoted sustainable development 'on the human scale'. He believed the best results in LEDCs come from small projects involving local people, as opposed to large projects controlled by international organisations which very often exploit people, and are unsustainable.)

3 Show photos of examples of self-help schemes in LEDCs, e.g. diguettes, afforestation, goats with kids, etc. Ask: What do these have in common?

Ideas for plenaries

1 Why couldn't people in eastern Ethiopia plant trees themselves ages ago? (Lack of knowledge/expertise, lack of tools, no spare money, etc.)

2 Many charities offer people in MEDCs like the UK ways to help those in LEDCs by buying gifts of sheep, goats, seeds and so on. Do you think this is a good idea? Why?/Why not?

3 Tell your neighbour the two key things you learned today.

4 Make a graffiti wall of what students have learned today.

Further class and homework activities

Many UK aid agencies work in LEDCs introducing self-help schemes. Find another example and prepare a class presentation.

Answers to the Activities

1 a Self-help schemes are where local people take part in activities (often introduced by aid agencies) to help themselves and their communities.

   b Sustainable development means improving people's lives without wasting resources or harming the environment.

2 a Answers will depend on students' knowledge, but could include examples such as self-help building schemes like those in Sao Paulo, Brazil, where groups of people are encouraged to help build their own homes, doing basic work such as digging trenches for water and sewage pipes. Local authorities provide building materials for the houses and the group provides the labour.

   b The example given in a improves people's lives by improving their living conditions, gives people new skills and helps to create a community spirit.

3 The mind map should include all the self-help schemes included in this Unit and any others that students are aware of. Students should explain the links between the ideas included on the mind map.

4 Appropriate technology meets the needs of the people using it. It is appropriate for them and the environment they live in.

 **Enquiry skills**

The big picture

These are the key ideas behind this chapter:

◆ Enquiry skills are important in geography. In terms of Standard Grade assessment, knowledge and understanding and enquiry skills are weighted 40%:60% respectively. Grades are based on external assessment and each paper assesses students in knowledge and understanding and enquiry skills.

◆ The syllabus specifies the following gathering techniques:
  – extracting information from maps
  – fieldsketching
  – measuring (rivers, weather)
  – recording observed information on a map (land-use, location, distributions)
  – observing and recording (traffic and pedestrian flows, environmental quality, buildings, services, weather)
  – compiling and using questionnaires and interviews.

◆ The syllabus specifies the following processing techniques:
  – classifying/tabulating/matrixing information
  – drawing graphs (bar, line, pie, scatter)
  – drawing maps (land-use, location, distributions)
  – drawing cross-sections/transects
  – annotating maps, graphs, and field sketches.

◆ Students may be encouraged to know that there are some educational uses that they can put their mobiles to – recording notes and data, taking photos, using them as voice reorders, etc. for fieldwork. Other new technologies such as the internet, PDAs, laptops, etc. can be used in a variety of ways to help with fieldwork.

Note that the students' version of the big picture is given in the students' chapter opener. Also note that some skills such as extracting information from maps and measuring the weather are included elsewhere in the students' book (e.g. Units 2.5 and 3.1 respectively), so they are not included in this chapter.

Chapter outline

Use this, and the students' chapter opener, to give students a mental roadmap for the chapter.

11 **Enquiry skills** As the students' chapter opener, this unit is an important part of the chapter; see page 11 of this book for notes about using chapter openers

11.1 **Questionnaires and interviews** When, and how, to use questionnaires and interviews, advantages and disadvantages of different types of questionnaire

11.2 **Surveys** Environmental quality surveys and traffic surveys

11.3 **Photography and field sketching** How and why photos, field sketches and sketch maps are used in geography fieldwork

11.4 **River fieldwork** What to measure, and how to measure it; drawing cross-sections

11.5 **New technologies for fieldwork enquiry** How to use the internet, mobile phones, PDAs and laptops to help with fieldwork

11.6 **Processing technique**s How to use different processing techniques in geography

Objectives and outcomes for this chapter

Objectives	Unit	Outcomes
Most students will understand:		Most students will be able to:
● The advantages and disadvantages of using different types of questionnaires.	11.1	List the advantages and disadvantages of face-to-face, postal and on-line questionnaires.
● When a questionnaire should be used, and when an interview should be used.	11.1	Explain that they would use questionnaires and interviews for seeking different types of information.
● What an environmental quality survey would be used for.	11.2	Explain what an environmental quality survey will tell you; give three examples of when an environmental quality survey could be used.
● The difference between a field sketch and a sketch map.	11.3	Explain what a field sketch is, and what a sketch map is. Draw a field sketch and a sketch map.
● That most river investigations will measure the channel characteristics and velocity of a river; how to draw a river cross-section.	11.4	Describe how to measure the width, depth and speed of a river, height of the bank and width of river channel; draw and label a river cross-section.
● How new technology can be used in geography fieldwork.	11.5	Give six ways that new technology can be used in geography fieldwork.
● How and why different processing techniques are used in geography.	11.6	Give five examples of different processing techniques. Justify their use and give examples of how they could be used.

These tie in with 'Your goals for this chapter' in the students' chapter opener, and with the opening lines in each unit, which give the purpose of the unit in a student-friendly style.

Using the chapter starter

The photo on page 194 of the *geog.SG* students' book shows a man taking a photo of the Red Rock Canyon in Nevada, USA.

Red Rock Canyon is a narrow valley that lies about 15 miles west of downtown Las Vegas, and is a National Conservation Area. This means that it is preserved and managed by the state in order to protect its unique geology, plants and animals.

Park Rangers are employed to look after these areas of special interest and beauty, and they have to learn the skills that will allow them to protect and improve their environment. These skills also allow them to gather and distribute scientific information to both visitors and researchers, so that everyone can learn more about what makes the land so unusual, and why it needs to be preserved.

Glossary

A

abrasion – the scratching and scraping of a river bed and banks by the stones and sand in the river

agribusiness – large-scale capital-intensive farming

agri-environment scheme – schemes which combine farming with looking after, and improving, the environment, such as Tir Gofal in Wales and Environmental Stewardship in England

air pressure – the weight of air pressing down on the Earth's surface. Low pressure means warm air is rising, so rain is on the way. (The rising air cools and its water vapour condenses.)

altitude – the height of the land above sea level

appropriate technology – meets the needs of local people and the environment they live in

arête – a sharp ridge between two corries (see corrie)

B

biome – a very large ecosystem. The rainforests are one biome. Hot deserts are another

birth rate – the number of live births in a country in a year, per 1000 people

C

CAP – Common Agricultural Policy set up by the EU and which subsidised farmers

CBD – central business district. It's the area at the centre of a town or city where you find the main shops and offices

cloud cover – how much of the sky is hidden by cloud. It is given in eighths (oktas).

cloud type – there are five main types of cloud: stratus, cumulus, nimbus, cumulonimbus and cirrus

commercial farming – outputs from the farm are sold to make a profit

corrie – a circular, armchair-shaped hollow cut into rock by ice during glaciation

counter-urbanisation – the movement of people out of cities to smaller towns and villages

crevasse – a vertical or wedge-shaped crack in a glacier

D

death rate – the number of deaths in a country in a year, per 1000 people

deforestation – clearing forest for another use. For example cutting down rainforest to make way for a motorway or cattle ranches

deindustrialisation – the decline in manufacturing (secondary) industry, and the growth in tertiary and quaternary industry

delta – a flat area at the mouth of a river, made of sediment deposited by the river

desertification – when soil in a savanna region gets worn out, dusty and useless

dietary energy supply – the number of calories per person available each day

distributaries – if sediment blocks a river it has to divide into small channels called distributaries

drainage basin – the land around a river, from which water drains into the river

drumlin – a smooth hill shaped by glaciers

E

economic migrants – people who move voluntarily for jobs and higher wages

ecosystem – a unit made up of living things and their non-living environment. For example a pond, a forest, a desert

employment structure – what % of workers are in the primary, secondary and tertiary sectors of the economy

esker – a long ridge of material deposited from streams flowing under glaciers

estuary – the mouth of a large river, which is affected by the tides. As the tide rises, sea water flows up into the estuary and mixes with the river water

extensive farming – has smaller inputs of labour, money or technology than intensive farming. Extensive farms are usually larger than intensive farms

F

factors – things which affect where industry, agriculture, settlements etc will locate

feedback – things are put back into the system – like profits which may be reinvested

fertility rate – the number of children, on average, a woman will have in her lifetime

floodplain – flat land around a river that gets flooded when the river overflows

footloose – an industry which is not tied to raw materials and so can choose where to locate

free trade – when goods and services can flow freely from country to country, without any taxes

freeze-thaw weathering – the weathering (breakdown) of rock by the action of water getting into cracks in the rock, freezing and thawing

G

GDP per capita (PPP) – GDP is gross domestic product. It is the total value of the goods and services produced in a country in a year. GDP per capita means the GDP divided by the population. PPP means purchasing power parity. GDP is adjusted because a dollar buys more in some countries than others

glacial trough – a steep-sided U-shaped valley caused by glaciers

global warming – the way temperatures around the world are rising. Scientists think we have made this happen by burning too much fossil fuel

green belt – an area of open land around a city, which is protected from development. This is to stop the city spreading further

greenhouse gases – gases like carbon dioxide and methane that trap heat around the Earth, leading to global warming

groundwater flow – the flow of groundwater through saturated rock or soil

H

hanging valley – a high-level tributary valley with a sharp fall to the main valley; a feature of glacial erosion

high-tech industry – an industry that develops and produces new and advanced products. For example new kinds of mobile phones or medical drugs

humidity – the % of water vapour in the air

I

impermeable – doesn't let water through

infant mortality – the number of babies out of every 1000 born alive, who die before their first birthday

infiltration – the soaking of rainwater into the ground

inputs – things that go into a system. They can be physical, or human and economic.

intensive farming – has large inputs of labour, money or technology to produce high outputs. Farms are usually quite small

interception – the capture of rainwater by leaves. Some evaporates again and the rest trickles to the ground

interlocking spurs – hills that stick out on alternate sides of a V-

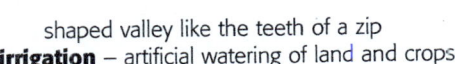
shaped valley like the teeth of a zip

irrigation – artificial watering of land and crops

K

kame – a mound or heap of material dropped from a glacier

L

latitude – distance north or south of the equator, expressed in degrees and minutes (for example, 51° 5′ North). Lines of latitude are imaginary circles drawn round the Earth parallel to the equator

levees – embankments built up on either side of a river channel

life expectancy – how many years a new baby can expect to live, on average. Life expectancy is higher for females than for males

M

materialism – wanting only belongings or comfort, and having no interest in morals

meander – a bend in a river

Mediterranean climate – a climate type that has hot dry summers and warm wet winters, named after the climate conditions found around the Mediterranean Sea

melt-water – the water produced when snow or ice melts. Glacial melt-water is produced by melting at the surface of the glacier and by pressure at the base of the glacier

misfit stream – a small stream in the bottom of glacial trough

moraine – material carried by a glacier

mouth – the end point of a river, where it enters the sea or a lake

N

natural increase – the birth rate minus the death rate for a place. It is always given as a % of the total population

NGO – non-governmental organisation. NGOs work to make life better, especially for the poor. Oxfam, the Red Cross and Greenpeace are all NGOs

O

ocean current – the movement of warm or cold ocean water, to a depth of about 100 metres. Cold ocean currents start in polar regions, warm ocean currents start in tropical waters

outputs - things that come out of the system (products)

ox-bow lake - a lake formed when a loop in a river is cut off by floods

P

percolation – the movement of water downwards through rock

plucking – when ice freezes on to rock, moves, and so plucks the rock away

population density – the average number of people per square kilometre

population growth rate – the number of people added to a population each year due to natural increase and net migration. It is given as a %

porous – lets water soak through

precipitation – water falling from the sky. It could fall as rain, hail, sleet or snow

prevailing wind – the one that blows most often; for the British Isles it's the south-west wind (blowing from the south-west)

primary industry – people extract raw materials from the land or sea. For example farming, fishing and mining

processes – things that happen in the middle of the system to turn inputs into outputs

pyramidal peak – the peak formed when three or more corries

form round a mountain (see corrie)

Q

quaternary industry – people are employed in industries providing information and expert help. For example IT consultants and researchers

quota – a limit on the amount of goods produced or purchased

R

refugee – a person who is forced to flee from danger (for example war or an earthquake) and seek refuge in another country

river channel – the bed and sides of a river form a river channel

river terraces – areas of flatter land above the floodplain

rural-urban fringe – the area where a town or city meets the countryside

S

secondary industry – people make, or manufacture, things. For example turning iron ore into steel, making cars and building houses

self-help scheme – a project or activity that helps people to help themselves; this term is often used about small-scale development schemes in LEDCs

set-aside land – land which isn't used for growing crops or keeping animals on; farmers are paid for this

source – the starting point of a river

sphere of influence – area around a settlement (or shop, or other service) where its effect is felt. London has a very large sphere of influence

SPS – Single Payment System, part of the CAP reform. Farmers now get one single payment a year instead of several different subsidy payments

striation – scratches in rock caused by abrasion in a glacier

subsistence farming – where farmers grow food to feed their families, rather than to sell

sunshine – in the study of weather and climate, sunshine is measured in hours

surface run-off – rainwater that runs across the surface of the ground and drains into the river

sustainable – can be carried on without doing any harm (to people, or other living things, or the environment)

sustainable development – development that will not lower our quality of life or harm the environment

sustainable management – meeting the needs of people now and in the future, and limiting harm to the environment

system – has inputs, processes and outputs. Industry and agriculture can be described as systems

T

tariff – a tax that a country places on goods being imported or exported

temperature – how hot or cold something is, usually measured in degrees Centigrade

tertiary industry – people are employed in providing a service. For example the health service (doctors, nurses, dentists) and education (teachers)

through-flow – the flow of rainwater sideways through the soil, towards the river

till – jumbled, unsorted material dropped by glaciers

trade balance – the difference between the value of imports and exports of a country

trade balance – the difference between the value of imports and exports a country

trade deficit – a country spends more on imports than it earns from exports

trade surplus – a country earns more money from exports than it spends on imports

trading bloc – a group of countries that have joined together to improve trade

TNC (transnational corporation) – a company with branches in many countries

tributary – a smaller river flowing into a larger river

truncated spurs – where a glacier has eroded and cut off interlocking spurs

tundra – the cold, treeless plains of northern Canada, Alaska, northern Europe and northern Russia

U

urban model – a simplified diagram of the way land is used in a city

urban redevelopment – clearance and rebuilding of old inner city areas

urban renewal – improving (without knocking down and clearing) old inner city areas

urban zones – areas of different land use in an urban area

urbanisation – an increase in the percentage of people living in towns and cities

U-shaped valley – see glacial trough

V

visibility – the greatest distance you can see, in km or m. On a foggy day it could be just 1 or 2 metres

V-shaped valley – a valley shaped like the letter V, carved out by a river

W

waterfall – where a river or stream flows over a steep drop

watershed – an imaginary line separating one drainage basin from the next

wind direction – the direction the wind blows from

wind strength (speed) – how fast the wind blows

World Trade Organisation – a body set up to help trade between countries

Klipp und Klar
Lösungen

Übungsgrammatik
Mittelstufe B2 / C1

Ernst Klett Sprachen GmbH
Stuttgart

Bildquellennachweis

130.1; 130.3 Fotolia.com (T. Michel), New York; **130.2** Fotolia.com (PictureP.), New York; **131.1; 131.6** Thinkstock (iStockphoto), München; **131.2; 131.8** shutterstock (Becky Stares), New York, NY; **131.3; 131.4; 131.7; 131.10** Fotolia.com (T. Michel), New York; **131.5; 131.9** URW, Hamburg; **Cover_4.1; Cover_4.2** Klett-Archiv, Stuttgart

Sollte es einmal nicht gelungen sein, den korrekten Rechteinhaber ausfindig zu machen, so werden berechtigte Ansprüche selbstverständlich im Rahmen der üblichen Regelungen abgegolten. Die Positionsangabe der Bilder erfolgt je Seite von oben nach unten, von links nach rechts.

Aufgrund der sprachlichen Vielfalt können die Lösungen nur eine Auswahl von richtigen Lösungsmöglichkeiten bieten.

In den Lösungen werden für Aufgaben zur freien Textproduktion Musterlösungen vorgeschlagen, diese sollen der Orientierung bzw. Hilfestellung dienen.

Auch für geschlossene Aufgabentypen gilt: Abweichende Wortstellung oder alternative Formulierungen können ebenso korrekt sein wie die hier vorgegebene Lösung.

1. Auflage   1 ⁵ ⁴ ³ ² ¹ | 2016 15 14 13 12

Redaktion: Eva Neustadt
Layoutkonzeption: Sandra Vrabec
Satz und Gestaltung: Eva Mokhlis, Swabianmedia, Stuttgart
Coverbild: shutterstock (stavklem), New York, NY
Illustrationen: Juan Pablo Amorocho, Leipzig
Umschlaggestaltung: Sandra Vrabec
Druck und Bindung: Medienhaus Plump GmbH, Rheinbreitbach
Printed in Germany

ISBN 978-3-12-675307-4

Inhaltsverzeichnis

1 Der Satz und seine Elemente

1.1 Das Verb und seine Ergänzungen

1.1

> Hi Kerstin,
>
> du **arbeitest** bestimmt noch an deiner Diplomarbeit und **hast** mir deshalb nicht **geantwortet**.
> Nur ganz kurz: **Kommst** du am Samstag zu Marios Party? Er **feiert** seinen Geburtstag wie letztes Jahr im Park. Wir wollen grillen, hoffentlich bleibt das Wetter bis zum Wochenende schön …
>
> Liebe Grüße aus Berlin
> Lena

> Hallo Lena,
>
> bitte entschuldige, dass ich dir erst jetzt **schreibe**. Ich **sitze** seit Monaten nur noch in der UB. Aber es hat sich gelohnt: Morgen **gebe** ich endlich meine Diplomarbeit **ab**! ☺ ☺ ☺
> Ich freu mich so auf den Moment.
>
> Danke, dass du mich an Marios Geburtstag **erinnerst**! Klar, ich komme auf jeden Fall und **bringe** einen Salat **mit**. Mein Nudelsalat **schmeckt** immer allen, oder?
> Wie alt wird Mario eigentlich? Und was **schenken** wir ihm denn? Hast du eine Idee?
>
> Ich ruf dich morgen mal an, viele Grüße
> Kerstin

1.2

Verb	Dativ-Ergänzung	Akkusativ-Ergänzung	Präpositional-Ergänzung
arbeiten			an deiner Diplomarbeit
antworten	**mir**		
kommen			**zu Marios Party**
feiern		**seinen Geburtstag**	
schreiben	**dir**		
sitzen			**in der UB**
abgeben		**meine Diplomarbeit**	
erinnern		**mich**	**an Marios Geburtstag**
mitbringen		**einen Salat**	
schmecken	**allen**		
schenken	**ihm**	**was**	

2 c > Die beiden Freundinnen **fragen den Verkäufer nach einem guten Buch.**
d > **Der Verkäufer berät seine Kundinnen.**
e > **Die beiden Kundinnen entscheiden sich für einen interessanten Titel.**
f > **Kerstin bezahlt das Geschenk mit einer Kreditkarte.**
g > **Später gibt Kerstin ihrem Freund das Buch.**
h > **Er freut sich über die schöne Überraschung.**

3 **b** Kerstin hat auf ihre E-Mail geantwortet. → **Präpositional-Ergänzung**
      beantworten: **Kerstin hat ihre E-Mail beantwortet.** → Akkusativ-Ergänzung

    c Lena kümmert sich um einen Grillabend. → **Präpositional-Ergänzung**
      planen: **Lena plant einen Grillabend.** → Akkusativ-Ergänzung

    d Mario und Thomas haben über den Vorschlag gesprochen. → **Präpositional-Ergänzung**
      besprechen: **Marion und Thomas haben den Vorschlag besprochen.** → **Akkusativ-Ergänzung**

    e Kerstin hilft ihrer Freundin beim Einkaufen. → **Dativ-Ergänzung**
      unterstützen: **Kerstin unterstützt ihre Freundin beim Einkaufen.** → **Akkusativ-Ergänzung**

    f Kerstin und Lena begegnen unterwegs einem Bekannten. → **Dativ-Ergänzung**
      treffen: **Kerstin und Lena treffen unterwegs einen Bekannten.** → **Akkusativ-Ergänzung**

4.1

Kleine Geschichte der Konsumgesellschaft

Das Buch von Wolfgang König befasst sich **mit** dem Konsum als Lebensform der Moderne. Nach einem einführenden Kapitel setzt sich der Autor **mit** den Voraussetzungen der Konsumgesellschaft und den verschiedenen Konsumfeldern auseinander. In drei weiteren Kapiteln geht es **um** „Konsumverstärker", „Individualisierung und Globalisierung als säkulare Prozesse" und um „Kritik und Grenzen der Konsumgesellschaft".
König geht in seinem Buch **auf** die Frage ein, wie es zur heutigen Konsumgesellschaft gekommen ist. Er verweist vor allem **auf** das Wechselverhältnis von Konsumtion und Produktion. Erst im 20. Jahrhundert aber könne man **von** einer Konsumgesellschaft sprechen, da erst in diesem Jahrhundert die Mehrheit der Bevölkerung zunehmend ein Einkommen erzielt habe, das deutlich über dem lag, was für die Befriedigung der Grundbedürfnisse notwendig war. Erst jetzt konnte die Mehrheit der Bevölkerung **an** neuen Konsumformen teilhaben, sodass dem Konsum herausragende kulturelle, soziale und ökonomische Bedeutung zukam ...

4.2 Die Studierenden befassen sich *mit* einem Problem. → Dativ
     Die Studierenden beschäftigen sich **mit** ein**em** Problem. → **Dativ**
     Die Studierenden setzen sich **mit** ein**em** Problem auseinander. → **Dativ**

     Es geht **um** ein**en** Vorschlag des Bildungsministeriums. → **Akkusativ**
     Sie gehen **auf** ein**en** Vorschlag des Bildungsministeriums ein. → **Akkusativ**
     In der Diskussion verweisen sie **auf** ein**en** Vorschlag des Bildungsministeriums. → **Akkusativ**

     Die Studierenden sprechen auch **von** ihr**em** Projekt. → **Dativ**
     Die Studierenden erzählen **von** ihr**em** Projekt. → **Dativ**
     Noch mehr Studierende sollen **an** dies**em** Projekt teilhaben. → **Dativ**

5 **b** Du kannst dich **auf mich** verlassen.
    c Ich kümmere mich **um den Fall**.
    d Du musst nicht **auf deinen Wunsch** verzichten.
    e Du kannst **bei deiner Idee** bleiben.
    f Du kannst **für dein Recht** kämpfen.
    g Ich glaube **an dich**.
    h Ich bemühe mich **um eine schnelle Lösung des Problems**.
    i Ich fange **mit der schlechten Nachricht** an.
    j Du musst dich schnell **an die neue Situation** anpassen.
    k Du solltest **über die Konsequenzen** nachdenken.

1.2 Angaben

1 b Auch in diesem Jahr will Mario mit Lena eine Party im Park organisieren.
      Wann?   **Mit wem?**   **Wo?**

  c Kerstin kommt jedes Jahr zu Marios Geburtstag für eine Woche aus Heidelberg.
          Wann?   **Wie lange?**   **Woher?**

  d Wegen ihrer Diplomarbeit verpasst sie dieses Jahr die Vorbereitungen.
      Warum?   **Wann?**

  e In Heidelberg hat sie in den letzten Monaten intensiv an ihrer Diplomarbeit gearbeitet.
     Wo?   **Wann?**   **Wie?**

  f Aber morgen fährt sie mit dem Zug nach Berlin.
     Wann?   **Womit?**

2.1 Beispiellösung:

  b Sie hat uns abgeholt. → Wo?
    > Sie hat uns **(mit ihrem Auto) am Bahnhof abgeholt.**

  c Wir sind zu ihr gefahren. → Womit?
    > Wir sind **(dann gleich) mit dem Auto zu ihr gefahren.**

  d Wir haben im Stau gestanden. → Warum? Wie lange?
    > Wir haben **wegen eines Unfalls fast eine Stunde im Stau gestanden.**

  e Wir haben uns die Stadt angesehen. → Wann? Mit wem?
    > Wir haben uns **am Abend mit meiner Schwester und ihrem Mann die Stadt angesehen.**

2.2 Beispiellösung:

Letzten Monat habe ich meine Schwester in München besucht. Sie hat uns am Bahnhof mit ihrem Auto abgeholt. Dann sind wir mit dem Auto zu ihr gefahren. Wegen eines Unfalls haben wir fast eine Stunde im Stau gestanden. Abends haben wir uns mit meiner Schwester und ihrem Mann die Stadt angesehen.

1.3 Funktionsverbgefüge

1

Reitverein Nordhessen
Protokoll der Vereinssitzung vom 22. Juni 2012

TOP 1: Begrüßung
Nach der Begrüßung und Eröffnung der Sitzung dankt die Vorstandsvorsitzende den Mitgliedern ausdrücklich für die freiwillig geleistete Arbeit im Verein.

TOP 2: Sinkende Mitgliederzahlen
Die Vorstandsvorsitzende berichtet vom weiteren Rückgang der Mitgliederzahlen. In der folgenden Diskussion **kommt** immer wieder zur Sprache, dass v.a. Kinder und Jugendliche im Verein fehlen. Um mehr Kinder für den Reitverein zu begeistern, bietet sich nach Ansicht von Frau Haller eine Zusammenarbeit mit der Käster-Schule an. Frau Haller, Grundschullehrerin an der Kästner-Schule, ist bereits mit dem Schuldirektor ins Gespräch **gekommen**: Die Schule **zieht** ein Reitangebot im Nachmittagsunterricht in Erwägung.

Frau Haller **erhält** vom Vorstand die Erlaubnis, im Namen des Reitvereins die Zusammenarbeit mit der Kästner-Schule zu besprechen. Sie wird den Vorstand über Neuigkeiten in Kenntnis **setzen**.

TOP 3: Renovierung Vereinsheim
Herr Sommer **bringt** erneut seine Sorge über den Zustand des Vereinsheims zum Ausdruck. Der Vorstand **fasst** den Entschluss, das Vereinsheim möglist bald zu renovieren, bisher wurde jedoch nichts unternommen.
Da im Moment kein Geld für die Renovierung zur Verfügung **steht**, ist der Verein auf freiwillige Helfer und Spenden angewiesen …

> Die Bedeutung der Verben ist abgeschwächt und eher abstrakt.

2

Zusammenfassung „Spracherwerbsforschung"

Die wichtigste These des Buches „Der Gegenstand der Fremdsprachenerwerbsforschung ist nicht der Lehrprozess *(teaching)*, sondern der Lernprozess *(learning)*" **bringt** der Autor auf verständliche Art und Weise zum Ausdruck.

Der Autor **stellt** seine These durch viele Beispiele unter Beweis und **nimmt** Bezug auf die psycholinguistische Perspektive des Lernens Er fragt, wie Lerner ein lin-guistisches System entwickeln und welche Faktoren Einfluss auf den Lernprozess **ausüben**. Er **zieht** die Beziehung zwischen dem sprachlichen Input und Output in Betracht.

Im Buch werden weitere spannende Fragen aufgeworfen und interessante Überlegungen **angestellt**. Einige der Thesen könnten in der Fachwelt auf Kritik **stoßen**.

1.4 Die Satzklammer im Hauptsatz

1

	Position 0	Vorfeld	linke Satzklammer	Mittelfeld	rechte Satzklammer	Nachfeld
	Position 0	Position 1	Position 2	Satzmitte	Satzende	
1		*Seine Semesterzahl*	*übersteigt*	*sein Lebensalter bei Weitem.*		
2		Rausschmeißen	kann	ihn die Uni nicht.		
3		**Das**	hat	**jemand für die Internetseite der medizinischen Fakultät der Uni Kiel**	geschrieben.	
4		**Mindestens bei einem Studenten**	**dürfte**	**eine intensive Begleitung der gesamten Ausbildung**	**ziemlich aufwendig sein.**	
5		**Bundesweit**	**nehmen**	**viele Hochschulen vor ihren Diplom- und Magisterstudenten**	**Abschied.**	
6		**Außerdem**	laufen	**nach und nach die Studienfristen**	**aus,**	die Alt-Studenten gewährt wurden.
7	Aber	**bis heute**	**konnte**	**der Kieler Marathon-Student nicht**	**zwangsexmatrikuliert werden.**	
8		**Wie**	ist	**das**	möglich?	
9			Gibt	**es keine Regelung dafür im Hochschulgesetz?**		

1.5 Das Vorfeld

1

Treffen mit Terézia Mora

Letzten Monat konnten wir Terézia Mora treffen. Die ungarische Schriftstellerin hat uns von ihrer Laufbahn erzählt. Sie hat ihren ersten Roman in deutscher Sprache vorgelegt. Dafür erhielt sie den mit 7000 Euro dotierten Adelbert-von-Chamiso-Förderpreis. Diesen Preis vergibt die Robert Bosch Stiftung jährlich für herausragende deutschsprachige Werke von Autoren, deren Muttersprache nicht Deutsch ist. Das war bis jetzt ihr größter Erfolg.

Den würde sie gerne in den kommenden Jahren wiederholen. Im Moment …

neu   neu

bekannt   bekannt   bekannt

2

Rahmen oder Ausgangspunkt der Handlung

Weiterführung und Spezifizierung zuvor genannter Information

Rahmen oder Ausgangspunkt der Handlung

[...] Vom Frühjahr 1917 an besuchte ich die Kantonsschule an der Rämistrasse. Sehr wichtig wurde der tägliche Schulweg dorthin und zurück. Zu Beginn dieses Weges, gleich nach der Überquerung der Ottikerstrasse, hatte ich immer dieselbe erste Begegnung, die sich mir einprägte. Ein Herr mit einem sehr schönen weissen Kopf ging da spazieren, aufrecht und abwesend, er ging ein kurzes Stück, blieb stehen, suchte nach etwas und wechselte die Richtung. Er hatte einen Bernhardi- ner, dem er öfters zurief: „Dschoddo komm zum Pápa!" Manchmal kam der Bernhardiner, manchmal lief er weiter weg, er war es, den der Pápa dann suchte. Aber kaum fand er ihn, vergass er ihn wieder und war so abwesend wie zuvor. [...]

Weiterführung und Spezifizierung zuvor genannter Information

bekannte Information

Kontrast zum vorherigen Thema

3 b Kennst du den älteren Herrn, der dort drüben winkt?
Ja, mit ihm habe ich mich vorhin länger unterhalten.
c Kennst du das Erasmus-Programm?
Ja, damit habe ich mein Auslandssemester finanziert.
d Die Sängerin hatte einen Auftritt in Berlin.
Dort war sie noch nie gewesen.

1.6 Die Satzklammer im Nebensatz

1

Hauptsatz		Nebensatz		
		Subjunktion	Mittelfeld	finites Verb / Verbalkomplex
1	Es gilt als allgemein anerkannt,	dass	Auslandsaufenthalte während des Studiums einen hohen Stellenwert für angehende Akademiker	haben.
2	Aktuelle Statistiken belegen allerdings,	**dass**	**die Zahl der deutschen Studierenden mit Auslandsaufenthalten seit 2000**	**zurückgeht,**
3		**obwohl**	**die Studienreform den Auslandsaufenthalt**	**erleichtern sollte.**
4	Die Zahlen der Studierenden mit einem Auslandssemester sinken,	**weil**	**sich die Universitäten und Studierenden durch die Hochschulreform erst einmal auf die neuen Bachelor- und Masterstudiengänge**	**haben umstellen müssen.**
5	Gegner der Studienreform führen an,	**dass**	**sie diese Entwicklung**	**haben kommen sehen.**

2 b Eine Kollegin hat gesagt, dass **er seine Zeit besser hätte planen sollen.**
c Die Kollegin meinte auch, dass **er alle lange auf sein Ergebnis hat warten lassen.**
d Er hat sich schlecht gefühlt, weil **er sein Versprechen gebrochen hat.**
e Die enttäuschte Kollegin hat gesagt, dass **sie das schon hat kommen sehen.**

1.7 Stellung von Ergänzungen und Angaben im Mittelfeld

1.7.1 Ergänzungen im Mittelfeld

2 b Das Davos-Organisationsteam **bietet den Besuchern die professionelle Organisation von Seminaren, Kongressen und Symposien.**
c Viele renommierte Professoren aus verschiedenen Bereichen der Medizin **präsentieren ihren Fachkollegen die neuesten Ergebnisse aus der Forschung.**
d Gerne **diskutieren die Kongressteilnehmer mit den Vortragenden über die angewandte Forschungsmethodik.**

3 **b** Mit Abstand hebt **sich** dieses Zentrum vom Durchschnitt ab. / Mit Abstand hebt dieses Zentrum **sich** vom Durchschnitt ab.

   c Vor Beginn eines Kongresses müssen **sich** die Vortragenden rechtzeitig anmelden. / Vor Beginn eines Kongresses müssen die Vortragenden **sich** rechtzeitig anmelden.

   d In bestimmten Sälen üben sie **sich** in freier Rede.

   e Im Back-Office bemüht **sich** die Organisation um jedes Detail. / Im Back-Office bemüht die Organisation **sich** um jedes Detail.

   f So können **sich** die Gäste komplett auf die Organisation verlassen. / So können die Gäste **sich** komplett auf die Organisation verlassen.

4 **b** Ein Stammgast **hat dem Hotelmanager ein Buch geschenkt.**

   c Der Hotelmanager **hat das Buch einem Mitarbeiter zum Aufbewaren gegeben.**

   d Der Mitarbeiter **hat ihnen den renovierten Swimmingpool gezeigt.**

   e Er **hat ihnen die Geschichte des Hotels erzählt.**

   f Er **hat sie ihnen schnell erzählt,** denn die Gäste brauchten ein Abendessen.

5 **b** Der Professor hat dem Studenten kein Gutachten geschrieben?

   > **Doch, er hat ihm eins geschrieben.**

   c Der Student hat der Auswahlkommission keine guten Antworten gegeben?

   > **Doch, er hat sie ihr gegeben.**

   d Die Auswahlkommission hat dem Studenten das Studium nicht ermöglicht?

   > **Doch, sie hat es ihm ermöglicht.**

1.7.2 Angaben im Mittelfeld

2 **b** Dagegen muss man etwas tun. Neulich habe ich gelesen, dass die Zahl der computersüchtigen Kinder **in den letzten Jahren in vielen Ländern rasant angestiegen ist.**

   c Warum hast du dich gestern nicht gemeldet?

   Ich konnte dich **den ganzen Tag wegen des schlechten Empfangs / wegen des schlechten Empfangs den ganzen Tag** nicht erreichen.

   d Endlich hat sich meine Schwester aus Brasilien gemeldet.

   Ich habe **seit Montag verzweifelt** auf den Anruf gewartet.

   e Gute Zeugnisse sind bei der Arbeitssuche von großer Bedeutung.

   Aber meine Freundin hat **trotz exzellenter Zeugnisse monatelang deutschlandweit** eine Arbeit gesucht.

3 **b** > **Sie nimmt dieses Jahr zum ersten Mal am Slalomwettbewerb teil.**

   c > **Ihre Mannschaft konnte heute wegen des Schneesturmes nicht so viel trainieren.**

   d > **Alle mussten sich gleich nach der ersten Abfahrt ganz schnell auf den Weg ins Hotel machen.**

   e > **Claudias Freunde treffen sich bei schlechtem Wetter am liebsten am neuen Swimmingpool.**

   f > **Auch Claudia kann sich dort in guter Gesellschaft stundenlang entspannen. / Auch Claudia kann sich in guter Gesellschaft dort stundenlang entspannen. / Auch Claudia kann sich dort stundenlang in guter Gesellschaft entspannen.**

4 **b** > **Der Trainer gibt den Spielern hoffentlich genaue taktische Anweisungen.**

   c > **Der Gegner stellt vermutlich die größte Herausforderung für die Abwehr dar.**

   d > **Die Mannschaft hat bedauerlicherweise das Spiel gestern hoch verloren. / Die Mannschaft hat das Spiel gestern bedauerlicherweise hoch verloren.**

   e > **In einer Woche findet glücklicherweise schon das Rückspiel statt.**

   f > **Es werden bestimmt viele Fans zum Heimspiel kommen.**

1.8 Das Nachfeld

1 Das Nachfeld

Vorfeld	linke Satzklammer	Mittelfeld	rechte Satzklammer	Nachfeld
Die Buchmesse	hat	bereits vor Jahren	angefangen,	diesen Sektor auszubauen.
Es	bleibt	der Messe auch keine Wahl,		**wenn sie nicht über kurz oder lang überflüssig werden will.**
Mit 7500 Ausstellern	hat	die Messe in etwa wieder so viel Fläche	vermietet	**wie im Vorjahr.**
Besonderes Interesse	werden	die neuen E-Book-Reader	wecken,	**die kürzlich auf den Markt gekommen sind.**

2   b > Die Ausstellungsfläche **war ungefähr so groß wie im letzten Jahr.**

     c > Jedoch **sind mehr Besucher gekommen als je zuvor.**

     e > Die neuen E-Book-Reader **waren für einige attraktiver als die entsprechenden Printmedien.**

     f > Die Auftritte der Autoren auf den Lesebühnen **waren so erfolgreich wie immer.**

     g > Die Fortbildungsseminare **haben eine wichtigere Rolle gespielt als im Vorjahr.**

2 Komplexe Sätze im Kontext

2.1 Konnektoren: Mittel der Textverbindung

1.1   > Konnektoren verbinden Aussagen und Sätze zu Texten. Sie werden deshalb auch als „Bindewörter" bezeichnet.

1.2 Konnektoren

Konnektoren

- sind **Bindewörter** und stellen Verbindungen zwischen Aussagen und Sätzen her.
- stellen inhaltliche **Zusammenhänge** im Text her.
- sorgen für Kohärenz im Text und sind Wegweiser für die **Interpretation**.

1.3

Funktion der Konnektoren	Beispiel
verdeutlichen die Argumentation	nicht nur ... sondern auch, **um ... zu, als, deshalb, kaum ... da, doch, stattdessen, bis, weil, nämlich, außerdem, schließlich**
haben vor allem grammatische Funktion	wie, **dass, und, die, wo**

1.4

Bedeutung	Konnektoren
zählt etwas auf	und, **nicht nur ... sondern auch, außerdem**
gibt einen Grund an	nämlich, **weil, deshalb**
drückt eine zeitliche Vorgabe aus	als, **kaum ... da, schließlich**
gibt einen Ort an	wo
drückt ein Ziel / einen Zweck aus	**um ... zu**
drückt die Art und Weise / einen Vergleich aus	**wie**
drückt einen Ersatz / Gegensatz aus	**stattdessen**

2.2 Konnektoren und Stellung im Satz

2.2.1 Konjunktionen und Verbindungsadverbien

1.1

Hauptsatz	Hauptsatz
II	Kennt ein junger Vorgesetzter das Problem selbst,	ist er **nämlich** meist sehr offen für Anfragen oder Vorschläge.

Verbindungsadverb im Mittelfeld des zweiten Hauptsatzes

Hauptsatz	Hauptsatz
III	Familienpolitik sei Sache des Staates.	**Deshalb** müsse sich jeder Einzelne persönlich kümmern.

Verbindungsadverb vor dem finiten Verb

Hauptsatz		Hauptsatz
IV	Andere nahmen das so hin,	**doch**	er gab nicht auf.

Verbindungsadverb zwischen den Hauptsätzen auf Position 0 des zweiten Hauptsatzes

1.3

Stellung von Konjunktionen und Verbindungsadverbien im komplexen Satz

- Konnektoren können an **verschiedenen** Stellen im komplexen Satz stehen: vor dem finiten Verb (Bsp. I und III), zwischen den Sätzen (Bsp. I und IV) und im Mittelfeld eines Teilsatzes (Bsp. II).
- Konjunktionen können Hauptsätze und Nebensätze verbinden (Bsp. V). Die Wortstellung ändert sich durch die Konjunktion im Satz nicht. Bei *und, aber, denn* kann das Subjekt im zweiten Satz **weggelassen** werden, wenn es identisch ist mit dem Subjekt im ersten Satz (Bsp. I und V).
- Konjunktionen verbinden auch Phrasen und **Wörter**: Die Vereinbarkeit von Karriere und Kindern ist nicht nur ein Frauenproblem.
- Verbindungsadverbien sind Satzglieder. Wenn das Adverb am **Satzanfang** steht, ändert sich die Wortstellung und das Subjekt rückt hinter das finite Verb (Bsp. I und III).
- Konjunktionen können auch Nebensätze miteinander verbinden. Wieder kann das **Subjekt** im zweiten Nebensatz weggelassen werden, wenn es sich um das **gleiche** Subjekt handelt.

2

Schuster: Ich rede jeden Tag mit den Bürgern. Auch in meiner Nachbarschaft gibt es Gegner von Stuttgart 21. Aber **deshalb** verabschiede ich mich nicht aus der Verantwortung für alle Bürger.

Sittler: Wo waren Sie denn am vorvergangenen Freitag, am Tag nach der gewaltsamen Polizeiaktion im Schlossgarten? Da demonstrierten 100000 Leute ausnahmslos friedlich, weil sie wollen, dass es der Stadt gut geht. […] Sie müssen doch die Sorgen der Bürger, die nicht gewalttätig werden, sondern friedlich demonstrieren, ernst nehmen!

Sittler: Die Stimmung ist katastrophal. Und die Kommunikationspolitik ist die schlechteste, die man sich vorstellen kann.

Schuster: **Deshalb** lade ich die Bürger zum Dialog ein. Noch mal: Machen Sie doch mit, Herr Sittler!

Sittler: Nur ohne Vorbedingungen.

Schuster: Aber Sie stellen doch Vorbedingungen! […]

Sittler: Wenn die Bundesregierung als Eigner sagt, wir stellen das Geld statt für Stuttgart 21 für die Renovierung des Gleisfelds und des Bahnhofs zur Verfügung, **dann** wird die Bahn das bauen. Die ist **nämlich** an Weisungen gebunden, Herr Schuster.

Bei K21 würden neue ICE-Gleise verlegt. Die Bürger, die dort wohnen, werden vielleicht auch für ihre Ruhe auf die Straße gehen, Herr Sittler.

Sittler: Na ja, man kann es nicht allen recht machen. **Außerdem** gibt es ja Lärmschutz.

Würde Ihr Positionswechsel einen Gesichtsverlust bedeuten, Herr Schuster?

Schuster: Mir geht es nicht um politische Ideologie und Rechthaberei. Viele Bürger sind, was die Fakten angeht, verunsichert. Das ist angesichts der Komplexität des Projekts nicht verwunderlich. **Deshalb** gibt es den großen Wunsch nach sachlicher Diskussion. Wir müssen auf jeden Fall in einen langfristig angelegten Dialog kommen.

Abschließend *fragen wir auch Sie, Herr Schuster: Was müsste passieren, dass Sie zu einem Gegner von Stuttgart 21 werden?*

Schuster: Wenn erneute Prüfungen ergeben würden, dass so, wie Herr Sittler sagt, alles Murks sei.

Sittler: Alles Murks ist übertrieben.

Schuster: Ach was.

Sittler: Na gut, ein bisschen übertrieben.

3

Aufzählung
1, **2, 3**, 4, 5, 7, 10, 12, **13**, 15, 18

Alternative
8

Einwand
17

Grund
6, 9

Kontrast nach Negation
11, **14, 16**

2.2.2 Subjunktionen

1

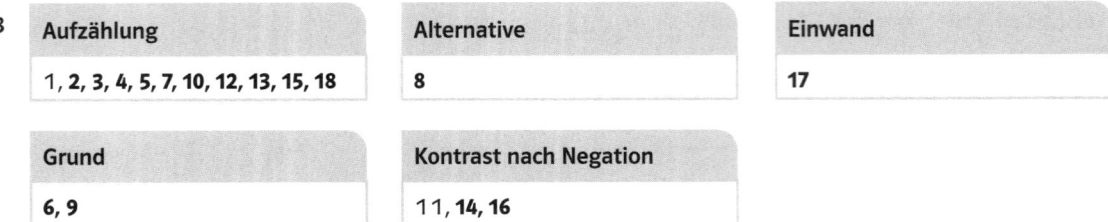

	Hauptsatz	Nebensatzklammer		
		Subjunktion	Mittelfeld	finites Verb
b	**In den Personalabteilungen ist die Botschaft inzwischen angekommen,**	**dass**	**Familienfreundlichkeit ein Unternehmen als Arbeitgeber attraktiv**	**macht.**

		Subjunktion	Mittelfeld	finites Verb
c	**Das sei aber auch verständlich, sagt Baisch,**	**weil**	**die mit ihren verbleibenden Mitarbeitern den Ausfall häufig**	**kompensieren müssten.**

Nebensatz auf Position 1			Hauptsatz	
Subjunktion	Mittelfeld	finites Verb	Position 2 = finites Verb	Mittelfeld
a Als	**vor acht Jahren seine erste Tochter zur Welt**	**kam,**	**nahm**	**sich der Leiter einer Jugendeinrichtung ein Jahr Elternzeit.**

Hauptsatz	Nebensatzklammer			Nebensatzklammer		
	Subjunktion	Mittelfeld	finites Verb	Subjunktion	Mittelfeld	finites Verb
d Wichtig sei,	**dass**	**dies im Unternehmen von oben**	**vorgelebt werde,**	**weil**	**somit ein gutes Arbeitsklima**	**geschaffen werden kann.**

2.1

Elke Heidenreich: Der Hund wird erschossen

[…] Ich weiß nicht, **ob** die Ehe meiner Eltern gut war. **Als** Kind denkt man über so etwas nicht nach, man kennt ja nichts anderes, man meint, so ist es eben und so muss es sein, das sind eben Eltern – erwachsen, langweilig, immer beschäftigt, unzufrieden. Ich habe nie gesehen, **dass** sie sich umarmt oder geküsst hätten, nur einmal gingen sie Arm in Arm, **und** das ist die Geschichte, die ich erzählen will. Streit gab es zu Hause eigentlich immer nur meinetwegen. Berti ist so schwierig, Berti ist so frech, ich werde mit Berti nicht mehr fertig, die Lehrer haben sich schon wieder über Berti beschwert, Berti ist unordentlich, Berti macht keine Schularbeiten, Berti treibt sich mit Jungens herum, Berti raucht heimlich – das waren so ungefähr die ständigen Klagen meiner Mutter, und sie seufzte, **wann** immer sie mich bloß sah und auch **wenn** ich gar nichts angestellt hatte[.] […] Von unten hörten wir unsere Eltern streiten. „Ich bin es leid", schrie Mutter, „ich kann machen, was ich will, wir kommen auf keinen grünen Zweig, und nun muss ich mir auch noch von dir vorwerfen lassen, ich wäre schuld daran." […] **Dann** knallte die Haustür, und kurz darauf kam meine Mutter laut heulend aus dem Wohnzimmer. Wir zogen uns schnell in unsere Zimmer zurück und hörten, **wie** im Schlafzimmer Schränke aufgerissen und wieder zugeschlagen wurden. Eine halbe Stunde später verließ unsere Mutter mit einem Koffer in der Hand und unter dem infernalischen Gebell von Molli das Haus und ging zur Bushaltestelle, **obwohl** doch in der Nacht dort gar kein Bus abfuhr. […]

2.2 Beispiellösung:

> Dann hörten wir ein weiteres Mal die Tür zuschlagen und wir nahmen an, dass es sich um unseren Vater handeln musste. Durch das offene Fenster hörten wir ihn zur Bushaltestelle rennen. Wir waren ganz verängstigt und hofften, dass sie bald wieder zurückkehren würden. Nach ungefähr 10 Minuten ging die Tür auf. Schließlich waren wir erleichtert, als wir ihre beiden Stimmen hörten. Kurz darauf betraten sie Arm in Arm das Wohnzimmer. Sie hatten sich beruhigt und wieder vertragen.

2.2.3 Präpositionen mit ähnlicher Bedeutung wie Konnektoren

1

Geisslers Plan als Test für die deutsche Streitkultur – Sind die Kontrahenten von «Stuttgart 21» zum Kompromiss fähig?

5 Als die Situation am verfahrensten war, hat der Schlichter im Streit um das Bahn-Großprojekt «Stuttgart 21» einen Kompromiss vorgeschlagen, bei dem beide Parteien ihr Recht bekom-

10 men, aber auch Abstriche machen müssen. In der Schweiz wäre so eine Lösung nahe liegend. Im Nachbarland ist es aber eine Herausforderung. **Bei** der Vorlage von Geisslers Joker wa-

15 ren die meisten Beteiligten im epischen Bahnhofsstreit von Stuttgart wie vor den Kopf gestoßen. Die Beobachtung **durch** den altgedienten CDU-Politiker aus nächster Nähe zeigte, wie

20 sich die beiden Lager immer mehr in ihre Argumente verbissen und so wagte er den Befreiungsschlag.

Auf Teufel komm raus

Die Bahn, die Bundesregierung und
25 auch die großen Parteien in Deutschland wollten das Großprojekt mit einem Stuttgarter Hauptbahnhof unter der Erde sowie einem Strecken-Neu-

bau für schnellere Fernverbindungen
30 auf Teufel komm raus durchziehen. Aber genauso entschlossen waren auf der anderen Seite die Ablehnung dieses Umbaus und die Forderung nach einem runderneuerten Bahnhof an al-
35 ter Stelle **durch** einen beträchtlichen Teil der Stuttgarter und eines heterogenen Bündnisses verschiedenster Bürgergruppen im Lande.
Nach der Entscheidung am Freitag soll-
40 ten die Experten schließlich einen Stresstest bringen. [...]

Niemand vorbereitet

[...] Über Wochen hatte der Schlichter im letzten Herbst geduldig und fair die
45 Konfrontation auf eine sachliche Ebene herunter geholt. Allenthalben lobte man die neue Streitkultur, die sich hier entwickelt habe. Doch was am Ende dieses Prozesses stehen sollte, wurde
50 nie gesagt. Beide Seiten konnten sich nichts anderes vorstellen, als schließlich auf ganzer Linie Recht zu bekommen.

2

Konjunktionen	Subjunktionen	Verbindungs-adverbien	Präpositionen mit ähnlicher Bedeutung
aber, **denn, und, sondern, oder** ...	**als, dass, weil, wenn, da, obwohl, nachdem** ...	**deshalb, nämlich, außerdem, dann, abschließend, darum** ...	**bei, nach, durch** ...

3 Partner 1:
Ich bin der Meinung, dass ...
Deinen/Ihren Standpunkt finde ich ...
Meiner Meinung nach ...
Es ist anzunehmen, dass ...

Partner 2:
Ich wiederum finde es aber nicht
angebracht, dass ...
Da muss ich dir/Ihnen widersprechen, da ...
Es erscheint naheliegend...
Meines Wissens ...

2.2.4 Zweiteilige Konnektoren

1.1

Eltern und Kinder haben ein Recht auf Krippenplätze

Ab 2013 gilt **zwar** per Gesetz der Rechtsanspruch auf einen Krippenplatz, **aber** es wird befürchtet, dass das Geld für den Ausbau der Betreuungseinrichtungen nicht reichen wird.

Mehr als 35 Prozent der Eltern wollen **entweder** ihre Kleinen in Krippen **oder** bei Tagesmüttern unterbringen. Das zumindest ist das Ergebnis einer neuen Umfrage bei jungen Frauen, die allerdings selbst noch kinderlos sind. Aber auch Wissenschaftler gehen davon aus, dass mehr Mütter und Väter auf ihren Anspruch pochen werden. Kann eine Gemeinde ihnen dann keine Krippenplätze anbieten, könnten die Eltern für ihren Krippenplatz klagen.

Es wird sich **zwar** immer noch darum gestritten, ob das Hausmütterchen oder das Karriereweib die schlechtere Mutter ist, **aber** die Streitgespräche sind längst nicht mehr so erhitzt und so unversöhnlich wie noch vor einigen Jahren. Die Stimmungslage hat sich verändert: Arbeitende Mütter, auch mit kleineren Kindern, sind selbstverständlich geworden. Das von der CSU erkämpfte Betreuungsgeld ist das eindeutigste Zeichen dafür, dass jede Frau (und mittlerweile auch jeder Mann) es sich heute gut überlegen muss, ob und wenn ja, für wie lange sie aus dem Beruf aussteigen kann, um ihre Kinder zu betreuen.

Nach der Änderung des Unterhaltsrechts ist der Versorger der Familie im Falle eines Scheiterns der Ehe nur noch sehr kurz und eingeschränkt unterhaltspflichtig. Es wird immer schwieriger, je länger die Frau (meistens ist es sie) aus dem Beruf ausgestiegen ist, wieder eine angemessene Arbeit zu finden. Somit haben **nicht nur** emanzipierte Frauen, **sondern auch** die Politik Fakten geschaffen, die ein Recht auf Kindergarten- und Krippenplätze unverzichtbar machen.

1.2 Zweiteilige Konnektoren: Bedeutung

- Gegenüberstellung: **entweder ... oder**
- Aufzählung: **sowohl ... als auch**
- Aufzählung mit Betonung des zweiten Teils: **nicht nur ... sondern auch**
- Einwand mit Betonung des zweiten Teils: **zwar ... aber**

2.1

Unproduktive Polemik

Mir scheint die Gegenüberstellung von Bemuttern hier und Selbstverwirklichung dort nicht zu greifen. **Weder** ist jede Mutter, die zu Hause bleibt, zwangsläufig eine Glucke, **noch** verwirklicht sich jede Frau, die außer Haus arbeitet, automatisch selbst.[1] Was mir in der Diskussion über die Forderung nach Krippenplätzen immer fehlt, ist die Forderung **sowohl** nach flexibleren **als auch** familienfreundlicheren Arbeitszeiten.[2] Wir sitzen vollkommen der Ideologie auf, dass eine maximale Präsenz am Arbeitsplatz auch maximale Effizienz bedeutet, was nicht der Fall ist, und deshalb muss nun auch schon das Leben der Kleinsten von Anfang an vollkommen durchorganisiert werden. **Zwar** sind Kinderkrippen unbestritten ein gutes Instrument, um sozial benachteiligte Kinder zu fördern, **aber** ein allgemeines Kinderrecht auf einen Krippenplatz unterschlägt zu stark die Defizite einer kinderunfreundlichen Arbeitswelt.[3] Wir sollten das Selbstbewusstsein aufbringen, nicht entscheiden zu müssen, ob wir **entweder** einen Acht-Stunden-Plus-Tag ab der Wiege **oder** häusliche Betreuung wünschen.[4]

Lisa Heuser aus Bonn

2.2 1 *Weder* im Mittelfeld: **Jede Mutter, die zu Hause bleibt, ist zwangsläufig weder eine Glucke, noch verwirklicht sich jede Frau, die außer Haus arbeitet, selbst.**

2 *Sowohl* innerhalb des Subjekts: **Sowohl die Forderung nach flexiblen als auch familienfreundlicheren Arbeitszeiten ist das, was mir in der Diskussion über die Forderung nach Krippenplätzen immer fehlt.**

3 *Zwar* im Mittelfeld: **Kinderkrippen sind zwar unbestritten ein gutes Instrument, um sozial benachteiligte Kinder zu fördern, aber ein allgemeines Kinderrecht auf einen Krippenplatz unterschlägt zu stark die Defizite einer kinderunfreundlichen Arbeitswelt.**

4 *Entweder* vor dem finiten Verb: **Entweder wünschen wir einen Acht-Stunden-Plus-Tag ab der Wiege oder häusliche Betreuung.**

3 *Lisa und Peter wollen einen Kindergartenplatz.*
Sowohl Peter als auch **Lisa wollen einen Kindergartenplatz.**
Sowohl Peter will einen Kindergartenplatz **als auch Lisa.**
Einen Kindergartenplatz wollen sowohl **Peter als auch Lisa.**
Es ist klar, dass sowohl **Peter als auch Lisa einen Kindergartenplatz wollen.**

4 Beispiellösung:

Ich habe weder **Zeit noch Lust die Kinderbetreuung allein zu bewerkstelligen und auf meinen Beruf zu verzichten.**

Zwar finde ich eine Kinderbetreuung **wünschenswert, aber ich würde mein Kind auch zu Hause erziehen.**

Entweder geht der Mann **oder die Frau in die Elternzeit.**

Weder die Frau **noch der Mann müssen auf die Elternzeit verzichten.**

Man kann sich entweder **selbst um die Kinderbetreuung kümmern oder einen Kindergartenplatz suchen.**

Ich finde es nicht nur **wünschenswert, sondern auch sehr wichtig, dass Kinder in einer Kindertagesstätte betreut werden.**

2.3 Komplexe Sätze nach semantischen Relationen

2.3.1 additiv: Aufzählung, Reihung, Ergänzung

1

> **Videokonferenz als interaktive Lernumgebung – am Beispiel eines Kooperationsprojekts zwischen japanischen Deutschlernenden und deutschen DaF-Studierenden (Makiko Hoshii und Nicole Schumacher)**
>
> Unsere Videokonferenz Waseda-Humboldt, eine Lernumgebung des Typs „Mehrere-zu-mehrere, technisch vermittelt", vereint in sich viele der Charakteristika der schon stärker etablierteren Lernumgebungen. Es gibt zwei Interaktionsräume (einen an der Waseda-Universität, einen an der Humboldt-Universität) **und** mehrere Interaktionssphären. Zunächst einmal gibt es die technisch vermittelte Gesamtsphäre. **Zudem** gibt es die technisch vermittelte Sphäre zwischen einer Lehrperson **und** einzelnen Lernenden oder der Lerngruppe. **Außerdem** gibt es mehrere Face-to-face-Interaktionssphären in beiden Räumen: **Sowohl** in Tokio **als auch** in Berlin können die Teilnehmer untereinander interagieren. Wie beim klassischen Fremdsprachenunterricht **und** beim Teamteaching können die Lernenden einander in Produktion **und** Rezeption unterstützen. Wie beim Teamteaching können **zudem** die angehenden Lehrenden miteinander interagieren.

Additive Satzverbindungen	
Konjunktionen	**Verbindungsadverbien**
sowie, **und**, **sowohl ... als auch**	auch, **außerdem**, daneben, darüber hinaus, dazu, des Weiteren, ebenfalls, ebenso, erstens ... zweitens ..., ferner, fernerhin, gleichfalls, noch dazu, obendrein, überdies, weiter, weiterhin, **zudem**, zusätzlich

2

Eine Möglichkeit, wie man Lernende in sinnvolle zielsprachliche Interaktionen involvieren (vgl. Rösler 2000: 129), ihnen ‚kommunikative Ernstfälle' bieten kann (vgl. Schlickau 2000: 2), besteht in der Einbindung digitaler Medien, was **sowohl** Lernende **als auch** Lehrende und damit auch die Lehrerausbildung vor neue Aufgaben stellt (vgl. Schneider & Würffel 2007). Ausgehend von diesen Vorüberlegungen ist 2004 ein Kooperationsprojekt zwischen der Waseda Universität Tokio **und** der Humboldt-Universität zu Berlin entstanden, in dem Studierende der beiden Universitäten gemeinsam per Videokonferenz Deutsch lernen bzw. lernen, Deutsch zu lehren (vgl. Mewes 2005; Hoshii & Niederhaus 2008). [...] Ziel dieser Vorstudie war es, die für unsere Videokonferenzen typischen Lernerfragen **und** die sich daran anschließenden Interaktionsmuster zwischen Lernenden **und / sowie** zwischen Lernenden **und / sowie** angehenden Lehrenden zu beschreiben.

3.1 Beispiellösung:

Lena Meier absolvierte im Jahr 2010 ihr Abitur und darüber hinaus ein Praktikum in einer Buchbinderei. 2010 begann sie außerdem ihr Studium im Fach Buchhandel / Verlagswirtschaft an der Hochschule für Technik, Wirtschaft und Kultur in Leipzig. Weiterhin spricht sie mehrere Fremdsprachen. Erstens Polnisch auf den Niveaustufen B1 bis B2 und zweitens Französisch auf den Niveaustufen A1 bis A2. Zudem hat sie sehr gute Kenntnisse in MS-Office-Anwendungen. Überdies ist sie aktives Mitglied im Kunstverein Leipzig e.V. und sie spielt Klavier.

3.2 Ich bin am ... geboren. Im Jahr ... absolvierte ich ...
Zudem habe / war ich von ... bis ...
Ich spreche außerdem verschiedene Sprachen:
In meiner Freizeit ...

4 Beispiellösung:

- Bitte beschreiben Sie kurz Ihren Ausbildungsweg.

   Nachdem ich mein Abitur absolviert hatte, ging ich für ein Jahr als Au-Pair nach Frankreich und betreute dort in einer Familie zwei Kinder. Danach habe ich mein Studium an der Goethe-Universität in Frankfurt begonnen. Überdies konnte ich weitere Auslandserfahrungen während meines ERASMUS-Austauschs in Schweden sammeln.

- Welche Tätigkeiten haben Sie bisher ausgeübt und welche Erfahrungen haben Sie gesammelt?
   In meinem Schulpraktikum konnte ich erste Erfahrungen in der Arbeit als Apothekerin sammeln. Zudem war es mir möglich, mir in den Ferien weitere Kenntnisse im Verkauf anzueignen.

- Auf welche Leistungen sind Sie in Ihrem bisherigen Lebensweg besonders stolz?
   Ganz besonders stolz bin ich auf meinen Fernstudienabschluss in Betriebswirtschaftslehre. Ich interessiere mich außerdem für zeitgenössische Kunst. Deshalb konnte ich bereits in einem Museum arbeiten und dort Führungen anbieten.

- Wie stellen Sie sich Ihre Tätigkeit in unserem Unternehmen vor?
   Ich denke, dass die Arbeit sehr anspruchsvoll werden wird. Ich freue mich aber darauf, in einem Team zu arbeiten.

2.3.2 temporal: Zeit

1

HEINRICH:	Wir möchten nächstes Jahr über die Osterfeiertage nach Leipzig fahren und dafür eine Reise buchen. Können Sie uns da vielleicht einen Vorschlag machen?
BERATERIN:	Für Sie beide oder noch jemand anders?
HEINRICH:	Für meine Frau und mich.
BERATERIN:	Und wie lange soll die Reise dauern?
KÄTHE:	Insgesamt 5 Tage.
BERATERIN:	**Seit** Mitte August haben wir ein Topangebot. Sie starten am Gründonnerstag und wären **dann** am Ostermontag gegen 18.00 Uhr zurück in Stuttgart. Auf dem Weg nach Leipzig würden Sie einen Zwischenstopp in Dessau machen. Das heißt, **bevor** Sie nach Leipzig fahren, haben Sie sogar noch die Möglichkeit das Bauhaus und die Innenstadt von Dessau zu besichtigen.
KÄTHE:	Und wann kämen wir dann in Leipzig an?
BERATERIN:	**Nach** einer Übernachtung in Dessau würden Sie Karfreitag gegen Mittag in Leipzig ankommen. Ihr Hotel würde direkt in der Ritterstraße liegen. Von dort aus starten jeden Tag verschiedene Ausflüge. Das Programm ist sehr vielseitig. Hier, ich gebe Ihnen die Broschüre, damit Sie sich das Programm näher anschauen können.
KÄTHE:	Das klingt wunderbar. Ich würde vorschlagen, dass wir uns nochmal melden, **nachdem** wir uns einen Überblick verschafft haben.
BERATERIN:	Alles klar, dann bis dahin. Falls Sie Fragen haben, rufen Sie mich doch einfach an.
KÄTHE UND HEINRICH:	Auf Wiedersehen.
BERATERIN:	Auf Wiedersehen.

Verbindungsadverb

Präposition

Subjunktion

2.1

Liebe Hildegard,

wir senden liebe Urlaubsgrüße aus Leipzig! Die Zugfahrt nach Leipzig ist problemlos verlaufen. **Zuvor** haben wir einen Stopp in Dessau eingelegt. Das hat sich wirklich gelohnt! **Während** Heinrich sich das bekannte Bauhaus angeschaut hat, bin ich an der Mulde spazieren gegangen. **Nach** dem Kurzaufenthalt sind wir dann einen Tag später nach Leipzig weitergefahren. Gleich als erstes haben wir an einer Führung durch das neue Museum der Bildenden Künste teilgenommen. Anschließend sind wir durch die wunderschöne Innenstadt geschlendert, **bis** es Abend wurde. **Nachdem** wir uns in Auerbachs Keller (Goethes Faust!) gestärkt hatten, waren wir im Gewandhaus bei Wladimir Kaminer. Wir haben herzlich gelacht! **Als** die Lesung vorbei war, sind wir im Barfußgäßchen in eine Jazz-Bar eingekehrt und so lange geblieben, **bis** die Musiker ihre Instrumente eingepackt haben. **Wenn** das Wetter so regnerisch bleibt, dann fällt unser für morgen geplanter Radausflug wortwörtlich ins Wasser. Drückt uns die Daumen, dass es besser wird!
Viele Grüße senden
Käthe und Heinrich

2.2

Stopp in Dessau	Heinrich besichtigt das Bauhaus	Ankunft in Leipzig	Führung durchs Museum	Spaziergang durch die Innenstadt	es wird Abend	Auerbachs Keller	Gewandhaus	Jazz-Bar	Musiker packen Instrumente ein
	Käthe spaziert an der Mulde								

2.3 Temporale Satzverbindungen

	Subjunktionen (Nebensatz)	Verbindungsadverbien	Präpositionen
vorzeitig	sobald, **nachdem**, als	vorher, bis dahin, **zuvor**	vor (+ D), **bis (+A)**, bis zu (+ D)
gleichzeitig	als, **wenn**, **während**, solange	währenddessen, inzwischen	bei (+ D), während (+G)
nachzeitig	**bis**, bevor, ehe	(gleich) danach, dann, seitdem	seit (+ D), **nach (+D)**

3   Beispiellösung:

Als ich am Morgen **das Bauhaus besichtigte, schien noch die Sonne.**
Wenn ich morgens aufstehe, **trinke ich normalerweise einen schwarzen Kaffee.**
Solange mein Freund / meine Freundin telefoniert, **schreibe ich eine E-Mail an meine Eltern.**
Während mein Freund / meine Freundin schlief, **habe ich meinen Roman zu Ende gelesen.**

4.1

Ein Boot aus Ästen und einer wasserdichten Plastik-Folie ist schnell gebastelt. *Zuvor* benötigt man drei ca. 4 Meter lange gerade und astfreie Äste von der Stärke eines Besenstiels (zwei kurze können auch zu einem langen zusammengebunden werden). Außerdem benötigt man etwa 15 Stöcke á 150 bis 200 cm. **Als erstes** wird Ast a in die Erde gerammt. Er bestimmt Länge und Höhe des Bootes. Darüber spannt man b. Er bestimmt die breiteste Stelle des Bootes. **Dann** folgen zweimal c. a und 2 x c werden an den beiden Kreuzungspunkten gut zusammengebunden.
Danach folgen zwei Ruderbänke d die aber nicht zum Sitzen gedacht sind! Sie sichern das Boot **später** gegen das Wieder-Flach-Werden. Nun folgen etwa acht fingerdicke Spanten, die wie b über den Kiel gebogen werden. Alle Berührungspunkte der Äste werden stramm mit Bindfaden verbunden. **Dabei** legt man die Bindfäden am besten doppelt – das erspart die halbe Arbeit – und schließlich dreimal stramm eine Taille (e). Die weiteren Äste in Längsrichtung können durch die Spanten hindurch-geflochten oder, wie die Zeichnung es zeigt, ebenfalls gebunden werden. Wichtig ist, auf Festigkeit der Bindungen zu achten. **Nach** zwei Tagen schrumpfen die Äste.

Dann besteht die Gefahr, dass die Bindungen sich lockern. **Nachdem** das Gerippe aus dem Boden gezogen wurde, werden alle überstehenden Ast-Enden auf 5 cm Länge gestutzt. Dann wird es, Öffnung nach oben, auf die auf der Erde liegende Folie (oder Rinderhaut) gelegt. Die Folie wird von außen über die Bootsränder nach innen ins Boot eingeschlagen und mit kurzen Bindfäden um die überstehenden Stutzen und an den Spanten verzurrt (f). Diese Bindung muss nicht mehr so stramm sein. Hat man reichlich Folie, genügt ein einfaches Nach-innen-Einschlagen. Ohne Bindungen. **Bevor** man lospaddeln kann, wird das Boot, zum bequemeren Sitzen, halb gefüllt mit Gras o. ä. Polstermaterial. **Inzwischen** werden die Paddel aus langen Holzstangen besorgt.
Während das Boot ideal auf strömendem Wasser ist, kann es, umgedreht, auch als Zelt dienen. Oder, schräg, als Windschutz. Oder, schräg, als Hitzereflektor, wenn man sich zwischen Feuer und Boot niederlegt oder als Hängematte. Oder zum Transport eines Verletzten mit zwei Trägern über Land.

Dieses Boot ist ein Muss für jedes Survival-Training!

4.2   2. Nach (+ Dativ): **Nach dem Schrumpfen der Äste besteht die Gefahr, dass die Bindungen sich lockern.**

3. Vor (+ Dativ): **Vor dem Lospaddeln wird das Boot, zum bequemeren Sitzen, halb gefüllt mit Gras o.ä. Polstermaterial.**

5   Beispiellösung:

Hallo ihr Lieben,

insgesamt waren wir zwei komplette Tage unterwegs, bevor wir in Australien ankamen. Während sich die Zugfahrt verzögerte, ist der Flug ohne Probleme verlaufen. Wir konnten sogar den beeindruckenden Burj Khalifa in der Ferne sehen, als wir in Dubai zwischengelandet sind. Bevor wir dann nach Melbourne gefahren sind, haben wir die erste Zeit an der weltberühmten Great Ocean Road verbracht, eine in den Zwanziger Jahren von ehemaligen Soldaten als Denkmal für die in Gallipoli Gefallenen errichtete Straße. Eine bekannte Sehenswürdigkeit sind die 12 Apostel. Dabei handelt es sich um acht Felsen, die vom Meer aus der Küste herausgespült wurden. Dann ging es weiter in den Osten Australiens.

Inzwischen ist es für uns auch angenehm warm, nachdem es gestern eine Art Temperatursturz von dreißig auf zwanzig Grad gab. Solange es so bleibt, fühlen wir uns pudelwohl.

Bis bald und herzliche Urlaubsgrüße aus Australien
Claudia und Martin

2.3.3 konditional: Bedingung

1.1

Die Bedingung steht im ...	Hauptsatz: **3, 5 , 7, 8**
	Nebensatz: **1, 2, 4, 6, 9, 10**

1.2

Eine Annahme oder sichere Vermutung wird ausgedrückt:	**Die Geschichtskundigen dürften sofort wissen, worum es dabei geht – andernfalls googeln.**
Wenn die Bedingung nicht erfüllt ist, entstehen Probleme:	**Ich sollte dann zum Abendessen pünktlich erscheinen, sonst ärgern sich meine Freundin und mein Sohn.**
Ein Befehl oder dringender Ratschlag wird ausgedrückt:	**Recherchieren Sie bloß gründlich, ansonsten merkt man sehr schnell, wenn sich ein Autor nur etwas ausgedacht hat.**

2.1

ROMAN RAUSCH: > **Ich muss viel Zeit einplanen, andernfalls werde ich keinen guten Thriller schreiben.**

KOLLEGIN: > **Wenn du ein interessantes Thema wählst, wird der Verlag ein neues Buch akzeptieren.**

ROMAN RAUSCH: > **Falls ich das brenzlige Thema aus Würzburg im Jahr 1628 wähle, wird der Verlag den Roman publizieren.**

KOLLEGIN: > **Wenn du in Eile bist, sprechen wir morgen.**

ROMAN RAUSCH: > **Ich muss rechtzeitig nach Hause, sonst bin ich nicht pünktlich beim Abendessen.**

2.2

Konditionale Satzverbindungen

Subjunktionen	Verbindungsadverbien	Präpositionen mit ähnlicher Bedeutung
falls, **wenn**, sofern	andernfalls, **sonst**, ansonsten	im Falle (+ Genitiv), **im Falle von (+ Dativ)**, **bei (+Dativ)**, ohne (+Akkusativ)

2.3   *Gesetzt den Fall*, **ich habe eine Schreibblockade, dann sorge ich dafür, dass die Muse mich küsst.**

      Angenommen, **dass ich eine Schreibblockade habe, sorge ich dafür, dass die Muse mich küsst.**

3.1

§ 4 Allgemeine Rechte und Pflichten der Benutzer

Falls Schäden und Verluste am Bibliotheksgut während der Benutzung entstanden sind, hat der Benutzer in angemessener Frist vollwertigen Ersatz zu leisten. **Andernfalls** bleibt es der Zentralen Hochschulbibliothek überlassen, einen Schadensersatzbetrag für die Wiederbeschaffung festzusetzen oder auf Kosten des Benutzers eine Reproduktion zu besorgen. **Bei** unersetzbaren Werken kann neben dem Ersatz der Kosten für die Herstellung der Reproduktion voller Wertersatz gefordert werden. **Sofern** ein beschädigtes Werk in Stand gesetzt werden kann, ersetzt der Benutzer die Kosten. **Im Falle** einer Instandsetzung eines beschädigten Werks, kann die Bibliothek auf einen vollwertigen Ersatz verzichten, **wenn** die Wertminderung des beschädigten Bibliotheksgutes durch Zahlung eines entsprechenden Beitrags ausgeglichen wird.

3.2   b   Die Bibliothek kann auf einen vollwertigen Ersatz verzichten, wenn die Wertminderung des beschädigten Bibliotheksgutes ausgeglichen wird.

       > Bei **Ausgleich der Wertminderung des beschädigten Bibliotheksgutes kann die Bibliothek auf einen vollwertigen Ersatz verzichten.**

    c   Sofern ein beschädigtes Werk in Stand gesetzt werden kann, ersetzt der Benutzer die Kosten.

       > Bei **Instandsetzung des beschädigten Werks ersetzt der Benutzer die Kosten.**

4 Beispiellösung 1:

AGB einer Videothek
§ 9 Mängelhaftung
Liegt ein Mangel an der DVD vor, kann der Kunde eine Beseitigung oder Ersatzlieferung von der Videothek verlangen.
Umformulierung: Gesetzt den Fall, dass ein Mangel an der DVD vorliegt, kann der Kunde eine Beseitigung oder Ersatzlieferung von der Videothek verlangen.
Ergänzung: Falls die Beseitigung der Mängel oder Ersatzlieferung nicht möglich sind, kann der Kunde vom Vertrag zurücktreten.
Änderung: Wenn ein Mangel an der DVD vorliegt, kann der Kunde von der Videothek eine Ersatzlieferung verlangen, es sei denn, eine Beseitigung des Mangels ist möglich.

Beispiellösung 2:

Anmeldung und Rücktritt von Modulprüfungen
Original: Wer zu einer Klausur nicht angemeldet ist, kann an der Klausur nicht teilnehmen.
Umformulierung: Falls Sie sich nicht zu einer Klausur angemeldet haben, können Sie nicht an der Klausur teilnehmen.
Ergänzung: Bei verpasster Anmeldefrist ist keine Anmeldung mehr möglich.
Änderung: Wer zu einer Klausur nicht angemeldet ist, kann an der Klausur nicht teilnehmen, ausgenommen, die Frist zur persönlichen Anmeldung konnte aufgrund eines Auslandsaufenthaltes nicht eingehalten werden.

2.3.4 kausal: Begründung

1.1 Beispiellösung:

Man macht Salz ins Kochwasser, damit das Wasser schneller kocht.
Man freut sich wie ein Schneekönig, weil man sich auch im tiefsten Winter freuen kann.
In südlichen Regionen gibt es mehr Gifttiere, weil die Tiere dort wärmere Temperaturen vertragen.

1.2

„Warum macht man Salz ins Kochwasser?
Schlaubi Schlümpfe werden jetzt sofort rufen: „Na **weil** das Salz den Siedepunkt des Wassers erhöht und **weil** das Gemüse dann schneller gar ist!" Schöne Idee, ist aber trotzdem falsch, **denn** wenn man 30 Gramm in ein' Liter Wasser gibt, dann erhöht sich die Siedetemperatur gerade mal um ein halbes Grad Celsius und die Kochzeit verkürzt sich um weniger als eine Sekunde. Andere werden jetzt sagen: „Das Salz gibt man ins Kochwasser, **damit** die Kartoffeln oder Nudeln den Salzgeschmack annehmen." Ja, nee, stimmt so nicht, wie ich auf einer schlauen Internetseite und in einem alten Schulbuch gefunden habe. Wenn Sie mal an ihren Biologieunterricht zurückdenken, dann erinnern Sie sich bestimmt daran, dass da auch mal das Wort Osmose gefallen ist. Das meint, dass sich Lösungen, wenn sie aufeinander treffen, in ihrer Konzentration ausgleichen. Das Salz sorgt nun dafür, dass das Gemüse nicht sein Aroma verliert, **denn** es verhindert, dass zu viel Wasser in das Gemüse wandert und all die Würze des Gemüses ins Wasser. Warum? Warum, Warum? Darum! Beim Berliner Rundfunk 91.4. Mit Simone Panteleit. "

„Warum freut man sich eigentlich wie ein Schneekönig?
Schneekönig ist eine andere Bezeichnung für den Zaunkönig, ein ziemlich unscheinbarer Singvogel, der im Winter nicht nach Süden zieht, sondern hier bei uns bleibt. Den umgangssprachlichen Namen Schneekönig hat er bekommen, **weil** er selbst im tiefsten Schnee und Eis noch fröhlich vor sich hin trällert, als wäre es bereits schönster Frühling und **deshalb** werden Menschen, die sehr gute Laune haben und sich freuen, ebenfalls als Schneekönige bezeichnet. Warum? Warum, Warum? Darum! Beim Berliner Rundfunk 91.4. Mit Simone Panteleit."

„Warum gibt es im Süden mehr Gifttiere als bei uns?
In Südeuropa, Südamerika und Afrika gibt es **wegen** der höheren Durchschnittstemperaturen grundsätzlich eine größere Artenvielfalt als hier bei uns. Die Konkurrenz um Lebensraum und Futter ist da **deshalb** auch deutlich größer. Ein gutes Mittel um störende Mitbewohner aus dem Weg zu räumen oder für eine schnelle Mahlzeit zu sorgen ist da Gift. **Deshalb** gibt es in den südlichen Ländern mehr Gifttiere als bspw. in Deutschland. Das könnte sich aber in naher Zukunft ändern. Durch den Klimawandel könnten **nämlich** viele gefährliche Tiere auch hier bei uns heimisch werden. [...]"

1.3

Hauptsatz und Nebensatz	Konjunktion auf Position 0	Nebensatz
Das Salz sorgt nun dafür, dass das Gemüse nicht sein Aroma verliert,	denn es verhindert,	dass zu viel Wasser in das Gemüse wandert und all die Würze des Gemüses ins Wasser.

Hauptsatz	Subjunktion	Nebensatz
Das Salz gibt man ins Kochwasser,	damit	die Kartoffeln oder Nudeln den Salzgeschmack annehmen.
Den umgangssprachlichen Namen Schneekönig hat er bekommen,	weil	er selbst im tiefsten Schnee und Eis noch fröhlich vor sich hin trällert.

Hauptsatz (Verbindungsadverb auf Position 1 oder im Mittelfeld)
Deshalb gibt es in den südlichen Ländern mehr Gifttiere als bspw. in Deutschland.
Die Konkurrenz um Lebensraum und Futter ist da deshalb auch deutlich größer.

1.4

Kausale Satzverbindungen

- *Weil* leitet einen Nebensatz ein, der den Inhalt des **Hauptsatzes** begründet.
- *Denn* (und *da*) gibt eher einen zusätzlichen **Grund** an oder eine Begründung für die Äußerungen an sich. *Denn* ist also meistens keine direkte Begründung für das im Hauptsatz Geäußerte.
- Der **Nebensatz** mit *weil/da* steht vor oder nach dem Hauptsatz.
- *Denn* steht immer im **zweiten** Hauptsatz.

1.5 Beispiellösung:

> Man sagt im Deutschen „Ach du grüne Neune!" aufgrund einer Übertragung der französischen Spielkarten auf das deutsche Blatt. Pik-Neun entspricht der Grünen Neun(e). Beim Kartenlegen bedeutet diese Karte etwas Schlechtes (Unannehmlichkeiten, Krankheit, Verluste).

> Man sagt „wie das Schwein ins Uhrwerk gucken", weil ein Schwein die Funktion eines Uhrwerks nicht begreift. Deshalb benutzt man diese Redewendung, um auszudrücken, dass man etwas überhaupt nicht versteht.

2

ANDREA:	Ich konnte letzte Nacht wirklich überhaupt nicht schlafen, **weil** die haben da drüben wirklich alle fünf Minuten rumgebrüllt.
MARTIN:	Ich wollte zur Abwechslung mal durchschlafen, **deswegen** hab ich bei Thorsten übernachtet. **Denn** ich hab mir schon gedacht, dass im Maximus wieder die Post abgeht!
ANDREA:	Ich halte das nicht mehr aus, wir müssen jetzt endlich was unternehmen.
MARTIN:	Aber nur **wegen** Lärm gleich die Polizei rufen? Das ist doch voll spießig ...
ANDREA:	Mir ist das mittlerweile echt egal. **Vor** lauter Schlafmangel hab ich ständig Kopfschmerzen. Wir sollten wirklich nicht länger zögern, nur **weil** es uns zu peinlich ist, uns zu beschweren.
MARTIN:	Na gut, dann lass uns gleich mal nen Brief schreiben. Kannst du mal deinen Bruder fragen, bei wem wir uns beschweren müssen, ich hab **nämlich** keine Ahnung, ob Polizei oder Ordnungsamt oder was weiß ich wo ...
ANDREA:	Okay, dann ruf ich ihn **deshalb** gleich mal an ...

3

Sehr geehrte/r Mitarbeiter/in des Ordnungsamtes,

gegenüber unserer Wohnung in der Hansedorfstraße 5 in 13362 Berlin befindet sich die Diskothek „Maximus". In der Nacht werden wir oft um den Schlaf gebracht, **da** durch die ständig offen stehende Tür die Musik zu hören ist. Zusätzlich ist unser nächtliches Wohlbefinden permanent gestört, **weil** die Diskothek eine extrem starke Reklamebeleuchtung hat, die ab 21 Uhr eingeschaltet wird. **Aufgrund** der chronischen Schlafstörungen ist unsere Gesundheit zunehmend strapaziert. **Daher** bitten wir um eine Prüfung, ob Dämmung sowie Leuchtanlage der Diskothek einer Wohngegend angemessen sind.

4 **Kausale Satzverbindungen**

Konjunktion	Subjunktionen	Verbindungsadverbien (Position 1 oder Mittelfeld)	Präpositionen
denn	**weil, da**	nämlich (nicht Position 1), aus diesem Grund, **daher, deshalb, deswegen**	wegen (+ G oder D), **aus (+D), vor (+D), aufgrund (+G)**

5

Ihre Adresse:

_____

_____

_____

Adresse des Ordnungsamtes:

_____

_____

_____

Ort, Datum

Betreff: _____

Sehr geehrte/r Mitarbeiter/in des Ordnungsamtes,

Mit freundlichen Grüßen

_____

2.3.5   final: Absicht

1.1 Um gesund zu bleiben (Z), benötigt man mehr als nur gute Ärzte (H). Die deutsche Gesundheitspolitik und das Gesundheitswesen nehmen immer mehr den Einzelnen in die Verantwortung (H), damit die Krankenkassen und Ärzte entlastet werden (Z).

Um den Erregern eine geringe Angriffsfläche zu bieten und das Immunsystem zu stärken (Z), ist eine gesunde Lebensweise mit vitaminreicher Ernährung und ausreichendem Schlaf besonders wichtig (H).

Damit man eine so genannte Winterdepression eindämmt (Z), die bei geringer Tageslichtdauer auftritt, kann man an sonnigen Tagen spazieren gehen, das Solarium besuchen oder pflanzliche Heilmittel wie Johanniskraut einnehmen (H).

1.2

Finale Satzverbindungen

- Wenn im Haupt- und Nebensatz das gleiche Subjekt verwendet wird, dann kann der Nebensatz mit **um... zu** gebildet werden.
- Wenn das Subjekt im Haupt- und Nebensatz unterschiedlich ist, dann muss der Nebensatz mit **damit** gebildet werden.
- Bei Nebensätzen mit *damit* kann unter Umständen das Modalverb **können** eingefügt werden.
- Bei Hauptsatz-Hauptsatz-Verbindungen mit Verbindungsadverb werden häufig *möchte, wollen, sollen* eingefügt, um die Absicht zu verdeutlichen.

2   Familiengespräch am Frühstückstisch

Tanja: Wozu muss ich mit Euch am Wochenende frühstücken?
Vater: **Damit** wir mal wieder zusammen sind und wir den anstehenden Geburtstag von Oma Hilde besprechen können.
Tanja: Aber das hat doch noch Zeit. Und wozu soll ich jetzt eigentlich schon wieder Klavier üben?
Vater: **Um** dich auf den Auftritt vorzubereiten, den du am Wochenende hast. Oder hast du schon geübt?
Tanja: Ich habe schon geübt. Dafür habe ich auch gerade sowieso keine Zeit.
Vater: Dann mach wenigstens deine Biologiehausaufgaben und vergiss nicht für die Klausur zu lernen.
Tanja: Wozu soll ich eigentlich noch Biologiehausaufgaben machen, wenn ich doch sowieso Mathematikerin werden möchte.
Vater: **Damit** du dein Allgemeinwissen verbesserst und das Abitur bestehst.
Tanja: Wisst ihr denn überhaupt, wie die Photosynthese funktioniert?
Vater: Ja, durch Photosynthese entsteht Energie, die die Pflanzen zum Wachsen benutzen. Aber ganz genau können wir dir das auch nicht erklären. **Zu diesem Zweck** haben wir aber das Lexikon.

Handlung	Ziel
am Wochenende frühstücken	**zusammen sein und Geburtstag besprechen**
Klavier üben	**Vorbereitung auf den Auftritt**
Hausaufgaben machen	**Verbesserung des Allgemeinwissens, Abitur bestehen**
Lexikon haben	Photosynthese erklären

3   **b** Um dich auf den Auftritt vorzubereiten, den du am Wochenende hast.
> Zur **Vorbereitung des Auftritts, den du am Wochenende hast, ...**
c Damit du dein Allgemeinwissen verbesserst.
> Zur **Verbesserung deines Allgemeinwissens ...**

4   Nicht an allen Urlaubsorten gibt es deutschsprachige Ärzte. **Damit** man auch im Ausland ärztlich gut versorgt ist, sollte man sich vor der Reise über Kliniken und Arztpraxen mit deutschsprachigem Personal informieren. **Zu diesem Zweck** empfiehlt es sich, Ärztelisten von den Webseiten der deutschen Botschaft im Urlaubsland herunterzuladen bzw. die Ratschläge des Gesundheitsdienstes des Auswärtigen Amtes auf deren Webseite anzuschauen. Wer verschreibungspflichtige Medikamente benötigt, sollte sich diese vor dem Urlaub in ausreichender Menge verschreiben lassen, **damit** man auch dann mit Medikamenten versorgt ist, wenn der Urlaub, z. B. wegen eines Streiks, kurzfristig verlängert werden muss. Lebenswichtige Medikamente gehören ins Handgepäck, denn das Klima im Gepäckraum ist nicht ideal für alle Medikamente, abgesehen von der Gefahr eines Gepäckverlusts.
Wegen der verschärften Sicherheitsbestimmungen an EU- und US-Flughäfen ist es ratsam, ein ärztliches Attest oder eine beglaubigte Rezeptkopie (auch in englischer Sprache) mitzuführen. **Für** Medikamente, die unter das Betäubungsmittelgesetz fallen, ist dies Pflicht!

5 Beispiellösung:

> *Spanien im Sommer:* Damit Sie Ihre Haut ideal vor der Sonne schützen, ist es empfehlenswert, Sonnencreme mit einem hohen Lichtschutzfaktor zu verwenden. **Um die Kopfhaut zu schützen, sollte man einen Hut tragen.**
> *Norwegen im Winter:* **Damit man sich nicht erkältet, sollte man warme Kleidung tragen und darauf achten, dass die Schuhe trocken bleiben.**
> *England im Herbst:* **Für Spaziergänge sollten Sie einen Regenschirm einpacken.**

2.3.6 konsekutiv: Folge

1.1

Herr Schmidt, vielen Dank, dass Sie sich für unser Interview Zeit nehmen.

W.S.: Guten Tag. Sehr gern.

Also lassen Sie uns anfangen mit der ersten Frage. Was würden Sie mir als Autofahrer als erstes raten, wenn ein Unfall passiert ist?

W.S.: Das erste Gebot: Anhalten! Das Gesetz verpflichtet jeden, dessen Verhalten zum Unfall beigetragen haben kann, zunächst am Unfallort zu bleiben. Ausnahmen gelten nur in Notfällen, z. B. wenn ein schwer Verletzter versorgt werden muss. Dann müssen Sie aber unverzüglich nachträglich die notwendigen Feststellungen ermöglichen.

Anhalten und Melden sind **also** das A und O. Was passiert ansonsten?

W.S.: Unfallflucht wird streng geahndet, **sodass** Sie das Führerschein und Versicherungsschutz kosten kann. Obendrein bringt sie Ihnen eine empfindliche Strafe ein. Nach § 142 des Strafgesetzbuchs macht sich grundsätzlich strafbar, wer sich als Unfallbeteiligter vom Unfallort entfernt, ohne die Feststellung seiner Personalien, seines Fahrzeugs oder der Art seiner Beteiligung zu ermöglichen.

Was muss ich denn bei einem Unfall alles angeben?

W.S.: Also, Sie müssen auf Verlangen Ihren Namen und Ihre Anschrift angeben, Führerschein und Fahrzeugschein vorweisen und nach bestem Wissen Angaben über Ihre Versicherung machen. Außerdem müssen Sie berichten, in welcher Weise Sie an dem Unfall beteiligt waren. Andernfalls machen Sie sich strafbar!

Also sollte ich mir den Unfallhergang möglichst gut einprägen. Ist denn bei der Schilderung des Unfallhergangs etwas Bestimmtes zu beachten?

W.S.: Ja, berichten Sie **so** genau, **dass** man den Tathergang gut nachvollziehen kann und der Bericht trotzdem leicht verständlich ist. Ist niemand an der Unfallstelle zu sehen, z. B. weil Sie gegen ein geparktes Auto gestoßen sind, **so** müssen Sie in jedem Fall eine angemessene Zeit warten.

Wie lange müsste ich am Unfallort warten?

W.S.: Das hängt von den Umständen ab, also etwa Tageszeit, Ort und Schwere des Unfalls. Kommt in dieser Zeit niemand, **so** dürfen Sie sich entfernen, haben aber unverzüglich dem Geschädigten oder einer nahe gelegenen Polizeidienststelle zu melden, dass Sie am Unfall beteiligt gewesen sind.

Muss ich eine solche Meldung auch machen, wenn ich mich berechtigt vom Unfallort entfernt habe, z. B. weil ich mich um einen Verletzten gekümmert habe?

W.S.: Wie gesagt, laut Gesetz müssen Sie als Beteiligter am Unfallort bleiben. Verlassen Sie einfach den Unfallort, machen Sie sich **demzufolge** strafbar!

Konsekutive Satzverbindungen

Subjunktionen	Verbindungsadverbien	Präpositionen mit ähnlicher Bedeutung
zu ..., als dass; sodass; so ..., dass	demnach, **also** , **demzufolge**, somit, folglich, infolgedessen	zu (+D)

1.2 Exkurs: Bedeutungen von *also*

Das Wort *also* ist nicht nur ein konsekutiver Konnektor, sondern hat verschiedene Bedeutungen:

- Einleitung eines neuen Themas oder Gesprächsabschnitts und Gewinnung von Planungszeit
  > *Also lassen Sie uns anfangen mit der ersten Frage.*
  > **Also, Sie müssen auf Verlangen Ihren Namen und Ihre Anschrift angeben, ...**
- Einleitung einer Erläuterung oder Präzisierung
  > **Das hängt von den Umständen ab, also etwa Tageszeit, Ort und Schwere des Unfalls.**

2.1

Bitte ausgefüllt zurücksenden an:

Polizeiwache 203/9
Hauptstädter Straße 30
44287 Dortmund

Zeugendaten:
Name: Kleinert, Thomas
Straße: Rahestr. 36
Ort: 44122 Dortmund

Zeugenbericht

Ereignisdaten:
Betrifft: Fahrradunfall auf der Landgrafenstraße
vom: 15.08.2011
Ort: Dortmund
Beteiligte: Sabine Strobel und Hannes Rath

Kurze Beschreibung des Beobachteten:
Ich befand mich zum Zeitpunkt des Unfalls auf der Landgrafenstraße, so dass ich den Verlauf gut beobachten konnte. Ich sah eine Frau mit dem Fahrrad auf mich zukommen. **Auf der Kölner Straße ist kein Fahrradweg in Richtung Hainallee vorhanden. Demzufolge musste die Frau auf dem Gehweg fahren. Weil in der oberen Hälfte der Landgrafenstraße gebaut wird, ist die Straße gesperrt. Somit bog ein männlicher Autofahrer vom Grundstück der Landgrafenstraße 3 nach rechts in die Landgrafenstraße in Fahrtrichtung Staufenstraße ab (s. beiliegende Skizze).**
Es ist davon auszugehen, dass die Radfahrerin infolge eines Schwächeanfalls nicht in der der Lage war, ihr Fahrrad sicher zu fahren. Infolgedessen prallte sie auf dem Gehweg gegen die rechte Fahrzeugseite und stürzte über die Motorhaube auf den Boden.

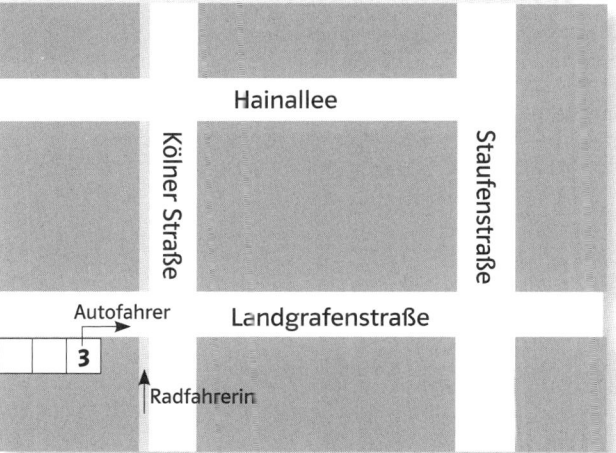

2.2 Beispiellösung:

Radfahrerin: Die Straße war extrem schmal, man kann als Radfahrer kaum überholt werden. Somit hatte ich keine andere Wahl als auf dem Gehweg zu fahren. **Die Kölner Straße ist auch immer so stark befahren und es gibt keinen Radweg. Mir war wegen der großen Hitze etwas unwohl, also fuhr ich langsamer. Die Landgrafenstraße wollte ich lieber mittig überqueren, denn das ist leichter. Infolgedessen bog ich in diese nach rechts ein, doch plötzlich wurde mir schwarz vor Augen. Erst im Krankenhaus kam ich wieder zu Bewusstsein.**

Autofahrer: Ich hatte angenommen, dass es an der Straße einen Radweg gibt. **Demzufolge hatte ich nicht damit gerechnet, dass jemand so schnell auf dem Gehweg heranfahren würde, als ich vorsichtig aus der Ausfahrt rollte. Plötzlich krachte mir etwas gegen die Seite. Der Radfahrerin schien es gut zu gehen, folglich hielt ich die Situation für ungefährlich und fuhr weiter.**

3 Beispiellösung:

Man sollte also immer umsichtig und rücksichtsvoll fahren.
Somit ist es besser, erst einmal stehen zu bleiben, wenn die Verkehrssituation unklar ist.
Man sollte also nicht unbedingt auf sein Recht bestehen.
Nur Kinder dürfen auf dem Gehweg Fahrrad fahren. Demnach sollten Erwachsenen mit dem Fahrrad immer auf dem Radweg oder auf der Straße fahren.

4.1

Fragen an den Anwalt

Vera Wahren braucht dringend Geld. Sie will deshalb ihr geliebtes Auto verkaufen. Sie einigt sich mit dem Autoliebhaber Anton Bettinger auf einen angemessenen Preis. Frau Wahren und Herr Bettinger vereinbaren, dass Herr Bettinger den Wagen schon am 1. Mai abholen darf, den Kaufpreis jedoch erst am 15. Mai zu zahlen braucht. Sie einigen sich weiterhin darauf, dass das Eigentum an dem Fahrzeug erst übergehen soll, wenn Herr Bettinger den Kaufpreis vollständig bezahlt hat. Herr Bettinger holt das Auto am 1. Mai ab. Am 2. Mai brennt das Fahrzeug bei einem Unfall aus. Das Auto wird dabei vollständig zerstört. Der Unfall war von Herrn Bettinger nicht zu vertreten. **Folglich** verweigert er die Kaufpreiszahlung. Herr Bettinger meint, Frau Wahren könne ihm das Eigentum an dem Wagen nicht mehr verschaffen, **also** sei er nicht mehr zur Zahlung verpflichtet. Kann Frau Wahren die Zahlung des Kaufpreises verlangen?

Aktivität / Folge

Das sagt der Anwalt

Frau Wahren könnte einen Anspruch auf Zahlung des vereinbarten Kaufpreises haben. Frau Wahren und Herr Bettinger hatten einen Kaufvertrag über das Auto geschlossen. **Demnach** wäre ein Anspruch auf den vereinbarten Kaufpreis zu bejahen. Frau Wahrens Anspruch könnte aber entfallen sein. Das wäre der Fall, wenn auch sie von ihrer Leistungspflicht befreit ist. Frau Wahren ist dazu verpflichtet, Herrn Bettinger das Eigentum an dem Fahrzeug zu verschaffen. Die Eigentumsverschaffung könnte **somit** unmöglich geworden sein. Die Gefahr des zufälligen Untergangs der verkauften Sache geht mit Übergabe auf den Käufer über. Übergabe ist die Übertragung des unmittelbaren Besitzes. Frau Wahren müsste die Sachherrschaft vollständig aufgegeben haben, und Herr Bettinger müsste sie vollständig erlangt haben. Beides kann im vorliegenden Fall unproblematisch bejaht werden. **Infolgedessen** kann Frau Wahren weiterhin die Zahlung des vereinbarten Kaufpreises verlangen.

2.3.7 adversativ: Einwand, Gegensatz

1.1

Logan
01.03.12
15:38

Hey Leute!
Zu meiner Mietwohnung gehört **zwar** ein Keller, **aber** dessen Wände sind sichtbar feucht. Ein, zwei Ziegel fehlen auch, damit das Grundwasser abfließen kann. Einen guten Schrank musste ich auch schon wegen Schimmelbefall entsorgen, sprich der Raum an sich ist eigentlich nicht nutzbar …
Der Vermieter sagte mir, dass ich selber keine baulichen Änderungen an den Wänden vornehmen darf. Sie selber würden **jedoch** auch nichts unternehmen, da das Grundwasser bei starkem Regenfall irgendwo abfließen müsse.
Meine Frage: Kann das Wasser nicht durch meinen Keller, **sondern** durch ein Abflussrohr fließen?
Was kann ich tun?
Vielen Dank im Voraus!

Max
02.03.12
19:21

Normalerweise sollte gar kein Wasser im Keller sein. **Während** dein Vermieter untätig bleibt, würde ich sofort aktiv werden. Denn wenn man gegen Wasser im Mauerwerk nichts unternimmt, dann wird dieses geschwächt. Während das Wasser vielleicht nicht so schnell nach oben zieht, kann sich rasch Schimmel an den Wänden bilden, was gesundheitliche Schäden zur Folge haben kann. Jeder Hausbesitzer würde sofort aktiv werden, dein Vermieter macht **hingegen** gar nichts. Die Wände werden schnell feucht, das Mauerwerk trocken zu legen dauert **dagegen** sehr lang. Ich würde an deiner Stelle so schnell wie möglich umziehen!

1.2

Hauptsatz	Konjunktion auf Position 0	Hauptsatz
Zu meiner Mietwohnung gehört zwar ein Keller,	aber	dessen Wände sind sichtbar feucht.
Kann das Wasser nicht durch meinen Keller,	sondern	durch ein Abflußrohr fließen?

Hauptsatz (mit Verbindungsadverb)	Nebensatz
Sie selber würden jedoch auch nichts unternehmen,	da das Grundwasser bei starkem Regenfall irgendwo abfließen müsse.

Hauptsatz	Hauptsatz (mit Verbindungsadverb)
Jeder Hausbesitzer würde sofort aktiv werden,	dein Vermieter macht hingegen gar nichts.
Die Wände werden schnell feucht,	das Mauerwerk trocken zu legen dauert dagegen sehr lang.

Subjunktion	Nebensatz	Hauptsatz
Während	dein Vermieter untätig bleibt,	würde ich sofort aktiv werden.

1.3 **Adversative Satzverbindungen: Wortstellung**

Konjunktionen	Subjunktion	Verbindungsadverbien	Präpositionen mit ähnlicher Bedeutung
(zwar) …, aber; **sondern**, doch	**während**	jedoch, **hingegen, dagegen,** allerdings	im Gegensatz zu (+D)

2

Darf bei einer Wohnung Wasser durch den Keller laufen?

Logan
03.03.12
10:38

Vielen Dank für die Infos! **Im Gegensatz zu** dir könnte ich leider nicht so schnell umziehen. Ich werde mich **aber** noch einmal mit meinen Nachbarn austauschen, ob sie ein ähnliches Problem haben. Letztens hatten wir auch einen Kurzschluss im Haus, der durch das Wasser im Keller verursacht wurde. Ich hatte deshalb die Hausverwaltung angeschrieben, **doch** sie hat bis jetzt nichts unternommen. Inzwischen bin ich **jedoch** davon überzeugt, dass der Vermieter etwas gegen den nassen Keller unternehmen muss. Das Haus fault ihm von unten nach oben weg, **aber** ihn scheint der sinkende Wert des Hauses bis jetzt noch nicht zu interessieren.

3 b Jeder Hausbesitzer würde sofort aktiv werden. Dein Vermieter unternimmt hingegen gar nichts. / Jeder Hausbesitzer würde sofort aktiv werden. Dein Vermieter hingegen unternimmt gar nichts.

c Die Wände werden schnell feucht. Das Mauerwerk trocken zu legen dauert dagegen sehr lang. / Die Wände werden schnell feucht. Das Mauerwerk dagegen trocken zu legen dauert sehr lang.

d Du würdest so schnell wie möglich umziehen. Ich könnte allerdings nicht so schnell umziehen. / Du würdest so schnell wie möglich umziehen. Ich allerdings könnte nicht so schnell umziehen.

e Das Wasser zieht vielleicht zunächst nicht so schnell nach oben. Es kann sich jedoch rasch Schimmel an den Wänden bilden, was gesundheitliche Schäden zur Folge haben kann.

4.1

LOGAN:	Guten Tag Frau Müller! Ich bin Ihr Nachbar von unten aus der ersten Etage.
FRAU UND HERR MÜLLER:	Guten Tag!
LOGAN:	Sie wohnen hier schon sehr lange. **Im Gegensatz zu** Ihnen bin ich erst vor einem Monat eingezogen. Deshalb möchte ich Sie fragen, ob Sie eventuell auch einen Wasserschaden im Keller bemerkt haben?
HERR MÜLLER:	Natürlich! Wir haben bereits mehrfach Beschwerden beim Vermieter eingereicht, **doch** dieser hat nicht viel unternommen. Wir hätten schon längst ausziehen sollen!
FRAU MÜLLER:	Die anderen sind ausgezogen. Wir sind **allerdings** hier wohnen geblieben.
LOGAN:	Und warum?
FRAU MÜLLER:	Wissen Sie, wir wohnen schon so lange hier. Mein Mann ist nun auch schon fast 80 Jahre alt.
HERR MÜLLER:	Die jungen Leute sind mobil und flexibel. Wir **jedoch** haben kein Auto und auch einfach keine Lust mehr umzuziehen.
LOGAN:	Ja, außerdem sind die Wohnlage und die Wohnung ja auch sehr schön. Ich bin mir einfach nicht sicher, was ich machen soll?
FRAU MÜLLER:	Sie zweifeln noch. Wir an Ihrer Stelle wären **dagegen** schon längst ausgezogen.
HERR MÜLLER:	Mit dem Vermieter haben Sie wirklich nichts als Ärger! Glauben Sie uns!

4.2

Element vor dem adversativen Adverb wird betont

Wir jedoch haben kein Auto und einfach keine Lust mehr umzuziehen.

Element vor dem adversativen Adverb wird nicht betont

Wir sind allerdings hier wohnen geblieben.

Wir an Ihrer Stelle wären dagegen schon längst ausgezogen.

4.3 b **Während wir bereits mehrfach Beschwerden beim Vermieter eingereicht haben, hat dieser nicht viel unternommen.**

c **Während die anderen ausgezogen sind, sind wir hier wohnen geblieben.**

d **Während die jungen Leute mobil und flexibel sind, haben wir kein Auto und auch einfach keine Lust mehr umzuziehen.**

e **Während Sie noch zweifeln, wären wir an Ihrer Stelle schon längst ausgezogen.**

5 Beispiellösung:

Im *Gegensatz zu* Logan würde ich **Mietminderung durchsetzen. / ... den Keller trocken legen lassen.**

Während Logan **untätig bleibt, würde ich mir unverzüglich Hilfe holen.**

Ich würde als Vermieter nicht **ausweichen,** *sondern* **den Schaden beheben lassen.**

Logan unternimmt nichts, *doch* **er könnte sich leicht Hilfe holen.**

6.1

SOZIOBAU AG
z.H. Herrn Panitzsch
Wilhelmsruher Damm 142

13435 Berlin

Berlin, 28.03.2012

Kündigung des Mietvertrages Nr. 11 00158

Sehr geehrter Herr Panitzsch,

leider muss ich feststellen, dass Sie **im Gegensatz zu** Ihren Ankündigungen den Wasserschaden im Keller noch nicht behoben haben.
Während Sie mich wegen meiner Mietminderung abmahnen, **doch** an den unerträglichen Zuständen nichts geändert haben, hat es bereits weitere Kurzschlüsse gegeben.

Auch hat die Anti-Schimmelpilzfarbe nichts bewirkt, **sondern** der Schimmel scheint sich weiter auszubreiten. Daher kündige ich den oben genannten Mietvertrag außerordentlich zum 31.10.2012.

Bitte bestätigen Sie den Erhalt des Kündigungsschreibens schriftlich. Sollte ich **hingegen** bis zum 15.10. nichts von Ihnen gehört haben, werde ich meinen Anwalt einschalten. Unsere zukünftige Anschrift teilen wir Ihnen so bald wie möglich mit. Für einen Termin zur Übergabe der Wohnung erreichen Sie mich **dagegen** schon jetzt unter der bekannten Telefonnummer.

Falls Sie weitere Fragen haben sollten, zögern Sie nicht, mich zu kontaktieren.

Mit freundlichen Grüßen

Logan Schmidt

2.3.8 konzessiv: unerwartete Konsequenz, Widerspruch

1.1

Fleisch ist mein Gemüse

Die Verfilmung von Heinz Strunks Bestseller feiert diese Woche Premiere. **Obwohl** *für den Film „Fleisch ist mein Gemüse" hervorragende Schauspieler, authentische Ausstattung und echte Kneipen als Kulissen gewählt wurden, berührt der Film nicht.*

Der „Jägerhof" in Hamburg-Harburg ist voll mit Leuten aus der Filmbranche. Die Meute ist in den Hamburger Vorort eingefallen, um die Premiere des Films „Fleisch ist mein Gemüse" stilgerecht zu feiern – mit Würstchen statt Salat.
Das 2004 erschienene Buch wurde überraschend zum Bestseller. Der Hamburger Künstler und Humorist Heinz Strunk erzählt darin von seiner verheerenden Jugend auf der falschen Seite der Elbe, in Harburg. **Zwar** erzählt Strunk gern von seiner fiktiven Jugend als Musiker, ansonsten regieren **aber** Ödnis und Trübseligkeit und selbst die Partys, die Strunk beschreibt, sind alles andere als hip. Es sind Dorffeste, Schützenfeste und Hochzeiten in der norddeutschen Provinz. So schrecklich das klingt: Für den 25-jährigen Strunk ist der Musiker-Job **trotzdem** der rettende Strohhalm.
Gleich zu Beginn des Films ist es mit der Ruhe **allerdings** vorbei: Die Mutter (Susanne Lothar) bricht zusammen und wird in eine Nervenklinik eingeliefert. Indessen muss sich Strunk mit seiner sagenhaft schlechten Band über Wasser halten. Die grellen Auftritte der Tiffanys mit Andreas Schmidt als Frontmann „Gurki" zählen zu den Highlights des Films: Stilechte Bühnen-Outfits aus den 80er Jahren, wie etwa knallenge Leggings, rote Angorapullis oder die rosafarbenen Glitzerjackets, katapultieren den Zuschauer direkt in das geschmacksfreie Jahrzehnt zurück. Maxim Mehmet in der Rolle des Heinz Strunk bleibt derweil neben Schauspielern wie Andreas Schmidt oder Livia Reinhardt etwas blass – **trotz** seiner erstaunlichen Ähnlichkeit mit dem echten Strunk.
Dennoch ist „Fleisch ist mein Gemüse" kein Film zum Schenkel klopfen. Er hinterlässt eher ein beklommenes Gefühl und wirkt seltsam unentschlossen zwischen der Absicht zu unterhalten und die dunklen Seiten des Lebens zu präsentieren.

1.2 b **Zwar erzählt Strunk gern von seiner fiktiven Jugend als Musiker, dennoch regieren ansonsten aber Ödnis und Trübseligkeit.**

c **Obwohl das so schrecklich klingt, für den 25-jährigen Strunk ist der Musiker-Job trotzdem der rettende Strohhalm.**

d **Obwohl er eine erstaunliche Ähnlichkeit mit dem echten Strunk hat, bleibt Maxim Mehmet in der Rolle des Heinz Strunk derweil neben Schauspielern wie Andreas Schmidt oder Livia Reinhardt etwas blass.**

1.3 Konzessive Satzverbindungen

- Konzessive Nebensätze werden bspw. durch *obwohl* eingeleitet und drücken einen Gegengrund oder Widerspruch aus.
- Im eingeleiteten Nebensatz mit **obwohl** werden die Umstände präsentiert, die gegen die im **Hauptsatz** folgende Handlung sprechen könnten, dies jedoch nicht tun.
- Verbindungsadverbien wie *trotzdem, dennoch, allerdings* etc. leiten die (unerwartete) **Handlung** ein.
- *Obgleich, gleichwohl* und *ungeachtet* (+G) werden vor allem in formellen Kontexten verwendet, im Mündlichen sind sie eher unüblich.

2

PETER:	Wie fandest du denn den Film?
SABINE:	**Obwohl** ich die Handlung ganz witzig fand, war er eher so lala!
PETER:	Wie bitte? Der Film war doch spitze! Also ich komme vom Land und kann nur sagen: Wie wahr!
SABINE:	Ja, das war ja auch nicht zu überhören, dass du dich amüsiert hast! Und ich gestehe dem Film auch einige lustige Szenen zu, **dennoch** war er insgesamt eher deprimierend!
PETER:	Deprimierend?! Kann es sein, dass wir **zwar** im selben Saal saßen, du **aber** einen anderen Film gesehen hast?
SABINE:	Nein, wir reden vom selben Film. **Allerdings** hat dieser nur deinen derben Humor angesprochen, mir war er viel zu flach.
PETER:	**Trotzdem** musst du zugeben, dass die Idee spitze war, auch wenn dir die Umsetzung nicht gefallen hat.
SABINE:	Ja, das stimmt. Auf jeden Fall besser als ein Hollywood-Blockbuster.

3 Konzessive Satzverbindungen

Konjunktion	Subjunktionen	Verbindungsadverbien	Präpositionen mit ähnlicher Bedeutung
zwar ..., aber	wenn ... auch, *obgleich*, **obwohl**	*gleichwohl*, **trotzdem**, **allerdings**, **dennoch**	*ungeachtet* (+ G), **trotz** (+G)

2.3.9 modal-substitutiv: Ersatz

1.2

„Die Führung der Deutschen durch Klose vom FC Bayern. Der lange Ball von Pogatetz kommt zwar präzise, aber was Royer draus macht, bleibt mir ein Rätsel. **Anstatt** den Ball **zu** sichern, lässt er ihn weit von der Brust abprallen. Der zweite Fehler: Der Innenverteidiger Schiemer kommt nicht zwischen Ball und Klose. So fällt das erste Tor für Deutschland gegen Österreich.

…

Bei Deutschland steht der erste Wechsel an. Joachim Löw wird wohl Schmelzer bringen. Oder es kommt mit Jerome Boateng der achte Bayern-Spieler. Höwedes muss nach 50 Minuten runter, kein schlechtes, aber auch kein auffälliges Spiel. Für ihn kommt nun doch Boateng **anstelle** von Schmelzer.

…

Da kommt Arnautovic. Freunde des österreichischen Fußballs aufgepasst! Immer noch Arnautovic. Tor! Ein fast schon historischer Moment – das erste Tor für Österreich gegen die Deutschen in dieser Qualifikation! Hummels und Badstuber greifen nicht ein. **Stattdessen** laufen sie nur nebenher wie zwei Jogger. So macht Arnautovic fast ungehindert den Anschlusstreffer."

2.1

Schönsein statt Sommermärchen

Die Leistung sollte im Mittelpunkt stehen. **Stattdessen** *entbrennt zur WM ein Wettstreit, welche Spielerin das schönste Glitzer-Girlie im DFB-Trikot ist. Warum konzentrieren wir uns nicht einfach auf den Sport? Der Frauenfußball hätte es verdient.*

Offiziell freuen sich alle Deutschen auf die Weltmeisterschaft. Aber das Aussehen der Spielerinnen steht im Mittelpunkt, **statt dass** über die Leistungen der Spielerinnen diskutiert wird. Eine Fußballspielerin muss scheinbar erst einmal beweisen, dass sie auch wirklich eine Frau ist.

«Da läuft etwas falsch», glaubt Andrea Schartner. **Statt** Misstrauen angesichts Fußball spielender Frauen **zu** hegen, sollten wir wieder den Sport in den Mittelpunkt stellen.

Dann würden wir sehen, dass die Randsportart Frauenfußball gerade dabei ist, aus ihrer Nische herauszutreten. Und dann hätte Deutschland auch die Chance auf ein neues Sommermärchen.

2.2. b **Wir hegen Misstrauen angesichts Fußball spielender Frauen. Stattdessen sollten wir wieder den Sport in den Mittelpunkt stellen.**

c **Er sichert den Ball nicht. Stattdessen lässt er ihn weit von der Brust abprallen.**

d **Schmelzer kommt nicht für ihn, stattdessen kommt Boateng.**

3.1 b **Statt dass Ballack (34) mit seinem 99. in den ungewollten Ruhestand verabschiedet wird, testet Löw Sven Bender. Anstelle einer Abschiedsgala findet ein knallharter Kandidaten-Check statt.**

c **Statt dass der Club Talente fördert, entlässt er den vielversprechenden Spieler Toko. Was denken Sie, liebe Leserinnen und Leser: Verliert der Club nicht die Glaubwürdigkeit, wenn er wieder und wieder seine Talente weggibt, anstatt etwas Konstantes aufzubauen?**

3.2

Modal-substitutive Satzverbindungen

- **Statt dass** wird verwendet, wenn der Haupt- und Nebensatz unterschiedliche Subjekte haben.
- Die Infinitivkonstruktion mit **statt / anstatt ... zu** wird verwendet, wenn es sich um das gleiche Subjekt im Hauptsatz und in der Infinitivkonstruktion handelt.

Konjunktion (nur zwischen Satzgliedern)	Subjunktionen	Verbindungsadverbien	Präpositionen mit ähnlicher Bedeutung
statt	(an)statt dass, (an)statt ... zu	stattdessen	anstelle von (+D), statt (+G)

4 **Beispiellösung:**

Spielfeld 2
Statt dass Spieler A den Ball übers rechte Mittelfeld nach vorn zu Spieler B passt, spielt er den Ball zurück zu Spieler C. / Spieler A soll den Ball übers Mittelfeld nach vorn spielen. Stattdessen passt er zurück zu Spieler C.

Spielfeld 3
Spieler A soll den Ball direkt zu Spieler C abspielen. Stattdessen dribbelt er ins Mittelfeld Richtung Tor. / Anstatt den Ball direkt zu Spieler C abzuspielen, dribbelt er ins Mittelfeld Richtung Tor.

Spielfeld 4
Anstatt den Ball nach vorn zu Spieler C zu passen, spielt B den Ball zurück zum Torwart. / Spieler B soll den Ball nach vorn zu Spieler C passen. Stattdessen spielt er den Ball zurück zum Torwart.

Spielfeld 5
Spieler A soll den Ball zu Spieler B passen. Stattdessen passt er ihn auf die linke Seite zu Spieler D. / Anstatt den Ball zu Spieler B zu passen, spielt er den Ball nach links zu Spieler D.

2.3.10 modal-instrumental: Art und Weise, Mittel

1.1

Netzwerk Angehörigenhilfe

Seit 2007 hat der ASB in Hamburg seine Aktivitäten für pflegende Angehörige verstärkt und im Netzwerk Angehörigenhilfe gebündelt. Dabei geht es zum einen darum, die pflegenden Angehörigen zu unterstützen, **indem** Wege aufgezeigt werden, wie die Pflege und Betreuung geleistet werden kann, **ohne dass** sie auf Kosten der eigenen Gesundheit gehen muss. Zum anderen sollen die Angehörigenhelfer entlastet werden, indem die Pflege und Betreuung zeitweilig und nach Möglichkeit kostengünstig – d. h. für breite Bevölkerungsschichten finanzierbar – durch andere übernommen wird.

Der ASB Hamburg versucht, hinreichend Informationen für die Angehörigen zur Verfügung zu stellen, **indem** geeignete Materalen zur Verfügung gestellt werden und diese über geeignete Wege die pflegen-

den Angehörigen erreichen. **Durch** 65.000 Mitglieder und damit jeden 25. Hamburger bestehen diese finanziellen Fördermöglichkeiten über Mitgliedsbeiträge.

So verfügt der ASB bewusst über eine hohe Zahl von 17 Sozialstationen im Stadtgebiet. Sie bilden die lokale Basis des Netzwerkes Angehörigenhilfe. **Dadurch** können die Einrichtungen von allen Angehörigen gut erreicht werden; vor Ort kann Information und Beratung umfassend geleistet werden. **Durch** die Verankerung vor Ort ist der ASB in Hamburg bekannt und gut mit allen Institutionen im Stadtteil – von den Ärzten, der bezirklichen Altenhilfe über die Kirchengemeinden bis hin zu den Krankenhaussozialdiensten – vernetzt.

1.2 Modal-instrumentale Satzverbindungen

Subjunktionen	Verbindungsadverbien	Präpositionen mit ähnlicher Bedeutung
soweit / soviel, dadurch, dass, **indem, ohne dass**	auf diese Weise, **so**, **dadurch**	mit (+ D), mit Hilfe von (+D), durch (+ A), wie, nach (+ D), ohne (+A)

2 **b** Dadurch, dass die Familie ein wichtiger Anker im Alter ist, kann man Geborgenheit und Sicherheit finden, wenn man Hilfe braucht. Man kann dadurch, dass die Familie ein wichtiger Anker im Alter ist, Geborgenheit und Sicherheit finden, wenn man Hilfe braucht.

c Dadurch, dass es durch berufliche Verpflichtungen nicht immer möglich ist, die Eltern zu pflegen, sind die Netzwerke Angehörigenhilfe eine gute Möglichkeit trotzdem die ausreichende Betreuung zu gewährleisten.

d Dadurch, dass jeder so früh wie möglich an seine Altersvorsorge denkt, kann man sehr zuversichtlich dem Alter entgegen sehen. Man kann dadurch, dass jeder so früh wie möglich an seine Altersvorsorge denkt, sehr zuversichtlich dem Alter entgegen sehen.

3 Alt werden, **ohne** die eigene Selbständigkeit **zu** verlieren, nimmt für die meisten Menschen einen hohen Stellenwert ein.

Eine Gemeinschaft von 7-9 Personen lebt, wie früher als Großfamilie, gemeinsam in einer großen Wohnung. Die zentrale Idee ist eine an der „Normalität" orientierte Organisation des Tagesablaufs, **ohne dass** es zu Isolation des Einzelnen kommt.

Ohne die Hilfe der Gruppe und des Pflegeteams könnte der / die Einzelne seine / ihre erlernten sozialen Verhaltensmuster nicht (wieder) finden.

Ihr / sein Leben erhält wieder einen Inhalt und sie / er einen neuen Platz in dieser Welt.

Hierbei kommt es darauf an, dass die täglichen Alltagsabläufe durch sie / ihn selber gemeistert werden können, **ohne dass** auf Anleitungen und Impulse des Pflegepersonals verzichtet werden muss.

Zur Pflege dementiell erkrankter Menschen werden Geduld und Ruhe benötigt, vor allem wenn die Abläufe nicht gleich koordiniert bzw. umgesetzt werden können.

Ohne ein ständiges Üben würden die alltäglichen Abläufe in Vergessenheit geraten. Aus diesem Grund werden die Bewohner in die täglichen Hausarbeiten, wie z. B. Beteiligung am Tischdecken, Essensvorbereitungen usw. einbezogen.

Jede/r Bewohner/in wird mit all ihren / seinen Stärken und Schwächen akzeptiert, **ohne** das Recht auf Privatheit gerade im Gemeinschaftsleben **zu** vernachlässigen. So werden z. B. der Wunsch nach Ruhe und Rückzug unbedingt erfüllt und respektiert.

4

Altersvorsorge: Je früher, desto besser

Im Interview mit „pressetext" spricht Walter Glanz, Pressesprecher der Deutschen Rentenversicherung Bund, über die Herausforderungen der finanziellen Altersvorsorge und geht auf die verschiedenen Formen der Alterssicherung ein. Sein Credo lautet: „Je früher, desto besser"

Altervorsorge ist wichtig. Aber ab wann sollte man damit beginnen? Welche grundlegenden Faustregeln gibt es hier?

Glanz: Eine Faustregel ist: Je früher man mit der Vorsorge beginnt, desto besser. **Indem** man bereits in jungen Jahren mit dem Aufbau einer zusätzlichen Altersvorsorge beginnt, kann man besonders vom Zinseszinseffekt profitieren. **Durch** eine Beratung bei der Deutschen Rentenversicherung kann man sich über die unterschiedlichen Modelle der zusätzlichen Altersvorsorge informieren.

Auf welchen Säulen baut die deutsche Alterssicherung auf?

Glanz: Die deutsche Alterssicherung beruht auf drei Säulen: Die erste Säule ist vor allem die Absicherung über die gesetzliche Rentenversicherung, die zweite Säule ist die betriebliche Absicherung und die dritte Säule ist die rein private Absicherung. **Dadurch, dass** die Menschen immer länger leben, viele früher als mit dem regulären Eintrittsalter in Rente gehen sowie einer sehr niedrigen Geburtenrate in Deutschland, ist die gesetzliche Rente in den letzten Jahren des Öfteren reformiert worden. So wurde die Absenkung des Rentenniveaus sowie die stufenweise Einführung der Rente mit 67 beschlossen. Damit ein Ausgleich der Niveauabsenkung erfolgen konnte, wurde mit der Reform 2001 die staatlich geförderte kapitalgedeckte Altersvorsorge, die sog. Riester-Rente, benannt nach dessen Erfinder und seinerzeitigen Sozialminister Walter Riester, eingeführt.

5   **Beispiellösung:**

- Was glauben Sie, wann sollte man mit der Altersvorsorge beginnen?
  So früh wie möglich, am besten mit Hilfe eines Experten und einer langfristigen Planung.
- Wie wollen Sie Ihre Rente ansparen?
  Indem ich eine zusätzliche Rentenversicherunger abschließe.
- Wie wichtig ist für Sie die Familie beim Thema Altersvorsorge?
  Die Altersvorsorge soll ohne die Hilfe meiner Familie gesichert sein.
- Haben Sie bereits etwas für das Alter geplant?
  Dadurch, dass ich noch kein festes Einkommen habe, konnte ich bisher noch nichts fürs Alter zurücklegen, aber sobald sich das ändert, möchte ich eine Rentenversicherung abschließen.
- Wie ist die Altersvorsorge in Ihrem Land organisiert?
  ...

2.3.11 Vergleich

1.1

So tun als ob

Verhältst du dich bewusst auf eine bestimmte Weise, verändert das deine Gefühls- und Gedankenwelt. Anders gesagt: mit dem Ausdruck kommt die Emotion.

In einer Studie forderte man eine Gruppe von Personen dazu auf, die Stirn zu runzeln, während man eine andere Gruppe dazu anhielt, ein leichtes Grinsen aufzusetzen. Unschwer zu erraten: Die „Grinser" fühlten sich so glücklich, wie wir gehofft haben. Die „Stirnrunzler" hingegen fühlten sich viel unglücklicher, als sie selbst vor Beginn der Studie erwartet haben.

Teilnehmer einer anderen Studie sollten verschiedene Produkte, die sich über einen Computer-Bildschirm bewegten, visuell fixieren und schließlich bewerten, ob ihnen die Artikel gefielen oder nicht. Manche Dinge wanderten dabei auf dem Bildschirm auf und ab (was beim Beobachten ein unbewusstes Nicken hervorrief), während andere sich horizontal bewegten (und eine Kopfbewegung von einer Seite zur anderen erforderten). Wie wir erwartet haben, bevorzugten die Probanden jene Produkte, deren Beobachtung eine „Ja"-Kopfbewegung provozierte. Das lässt uns schlussfolgern: Wenn du dich aufmuntern willst, dann verhalte dich nicht nur so, als ob du bereits glücklich wärst. Handle auch so, als wärst du glücklich!

1.2

	Nebensatz	Hauptsatz	Nebensatz	Nebensatz mit Verb an 2. Position
1		Die „Grinser" fühlten sich so glücklich,	wie wir gehofft haben.	
2		Die „Stirnrunzler" hingegen fühlten sich viel unglücklicher,	als sie selbst vor Beginn der Studie erwartet haben.	
3	Wie wir erwartet haben,	bevorzugten die Probanden jene Produkte,	deren Beobachtung eine „Ja"-Kopfbewegung provozierte.	
4	Wenn du dich aufmuntern willst,	dann verhalte dich so,	als ob du bereits glücklich wärst.	
5		Handle auch so,		als wärst du glücklich.

1.3 Reale Vergleichssätze

- Zu den realen Vergleichssätzen zählen Sätze 1, 2 und 3 mit (so) ... *wie* und der *Komparativ + als*.
- Bei Vergleichssätzen mit *so* (+ Adjektiv im Positiv) steht *wie* im Nebensatz.
- Bei Vergleichssätzen mit Komparativ steht *als* im Nebensatz.
- Der Nebensatz mit *wie* wird verwendet, wenn die Erwartung mit der Aussage im Hauptsatz **übereinstimmt**. Der Nebensatz mit *wie* kann **variabel**, d.h. vor oder nach dem Hauptsatz im komplexen Satz stehen.
- Der Nebensatz mit *als* wird verwendet, wenn die Erwartung mit der Aussage im Hauptsatz **nicht übereinstimmt**. Der Hauptsatz enthält dann einen **Komparativ**.

Fiktive Vergleichssätze

- In den Sätzen 4 und 5, die einen fiktiven Vergleich ausdrücken, wird der Nebensatz mit *als ob* und *als* eingeleitet.
- Das Verb steht bei fiktiven Vergleichssätzen im **Konjunktiv II**.
- Der **Nebensatz** mit *als ob* wird verwendet, wenn das Subjekt im Haupt- und Nebensatz gleich ist.
- Wenn *als* **in fiktiven Vergleichssätzen** allein den Nebensatz einleitet, steht das Verb **an zweiter Position**.

2

Auf dem Bild sieht der Mann so aus, als ob er sehr unglücklich / verzweifelt wäre.
Der Mann wirkt, als wäre er sehr unglücklich / verzweifelt.
Es sieht so aus, als wäre etwas Schlimmes passiert / als ob etwas Schlimmes passiert wäre.

Das Kind sieht so aus, als wäre es sehr unglücklich.
Das Kind sieht so aus, als ob es etwas nicht bekommen hätte / als hätte es etwas nicht bekommen.

Die Frau sieht so aus, als wäre sie sehr entspannt.
Die junge Frau sieht so entspannt aus, als hätte sie Urlaub / als ob sie Urlaub hätte.

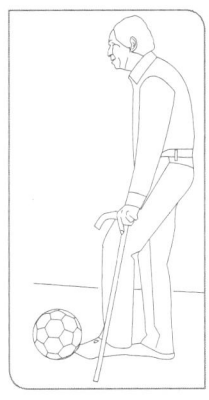

Der alte Mann sieht so aus, als wäre er sehr glücklich / zufrieden.
Der alte Mann sieht so aus, als hätte er Spaß.

3 Beispiellösung:

Wenn ich mir direkt vor einem Referat Tomatensoße über mein Hemd gekleckert hätte, würde ich so tun, als wäre mir kalt und meine Jacke anziehen.
Kurz vor meinem Vortrag habe ich mir Tomatensoße über meinen Kragen gekleckert. Ich habe dann so getan, als wäre ich krank und mir meinen Schal umgebunden.

---

4 **Reale Vergleichssätze: Wortstellung**

- Sätze mit *je ... desto* und *je ... umso* zählen zu den realen Vergleichssätzen.
- Sätze mit *je ... desto* bzw. *je ... umso* sind auffällig durch die Verwendung von **zwei Komparativen**. Der Satz mit *je* ist ein Nebensatz, d.h. das Verb steht **am Ende**. Der Satz mit *umso* bzw. *desto* ist ein **Hauptsatz**.

---

5 Beispiellösung:

Je näher wir unserer Unterkunft kamen, umso **mehr freute ich mich auf den Urlaub**.
Das Hotel war schöner, als ich **es mir ausgemalt hatte**.
Der Urlaub war so, wie ich **es mir erträumt hatte**.
Je näher die Abreise rückte, desto mehr genoss ich den Strand.

2.4 Weitere komplexe Sätze

2.4.1 Infinitivkonstruktionen und *dass*-Sätze

1.1

Ursula von der Leyen im Interview

„Lehrlinge werden mehr bekommen"

von EVA QUADBECK

(RP) Bundesarbeitsministerin Ursula von der Leyen spricht mit unserer Redaktion über den wachsenden Fachkräftemangel.

Der Fachkräftemangel macht sich schon bemerkbar: <u>Zurzeit stehen wir vor der ungewohnten Situation, dass es mehr Ausbildungsplätze als geeignete Bewerber gibt.</u> Was müssen Sie da unternehmen?

Von der Leyen: Die Jugendlichen haben in dieser Situation mehr Möglichkeiten bei der Auswahl eines Berufes. <u>Den Unternehmen muss klar werden, dass Azubis nicht mehr im Überangebot da sind.</u> <u>Sie müssen lernen, auch den auf den ersten Blick weniger geeigneten Bewerbern eine Chance zu geben.</u> Die Politik begleitet den Prozess, indem wir den Übergang von der Schule in die Ausbildung erleichtern. <u>Dazu gehört, dass kein Kind mehr die Schule ohne Abschluss verlässt.</u> Bereits in der Schule beraten wir über Berufe und begleiten zögernde Jugendliche bis in die ersten Monate der Ausbildung, damit sie Fuß fassen.

Müssen die Azubis auch besser bezahlt werden?

Von der Leyen: Das wird nicht die Politik entscheiden, sondern automatisch kommen. Wo Knappheit herrscht, steigen die Preise. <u>Die Unternehmen werden sich künftig mehr anstrengen, attraktivere Angebote zu machen. Wir unterstützen dabei mit Rat und Tat, und werden bei Interesse dabei helfen, größere Anreize für Auszubildende zu schaffen.</u>

1.2 <u>Sie</u> müssen lernen, dass <u>sie</u> auch auf den ersten Blick weniger geeigneten Bewerbern eine Chance geben.

<u>Die Unternehmen</u> werden sich künftig mehr anstrengen, dass sie attraktivere Angebote machen.

<u>Wir</u> unterstützen dabei mit Rat und Tat, und <u>wir</u> werden bei Interesse dabei helfen, dass <u>größere Anreize</u> für Auszubildende geschaffen werden.

1.3 Infinitivkonstruktionen und *dass*-Sätze

- Infinitivkonstruktionen ersetzen einen *dass*-Satz, um eine **Dopplung** des Subjekts zu vermeiden: *Sie müssen lernen, auch den auf den ersten Blick weniger geeigneten Bewerbern eine Chance zu geben.*

- Eine Infinitivkonstruktion kann auch folgen, wenn sich die Ergänzung im Einleitungssatz und das logische Subjekt in der Infinitivkonstruktion auf **dieselbe** Person beziehen: *Wir unterstützen dabei mit Rat und Tat und werden bei Interesse dabei helfen, größere Anreize für Auszubildende zu schaffen.*

- Wenn im Haupt- und *dass*-Satz **verschiedene** Subjekte stehen, kann der *dass*-Satz nicht durch einen Infinitiv ersetzt werden: *Den Unternehmen muss klar werden, dass Azubis nicht mehr im Überangebot da sind.*

- Der Infinitiv mit *zu* steht nach bestimmten Verben der **Erlaubnis** (*Es ist erlaubt, kurze private Telefonate am Arbeitsplatz zu führen.*), Absicht (*Wir haben vor, neue Maßnahmen einzuführen.*) oder des Gefühls (*Es ist großartig, mehr Lehrstellen anbieten zu können.*)

- *Dass*-Sätze stehen nach Verben des Sagens, der persönlichen Haltung, Verben mit festen Präpositionen oder unpersönlichen Ausdrücken.

2

> FRAU SCHMIDT: Frau Schulze ist nun endgültig von ihrem Chef gefeuert worden. Sie behauptet aber,
>   nicht gefeuert worden zu sein.
> FRAU MAIER: Ja, ich weiß! Sie hat mir erzählt, **den Job gekündigt zu haben**.
> FRAU SCHMIDT: Das verstehe ich nicht. Sie glaubt auch noch daran, **rasch einen neuen Job zu finden**.
> FRAU MAIER: Sie ist überzeugt davon, **sehr kompetent zu sein**.
> FRAU SCHMIDT: Wie sie meint. Also mir gefällt es hier – ich bin mir sicher, **in meinem Betrieb bleiben zu wollen**.
> FRAU MAIER: Ich weiß nicht. Ich denke, **irgendwann auch noch mal etwas anderes machen zu wollen**.

3.1 b Grundsätzlich gehe ich davon aus, **dass der Chef in einer klassischen hierarchischen
            Arbeitsstruktur die Aufgaben an seine Angestellten überträgt**.
   → **Ergänzung: Wovon gehe ich aus?**
   c Das Kernproblem ist doch aber, **dass heutzutage auf immer weniger Arbeitnehmer immer mehr
            Aufgaben zukommen**.
   → **Subjektsatz: Was ist doch aber das Kernproblem?**
   d Es ist nicht verwunderlich, **dass der Arbeitsverdichtung schnell Zeitmangel beim Arbeitnehmer folgt**.
   → **Subjektsatz: Was ist nicht verwunderlich?**
   e Es ist ratsam, **dass man dem Chef solche Überlastungssituationen mitteilt**.
   → **Subjektsatz: Was ist ratsam?**
   f Es lässt sich daher schlussfolgern, **dass das richtige und effektive Zeitmanagement für jeden
            Arbeitnehmer sehr bedeutend ist**.
   → **Subjektsatz: Was lässt sich daraus schlussfolgern?**

3.2 | ***dass*-Sätze**

- Der Subjektsatz übernimmt die Rolle des Subjekts im Hauptsatz, der **Ergänzungssatz** die Rolle einer Ergänzung
  im Hauptsatz.
- Steht ein **Subjektsatz** in Endstellung, wird er oft durch das Pronomen **es** angekündigt:
  Es ist ratsam, dass …
- Subjekt- und Ergänzungssätze werden durch *dass* oder auch *ob* eingeleitet:
  Ob die Aufgaben in der vorgegebenen Zeit erledigt werden können, ist zum heutigen Stand nicht sicher.
- Subjekt- und Ergänzungssätze können auch durch ein **Fragepronomen** eingeleitet werden:
  Wer die Aufgaben in der vorgegeben Zeit erledigen wird, ist bis zum heutigen Stand nicht sicher.

3.3 Dass der Arbeitsverdichtung schnell Zeitmangel beim Arbeitnehmer folgt, ist nicht
      verwunderlich.
      Dass man dem Chef solche Überlastungssituationen mitteilt, ist ratsam.
      Dass das richtige und effektive Zeitmanagement für jeden Arbeitnehmer sehr bedeutend ist,
      lässt sich daher schlussfolgern.

4 Beispiellösung:

Ich denke, dass **es schwerer ist, eine Arbeit aufzugeben, als eine neue zu beginnen**.
Ich bin davon überzeugt, **dass die Arbeit, die wir tun, anstrengender ist als die, die wir nicht tun**.
Dass man **auch mit einem vollen Terminkalender glücklich sein kann**, glaube ich sehr wohl.
Ich bin (nicht) der Meinung, dass **Denken die schwerste Arbeit ist**.
Es ist ratsam, **sich auch um die Anerkennung der eigenen Arbeit zu bemühen**.

2.4.2 Relativsätze

1.1

> KOMMISSAR: Können Sie bitte den Mann beschreiben, **den** Sie gestern
> beobachtet haben.
>
> ZEUGIN: Ja. Er trug dunkle Kleidung und Handschuhe sowie eine schwarze
> Mütze, **die** er tief ins Gesicht gezogen hatte.
>
> KOMMISSAR: Konnten Sie das Gesicht des Mannes erkennen?
>
> ZEUGIN: Ganz kurz. Alles ging so schnell. Er hatte ein schmales Gesicht und war
> rasiert. Ich erinnere mich an eine goldene Uhr, **die** kurz zwischen
> Handschuh und Ärmel zu sehen war.
>
> KOMMISSAR: Ich zeige Ihnen nun Fotos von den verschiedenen verdächtigen
> Personen. Zeigen Sie bitte auf denjenigen, **den** Sie von dem Überfall
> wiedererkennen.
>
> ZEUGIN: Moment, noch mal das letzte Foto, bitte. Ah, genau das ist der Mann,
> **den** Sie suchen. Ich erkenne ihn ganz genau, ich werde diesen kühlen
> Gesichtsausdruck nie vergessen.
>
> KOMMISSAR: Sind Sie sich ganz sicher, dass das die Person ist, **die** den Mann gestern
> überfallen hat?
>
> ZEUGIN: Ja, das ist der Mann, **der** auf den älteren Herrn eingeschlagen und ihn
> ausgeraubt hat.
>
> KOMMISSAR: Vielen Dank! Ich habe erstmal keine weiteren Fragen.

1.2

Detaillierte Täterbeschreibung führt zur Ergreifung eines Räubers

Die Osnabrücker Polizei konnte einen Räuber überführen, **der zuvor durch schnelle Fahndungsmaßnahmen festgenommen worden war.**

Der Räuber hatte zunächst vor den Augen von zwei Frauen auf einen Mann eingeschlagen, **der an einer Telefonzelle gestanden hatte.**

Danach hat er dem Opfer zwei Handys abgenommen, **mit denen er flüchtete.**

Danach war er mit einem Fahrrad, **das er einem Passanten entrissen hatte,** in Richtung Innenstadt geflohen.

Aufgrund der sehr detaillierten Personenbeschreibung von drei Zeugen konnte eine Polizeistreife den Mann, **der bereits polizeibekannt ist, schnell ausfindig machen und festnehmen.**

1.3

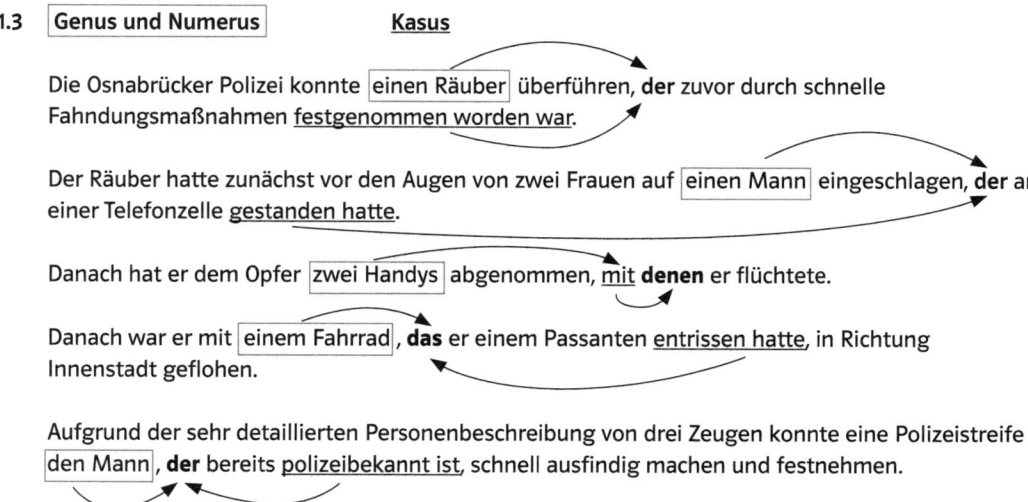

Genus und Numerus   Kasus

Die Osnabrücker Polizei konnte einen Räuber überführen, **der** zuvor durch schnelle Fahndungsmaßnahmen festgenommen worden war.

Der Räuber hatte zunächst vor den Augen von zwei Frauen auf einen Mann eingeschlagen, **der** an einer Telefonzelle gestanden hatte.

Danach hat er dem Opfer zwei Handys abgenommen, mit **denen** er flüchtete.

Danach war er mit einem Fahrrad, **das** er einem Passanten entrissen hatte, in Richtung Innenstadt geflohen.

Aufgrund der sehr detaillierten Personenbeschreibung von drei Zeugen konnte eine Polizeistreife den Mann, **der** bereits polizeibekannt ist, schnell ausfindig machen und festnehmen.

2.1

Das Rendezvous am Freitag

MARIA:	Hier sind die aussortierten Klamotten, **von denen** ich dir erzählt habe.
ANNA:	Ist das die Hose, **von der** du erzählt hast? Die sieht ja super aus! Bist du dir sicher, dass du die hergeben willst?
MARIA:	Ja! Das hab ich dir doch schon gesagt. Das ist die, **in die** ich nicht mehr reinpasse. Also nimm schon!
ANNA:	Darin sehe ich heute Abend bestimmt fabelhaft aus. Endlich kommt der Tag, **auf den** ich so lange gewartet habe!
MARIA:	Wo trefft ihr euch gleich nochmal?
ANNA:	Vor dem neuen indischen Restaurant, **von dem** ich dir schon mal erzählt habe.

2.2

Elternabend

PETER:	**Wer** ist das, **der** da drüben mit der Lehrerin spricht?
ANNA:	Das ist der Vater von Edith, **mit der** Malte in der Schule so gut befreundet ist.
PETER:	Könntest du bitte nicht so offensichtlich auf die beiden zeigen!
ANNA:	Ach, Peter!
PETER:	**Wer** das sieht, **der** denkt doch, dass wir schlecht über ihn reden.
ANNA:	Die meisten denken doch: **Was** ich nicht weiß, macht mich nicht heiß. Die kümmert es gar nicht, ob ich auf sie zeige oder nicht. Und **wem** ich meine Meinung sagen möchte, **dem** sage ich die schon persönlich!
PETER:	Jetzt sprich doch bitte nicht so laut! Da hat sich gerade jemand nach uns umgedreht.

2.3

Rückblick

PETER:	Es war die Geburt unserer Kinder, **worüber** ich mich am meisten gefreut habe. Unsere Kinder sind einfach das, **worauf** wir ganz besonders stolz sind.
ANNA:	Freundschaft ist das, **was** das Leben reicher macht. Mit Maria habe ich immer noch eine gute Freundschaft. Unserer Freundschaft verdanke ich vieles, **was** ich heute erreicht habe.
PETER:	Ich bin immer noch sehr glücklich mit Anna. Sie gibt mir alles, **wovon** ich schon immer geträumt habe. Wir reden über vieles, **was** uns im Alltag widerfährt. Das ist glaub ich das, **worauf** es ankommt.

3   Freundschaften hat jeder. Das Wichtigste, worauf **es wirklich ankommt, ist, dass man Freundschaften pflegt, damit sie nicht einschlafen.**

In der Freundschaft gibt es vieles, was **mal stören kann.**

Eine Beziehung / Ehe sollte wie eine Freundschaft sein. Das ist das gewisse Etwas, was **zwei Menschen lange zusammen leben lässt.**

2.4.3 Indirekte Fragesätze: Zeit, Grund, Art und Weise, Ort

1

> Daher möchte ich nachfragen, warum diese Erkenntnis bei Ihnen erst nach mehreren verlorenen Landtagswahlen kommt und nicht schon viel früher? Zudem interessiert mich, ob ich Ihnen noch vertrauen kann? Haben Sie nicht noch vor wenigen Wochen und Monaten eine ganz andere Meinung vertreten?
> Würden Sie einem Politiker einer anderen Partei Glaubwürdigkeit bescheinigen, wenn er sein Fähnchen so nach dem Wind stellt?
> Ich hätte gern gewusst, welche Ihrer Wahlkampfthemen nun über Bord geworfen werden sollen.
> Worauf ist Ihre Kurswende zurückzuführen und wie wollen sie weiterregieren?

2 Beispiellösung:

- Ich möchte gerne wissen, wen Sie bei der nächsten Bundestagswahl wählen würden?
  Ich würde … wählen, weil sie/er eine gute Umweltpolitik verfolgt.
- Was glauben Sie, welche Partei die kommenden Wahlen gewinnen wird?
  Ich denke, dass die … gewinnen wird, weil sie kommunal stark verankert ist.
- Mich interessiert, welche Politikerin / welchen Politiker Sie schätzen?
  Ganz besonders schätze ich …, weil er/sie sich für eine strenge Steuerpolitik einsetzt.
- Sind Sie der Meinung, dass …
  Ja, ich bin der Meinung, dass weniger Sicherungsverwahrungen verhängt werden sollen.
- Wie schätzen Sie die derzeitige Situation der Bundesregierung ein?

  …

2.5 Nominalisierung

2.5.1 Nominalisierung von Infinitivsätzen und *dass*-Sätzen

1 **Es ist wichtig, dass man bis Vierzig durchhält**

→ verbal

Ein Gespräch mit Bundesarbeitsministerin Ursula von der Leyen

→ **nominal**

Detaillierte Täterbeschreibung führt dazu, dass der Täter nach monatelanger Suche ergriffen wird

→ **verbal**

Polizei gelingt es, den Verdächtigen nach Raubüberfall schnell festzunehmen

→ **verbal**

ABSTURZ STATT KRÖNUNG

→ **nominal**

Das Schönsein zählt mehr als ein Sommermärchen

→ **nominal**

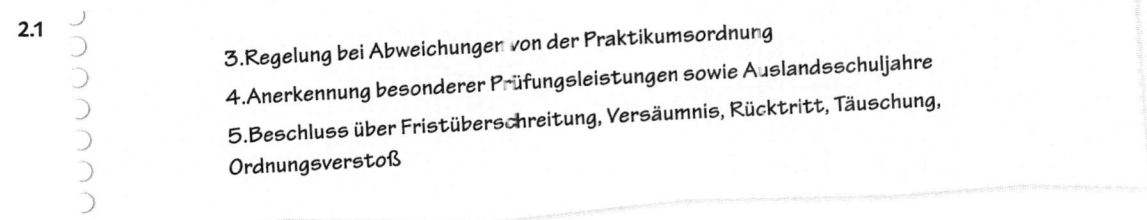

2.1

3. Regelung bei Abweichungen von der Praktikumsordnung

4. Anerkennung besonderer Prüfungsleistungen sowie Auslandsschuljahre

5. Beschluss über Fristüberschreitung, Versäumnis, Rücktritt, Täuschung, Ordnungsverstoß

> Er ist dafür zuständig, **über die Prüfungsteilnahmen bei Gasthörerschaft einzuwilligen.**
> Der Prüfungsausschuss ist dazu verpflichtet, **regelmäßig an den Fachbereichsrat zu berichten. / regelmäßig Bericht an den Fachbereichsrat zu erstatten.**
> Der Prüfungsausschuss ist dafür zuständig, **zur Reform von Studium und Prüfungen anzuregen.**

3 Beispiel-Exzerpt:

2. Voraussetzung für erfolgreiches Argumentieren:
   – wichtige Argumente erkennen und angemessener Umgang damit
   – Ziel: Formulierung eigener Argumente, Entkräftung von Gegenargumenten
   – Kenntnis von Argumentationstechniken und richtige Anwendung
3. Voraussetzung für wissenschaftliches Schreiben: Kenntnis des Aufbaus einer wissenschaftlichen Arbeit
   – Ziel: Sicheres Verfassen eigener Texte, gezielter Einsatz gelernter Methoden des wissenschaftlichen Schreibens
   – Aufbau und Gliederung wissenschaftlicher Texte, Beachtung der Formalia (z.B. Zitierregeln)
   – Strukturierung des Verfassens wissenschaftlicher Texte
4. Voraussetzung für interkulturelle Kommunikation: Auseinandersetzung mit anderen Kulturen
   – Ziel: Durchführung von Projekten im interkulturellen Kontext
   – Beachtung von Kulturstandards, typischer Rhetorik, unterschiedlicher Kommunikations-/Vortragsstile

2.5.2 Nominalisierung von weiteren Haupt- und Nebensätzen

1.1

Zur Erforschung der Veränderung hatten das Helmholtz-Zentrum Geesthacht und das Alfred-Wegener-Institut für Polar- und Meeresforschung 100.000 Kilometer arktischer Küstenlinie untersucht. (**final**) Am Zustandsbericht „Stand der Arktikküste 2010" waren zudem Forscher aus zehn Ländern beteiligt.
Statt aus Fels bestehen zwei Drittel der arktischen Küsten aus dauerhaft gefrorenem Boden, dem so genannten Permafrost. (**modal-substitutiv**) Dieser taut nun teilweise. Die Wissenschaftler warnen vor den großen Veränderungen für die küstennahen arktischen Ökosysteme und die dort lebende Bevölkerung. Trotz der dünnen Besiedelung der arktischen Landstriche sind auch im Hohen Norden die Küsten wichtige Achsen für das wirtschaftliche und gesellschaftliche Leben. (**konzessiv**)

Bisher seien die arktischen Küsten vor der erodierenden Kraft der Wellen durch ausgedehnte Meeresflächen geschützt, heißt es in der Untersuchung. Infolge des kontinuierlichen Rückgangs des Meereises sei dieser Schutz nun gefährdet. (**kausal**)
Seit dem Anwachsen des Bedarfs an globalen Energieressourcen und dem zunehmenden Tourismus und Gütertransport wird der menschliche Einfluss auf die Küstenregionen verstärkt. (**temporal**) Aufgrund der Zunahme der Erosion änderten sich die ökologischen Bedingungen für Wildtierbestände wie die großen Karibuherden des Nordens erheblich, warnen die Experten. (**kausal**)
Neue Messungen hatten gezeigt, dass manche Gletscher in Grönland schneller tauen. Im Falle einer Erhärtung dieses Verdachts würde sich der Anstieg der Meeresspiegel beschleunigen. (**konditional**)

1.2

Bedeutung	Präpositionen	Äquivalente (Subjunktionen, Konjunktionen, Verbindungsadverbien)
modal-instrumental	durch	indem, **dadurch, dass**
final	**zu**	um … zu
adversativ		während, **dagegen, sondern**
konzessiv	**trotz**	**obwohl**
konsekutiv		**folglich**, infolgedessen, **also, demnach, somit**
temporal	**seit**	**seitdem**
kausal	**infolge, aufgrund**	**weil, darum, nämlich**
konditional	**im Falle**	**falls**
modal-substitutiv	statt	**stattdessen, anstatt**

1.3   **b** > Um die Veränderung zu erforschen, hatten das Helmholtz-Zentrum Geesthacht und das Alfred-
Wegener-Institut für Polar- und Meeresforschung 100.000 Kilometern arktischer Küstenlinie
untersucht.

c > Obwohl der Landstrich dünn besiedelt ist, sind auch im Hohen Norden die Küsten wichtige Achsen
für das wirtschaftliche und gesellschaftliche Leben.

d > Der Schutz ist gefährdet, weil das Meereis kontinuierlich zurückgeht.

e > Seitdem der Bedarf an globalen Energieressourcen anwächst und Tourismus und Gütertransport
zunehmen, wird der menschliche Einfluss auf die Küstenregionen verstärkt.

f > Die Erosion nimmt zu, darum änderten sich die ökologischen Bedingungen für Wildtierbestände wie
die großen Karibuherden des Nordens erheblich, warnen die Experten.

g > Falls sich der Verdacht erhärtet, würde sich der Meeresspiegel schneller heben.

2   **Erosion** (lat.), Vorgänge, die **durch** Abtragung von Gestein u. Boden **zur** Bildung der Oberflächen-
formen der Erde beitragen: a) *Fluss-E.*, ausfurchende und einschneidende Arbeit des fließenden
Wassers (Täler, Schluchten); b) *Wind-E.* (Deflation), Abtragung **durch** Windeinwirkung (bes. in
Gebieten mit spärl. Vegetation); c) *Glazial-E.*, ausräumende Wirkung der Gletscher; d) *marine E.*
(Abrasion), abtragende Wirkung von Brandung, Gezeiten u. Meeresströmungen.

Ökosystem, funktionelle Einheit von Lebensraum (Biotop) u. Lebensgemeinschaft (Biozönose);
ohne störende äußere Einflüsse **durch** stete Selbstregulierung im biol. Gleichgewicht.

Polarstern (Nordstern), Stern 2. Größe im kleinen Bären. Dient **wegen** seines geringen Abstandes
vom Himmelspol (ca. 1) **zur** Bestimmung der Himmelsrichtungen u. der geograph. Breite.

3   Nominale Gruppen

3.1   Nominale und präpositionale Gruppen erkennen

1.1 **b** Was nahm er auf die Reise mit?   Er reiste **ohne Gepäck**.

c Warum reiste der Mann weiter nach Kopenhagen?
Er fand in England **keine baldige Rückfahrgelegenheit**.

d Welches Verkehrsmittel wohin benutzte er von Kopenhagen aus?
Er benutzte **den Postdampfer nach Irland**.

e Welche Reise konnte er wegen der rauhen Jahreszeit nicht mehr antreten?
Er konnte **die lange Landreise in den Norden** nicht mehr antreten.

f Wann kehrte der Mann von seinem Sonntagsausflug nach Hause zurück?
Er kehrte **im folgenden Sommer / ein Jahr nach dem Aufbruch** nach Hause zurück.

g Woher genau kommt der Mann, dessen Reise hier beschrieben wird?
Er kommt **aus dem Norden von Island**.

h Warum handelt es sich um „einen Fall von gewaltiger Zeitverschwendung"?
Ein Mann will am Sonntag **einen Bekannten** besuchen, benötigt aber für den Besuch und die Rückkehr nach Hause ein ganzes Jahr.

1.2

Soeben ist mir **ein** Fall von gewaltiger Zeitverschwendung zu Ohren gekommen. In Island passierte es **einem** Mann, der über Sonntag einen Bekannten besuchen wollte und **ein** in der Gegend verkehrendes Schiff bestieg, dass **dieses** Schiff infolge der ungewöhnlichen Strömungsverhältnisse nicht an der beabsichtigten Stelle anlegen konnte. Der Mann, der ohne Gepäck reiste, musste nun weiter nach England fahren, und da er dort **keine** baldige Rückfahrgelegenheit vorfand, reiste er weiter nach Kopenhagen, um von dort aus **den** vier Wochen später fälligen Postdampfer nach Island benutzen zu können. Er kam auch glücklich an, aber die Jahreszeit war inzwischen so rauh geworden, dass er **die** lange Landreise in den Norden der Insel, aus dem er gekommen war, nicht mehr antreten konnte. Deshalb kehrte der Mann von **seinem** Sonntagsausflug erst im folgenden Sommer, also **ein** Jahr nach dem Aufbruch, zurück.

1.3   In der folgenden Tabelle finden Sie in der linken Spalte nominale Gruppen mit verschiedenen Artikelwörtern und Erweiterungen (Attributen). Ergänzen Sie in der rechten Spalte Beispiele aus dem Text, die die gleiche Form haben.

Zeit	Gepäck / Island / Sonntag / **England / Kopenhagen**
die Zeit	die Gegend / **der Mann / die Jahreszeit**
eine Zeit	ein Fall / **ein Mann / ein Jahr**
diese Zeit	dieses **Schiff**
deine Zeit	sein **Sonntagsausflug**
die Zeit des Wartens	der Norden **der Insel**
die Zeit, die mir fehlt	der Mann, der **ohne Gepäck reiste** ein Mann, **der über Sonntag einen Bekannten besuchen wollte**
keine anstrengende Zeit	**keine baldige Rückfahrgelegenheit**
die schlimmste Zeit	die **ungewöhnlichen** Strömungsverhältnisse / **die lange Landreise**
eine Zeit nach dem Studium	ein Fall **von gewaltiger Zeitverschwendung**

2.1   links vom Nomen / rechts vom Nomen

Bei der industriellen Nutzung der Kernenergie war es von Beginn an zu Unfällen mit zum Teil schweren Umweltbelastungen gekommen. Doch erst die Reaktorkatastrophen von Harrisburg 1979 und Tschernobyl 1986 sowie das nach wie vor ungeklärte Problem der sicheren Entsorgung nuklearer Abfälle führten weltweit auch in Bezug auf die Kernenergie zu einem Umdenken. Die immer wieder als Energiequelle der Zukunft angeführte Kernfusion, nach deren technischer Ausnutzung bereits über ein halbes Jahrhundert geforscht wird, hat die Hoffnungen, die in sie gesetzt wurden, bisher nicht erfüllen können.

Im Bereich der traditionellen fossilen Energieträger wie Steinkohle, Braunkohle und Erdöl war es vor allem der Anstieg an klimagefährdenden Emissionen, der auch hier zu einer verstärkten Suche nach Alternativen geführt hat. In letzter Zeit gewinnt dabei neben der Verwendung des emissionsärmeren Erdgases und Anstrengungen zur Energieeinsparung besonders die Rückkehr zu regenerativen Energieträgern wie Wind, Sonne, Erdwärme und Biogas an Bedeutung.

2.2

Erweiterungen rechts vom Nomen-Kern	
Unfälle mit zum Teil schweren Umweltbelastungen	Präpositionalattribut
die **Reaktorkatastrophen von Harrisburg 1979 und Tschernobyl 1986**	
der **Anstieg an klimagefährdenden Emissionen**	
eine **Suche nach Alternativen**	
die **Rückkehr zu regenerativen Energieträgern wie Wind, Sonne Erdwärme und Biogas**	
das **Problem der sicheren Entsorgung nuklearer Abfälle**	Genitivattribut
die **Kernfusion, nach deren technischer Ausnutzung bereits über ein halbes Jh. geforscht wird**	Relativsatz
Hoffnungen, die in sie gesetzt wurden	

	Erweiterungen links vom Nomen-Kern
Adjektivattribut / Partizipialattribut	das nach wie vor ungeklärte **Problem**
	eine verstärkte Suche
	die immer wieder als Energiequelle der Zukunft angeführte Kernfusion

3.1

Ernährung

Die Empfehlungen für die tägliche Aufnahme der wichtigsten Nährstoffe wurden im frühen 20. Jahrhundert ursprünglich im militärischen Kontext entwickelt. Wenn sich **ein** einzelner Mensch falsch ernährt, ist das sein Privatproblem. Hat man aber **eine große** Menge Menschen zu ernähren, also etwa Soldaten, Flüchtlinge oder die Bewohner **eines** Staats, in dem Lebensmittelrationierung herrscht, dann ist es nützlich zu wissen, **welche** Nährstoffe diese Menschen brauchen.

Das herauszufinden ist aber keine triviale Aufgabe. Der einzige Mensch, dessen Nährstoffbedarf relativ genau bekannt ist, ist **ein** Säugling, der von einer gesunden Mutter gestillt wird. Da die Natur in solchen Dingen üblicherweise weiß, was sie tut, kann man davon ausgehen, dass die Muttermilch nicht mehr und nicht weniger als **die** für den Säugling erforderlichen Nährstoffe enthält. Sobald der Mensch aber anfängt, an Bananen, Brezelstücken und Tischbeinen zu kauen, wird es kompliziert.

So beruht **die** Empfehlung, täglich mindestens 130 Gramm Kohlenhydrate zu verzehren, auf **dem** errechenbaren Energiebedarf des Gehirns und **der** Annahme, dass das Gehirn diesen Energiebedarf am liebsten aus Stärke oder Zucker deckt.

3.2

Erweiterungen rechts vom Nomen-Kern	
die **Empfehlung**, täglich mindestens 130 Gramm Kohlenhydrate zu verzehren, …	Infinitivkonstruktion
Die **Annahme, dass das Gehirn diesen Energiebedarf am liebsten aus Stärke oder Zucker deckt.**	*dass*-Satz

3.3 a ein Staat, **in dem Lebensmittelrationierung herrscht**
b ein Säugling, **der von einer gesunden Mutter gestillt wird**
c die Nährstoffe, **die für den Säugling erforderlich sind**
d der Energiebedarf, **der errechenbar ist**

3.2 Artikelwörter

1.1

Herr Jensen steigt aus

Am nächsten Morgen wachte Herr Jensen pünktlich in **seinem** Sessel auf. Pünktlich wozu, dachte er frustriert. Lächerlicherweise plagten ihn Kopfschmerzen, obwohl er am Vorabend **keinen** Alkohol getrunken hatte. Spielend hätte er es nun zur Arbeit geschafft, die er nicht mehr hatte, und das, wo es ihm in **allen** Jahren, in denen er **diese** Arbeit noch hatte, immer so schwer gefallen war, pünktlich aufzustehen, und er sich **jeden** Morgen gewünscht hatte, noch liegenbleiben zu können. – Herr Jensen stand ächzend auf und probierte durch vorsichtige Bewegungen aus, welches Gelenk am schlimmsten schmerzte.

1.2

Artikelwörter

	Deklinationsgruppe	Beispiel	Variation
1	der / die / das dies-, jen-, welch-, manch-, jed-	Er probierte aus, welches Gelenk schmerzte. diese Arbeit jeden Morgen	Er merkte, dass jedes Gelenk schmerzte. jene Arbeit manchen Morgen
2	ein / eine / ein mein-, … irgendein-, was für ein-, kein-	Er wachte in seinem Sessel auf. **Er hatte keinen Alkohol getrunken.**	**Er wachte an irgendeinem Ort auf.** Was für einen Kaffee trinkst du?
3	all-, einig-, irgendwelch-, etlich-, mehrer- (solch-)	**in allen Jahren**	an einigen Tagen

2   › Ein solcher Schritt: „solch-" wird wie ein Adjektiv verwendet und bezieht sich auf den Satz zuvor.

3   a   Schon **wenige** Worte Spanisch helfen beim Ferienaufenthalt. Der Kurs im August richtet sich an Personen, die bereits **geringe** Spanischkenntnisse haben.

b   Obst und Gemüse haben **wenig** Fett, eine **geringe** Energiedichte, einen **hohen** Vitamingehalt und sind schmackhaft. Deshalb soll man **viel** frisches Obst und Gemüse essen.

c   Zu **wenig** Wähler gingen zur Wahl. Es habe eine **viel** zu **geringe** Wahlbeteiligung gegeben, so das Ergebnis einer Wahlanalyse.

4

Kleine Abfallgeschichte(n)

Arte 19.30 Neue Reihe: Umwelt-Doku „Paris"

Jeden Tag produziert **die** Menschheit gewaltige Mengen Abfall – doch wohin mit **dem** Müll? Noch vor wenigen hundert Jahren lief man in **den** Städten durch Müll und Exkremente. Erst im 19. Jahrhundert entwickelten sich in Ballungsgebieten Abfallentsorgung, Abwassersysteme und auch **die** erste Mülltrennung.

Die Doku-Reihe von Nick Quinn zeigt, wie dies **den** Städtebau veränderte und sogar Revolutionen auslöste.

	Nomen ohne Artikelwort
Plural unbestimmt	Mengen, **Exkremente**, **Ballungsgebiete**, **Revolutionen**
abstrakte Nomen	Intelligenz, Mut
Substanznomen	Wasser, Obst, Geld, **Abfall**, **Müll**
Institutionen oder abstrakte Vorgänge	Abfallentsorgung
Namen	Nick Quinn

5

Fahrer eines Rettungswagens verursacht Unfall

Helmstedt. Bei einem Verkehrsunfall auf der Landesstraße 292 wurden gestern Mittag drei Menschen leicht verletzt. Dazu zählt der 35 Jahre alte Fahrer eines Rettungswagens aus Helmstedt.Zum Hergang machte die Polizei folgende Angaben: Der Helmstedter erlitt während der Fahrt einen Schwächeanfall und geriet dadurch mit seinem Fahrzeug auf die Gegenfahrbahn. Eine entgegenkommende 65-jährige Fahrerin eines Pkws konnte noch ausweichen, trotzdem stießen beide Fahrzeuge zusammen. Während der 35-jährige Krankenwagenfahrer bei dem Aufprall **einen** Schock erlitt, wurden **die** Pkw-Fahrerin und ihre Beifahrerin leicht verletzt. **Ein** zum Unfallort bestellter Rettungssanitäter brachte die drei Leichtverletzten ins Wolfsburger Klinikum. **Dem** Patienten im Rettungswagen sei laut Polizeiangaben nichts passiert

Bestimmter und unbestimmter Artikel

bestimmter Artikel	die Person oder Sache ist persönlich oder allgemein bekannt	*die Polizei*
	eine Sache / Person / ein Ereignis wurde im Text bereits eingeführt	*der Helmstedter* **bei dem Aufprall**
	etwas ist einmalig (Unikat: z. B. *die Erde*)	**auf der Landstraße 292**
	etwas gehört zu einer Sache oder Person, die aus dem Kontext bereits bekannt ist	**die Gegenfahrbahn**
	die Sache oder Person wird im Text genauer bestimmt, z. B. durch einen Relativsatz oder ein anderes Attribut	**der 35 Jahre alte Krankenwagenfahrer aus Helmstedt**
unbestimmter Artikel	eine Sache oder Person wird neu eingeführt	**bei einem Verkehrsunfall**
	man bezieht sich auf eine beliebige Sache / eine beliebige Person	*Ein Pkw ist ein Auto.*

6

Weisheit der Vielen

In **einer/der** Gruppe treffen Menschen klügere Entscheidungen als allein, wie Experimente schon mehrfach nachgewiesen haben. So lässt sich beispielsweise **das** Gewicht eines Rindes relativ präzise bestimmen, wenn man **den** Mittelwert vieler individueller Schätzungen bildet. Auf **diese** Weise kann der Einzelne von der sogenannten „Weisheit der Vielen" profitieren.

7 Gestern kam es um 08.20 Uhr zwischen Peine-Ost und Hämelerwald **zu einem Zusammenstoß von fünf Autos.** Laut Polizeiangaben wollte ein Fahrer die Spur wechseln und übersah dabei ein anderes Auto. Die übrigen drei Fahrzeuge konnten wegen zu hoher Geschwindigkeit nicht rechtzeitig bremsen. Wie die Polizei sagte, gab es keine Verletzten, alle Beteiligten hatten großes Glück. Während der Unfallaufnahme und der Bergung der Fahrzeuge wurde der Verkehr beeinträchtigt.

3.3 Komplexe nominale Gruppen

3.3.1 Adjektivattribute

1.1

> ### Neugier
>
> Neugier ist ein Glücksfaktor. Mit ihr suchen wir nach **neuen** Genüssen, probieren **fremde** Speisen, reisen in **unbekannte** Länder, lesen **neue** Bücher, erweitern ständig unser Wissen, trainieren eine **neue** Sportart, verfolgen die Nachrichten, lesen Tageszeitungen und Journale, schauen uns **neue** Filme an, lassen uns im Zirkus (ohne Tiere) von Artisten verblüffen* oder im Theater durch **frische** Inszenierungen von **altem** Stoff neu unterhalten und inspirieren. **Kleinen** Kindern sehen wir schon im Gesicht an, wie Neugier ihnen Freude bereitet, wie sie strahlen, wenn wir sie mit einer Kleinigkeit überraschen. Sie sagen uns deutlich genug, was sie nicht möchten, dass wir ihnen ihre Neugier abgewöhnen.

1.2   Aus Neugier suchen wir nach einem neuen Genuss, probieren **eine fremde** Speise, reisen in **ein unbekanntes** Land, lesen **ein neues** Buch, trainieren **eine neue** Sportart, lesen **eine** Tageszeitung und schauen uns **einen neuen** Film an. **Einem kleinen** Kind sehen wir schon im Gesicht an, wie Neugier ihm Freude bereitet.

1.3

Aus Neugier probieren wir ein**e** frem**de** Speise,	Aus Neugier probiert er d**ese** frem**de** Speise,
reisen in ein unbekannt**es** Land,	reist in jen**es** unbekannt**e** Land,
lesen ein neu**es** Buch,	liest dies**es** neu**e** Buch,
trainieren ein**e** neu**e** Sportart,	trainiert jen**e** neu**e** Sportart,
lesen ein**e** Tageszeitung und	liest **die** Tageszeitung und
schauen uns ein**en** neu**en** Film an.	schaut sich jed**en** neu**en** Film an.
Ein**em** klein**en** Kind sehen wir schon im Gesicht an, wie Neugier ihm Freude bereitet.	Manch**em** klein**en** Kind sieht er schon im Gesicht an, wie Neugier ihm Freude bereitet.

> Die Markierung für Kasus, Genus und Numerus trägt entweder die Adjektivendung oder das Artikelwort.

2

> ### Komplementär-Kontrast
>
> Ein **weiteres** Stilmittel ist der Komplementär-Kontrast, bei dem ein **im Farbkreis komplementäres** Farbenpaar gegenübergestellt wird. Dies ist die **stärkste** Kontrast-möglichkeit in der Malerei. Beim Auftragen **reiner, voll gesättigter** Farben kann ein derartiger Kontrast eine **höchst intensive, oft das Auge des Betrachters schmerzende** Leuchtkraft entwickeln. Dennoch ergibt sich aus der Gesetzmäßigkeit **im Farbkreis gegenüber-liegender** Farben eine **harmonische** Bildgestaltung.

> Die markierten Adjektive beziehen sich auf das folgende Adjektiv, nicht auf das Nomen. Die vorangestellten Adjektive werden nicht flektiert.

3.3.2 Partizipialattribute

1.1

> **Sprit statt Brot**
>
> Vor ein paar Jahren dachten viele, mit Biosprit habe man eine saubere Alternative zum **umweltverschmutzenden** Öl gefunden. Als im Frühjahr 2008 eine Ernährungskrise ausbrach, fand man als Grund für diese Entwicklung schnell die zur Gewinnung von Energie **angebauten** Pflanzen. Aufgrund des staatlich **geförderten** Anbaus hatten sich viele Farmer und Bauern bereits gegen die Erzeugung von Nahrungspflanzen entschieden.

> **Empfehlungskatalog zum Meeresschutz vorgelegt**
>
> Die **zunehmenden** Konzentrationen des Treibhausgases Kohlendioxid in der Atmosphäre und der dadurch **ausgelöste** Klimawandel mit **steigenden** Temperaturen bedrohen die Weltmeere gleich zweifach: Die Ozeane erwärmen sich, außerdem löst sich mehr CO_2 im Wasser und macht es dadurch saurer. Deshalb empfiehlt der Wissenschaftliche Beirat der Bundesregierung für Globale Umweltveränderungen (WBGU) **weit reichende** Schutzanstrengungen für die Meere sowie Anpassungsmaßnahmen für die Küstenbewohner.

1.2

umweltverschmutzendes Öl	Öl verschmutzt die Umwelt.
zur Gewinnung von Energie angebaute Pflanzen	Pflanzen **werden** zur Gewinnung von Energie **angebaut**.
der staatlich geförderte Anbau	Der Anbau **wird staatlich gefördert**.
die zunehmenden Konzentrationen des Treibhausgases Kohlendioxid in der Atmosphäre	Die Konzentrationen des Treibhausgases Kohlendioxid in der Atmosphäre **nimmt zu**.
der ausgelöste Klimawandel	Der Klimawandel wurde ausgelöst.
die steigenden Temperaturen	**Die Temparaturen steigen**.
weit reichende Schutzanstrengungen für die Meere	**Die Schutzanstrengungen für die Meere reichen weit.**

2.1

Partizipialattribut	Verb	Satz-Entsprechung
der **lesende** Student	lesen	Der Student liest.
das **gelesene** Buch	lesen	Das Buch wurde gelesen.
die **bestandene** Prüfung	bestehen	**Die Prüfung wurde bestanden.**
der **bezahlende** Kunde	bezahlen	**Der Kunde bezahlt.**
das **empfohlene** Buch	empfehlen	Das Buch wurde empfohlen.
der **rechnende** Computer	rechnen	**Der Computer rechnet.**
der **geschriebene** Brief	schreiben	**Der Brief wurde geschrieben.**
der **produzierende** Betrieb	produzieren	Der Betrieb produziert.
die **lernende** Studentin	lernen	**Die Studentin lernt.**
die **beantwortete** Frage	beantworten	**Die Frage wurde beantwortet.**

2.2   a   Der **im April 2008** auf ein Rekordniveau *gestiegene* Weizenpreis führte in Frankreich zu einem Baguettepreis von über einem Euro.

      b   Wir sollten uns nicht von den **seit Herbst 2008** kaum **gestiegenen** Preisen für Agrarrohstoffe und Lebensmittel irritieren lassen. Die Ruhe an der Preisfront ist nur vorübergehend.

      c   **Weiter** steigende Rohölpreise werden in diesem Jahrzehnt zu höheren Nahrungsmittelpreisen führen.

      d   Die **im Jahr 2011** um zehn Prozent **gestiegene** Produktion konnte den Bedarf trotzdem nicht decken.

      e   Wir rechnen **in den nächsten Tagen** mit **steigenden** Milchpreisen.

3.1

Hauptmahlzeit **der Deutschen** war üblicherweise das **von der Hausfrau liebevoll bereitete warme** Mittagessen, jedenfalls solange der Mann in der Nähe des heimischen Herdes arbeitete. Während morgens und abends das **gesunde dunkle** Brot nebst Aufstrich – morgens süß, abends deftig – auf den Tisch kam, gab es mittags jahrhundertelang Eintopf: Suppe, Brei und Grütze in verschiedenen Formen. [...] Als Grundnahrungsmittel dienten lange Zeit die **in Süddeutschland und Österreich beliebten** Mehlspeisen. [...]

Während die Spanier in der Neuen Welt den Weizen einführten, brachten sie von dort die **vielfach als „Erdapfel" bezeichnete** Kartoffel nach Europa.

3.2 **a** Hauptmahlzeit der Deutschen war das warme Mittagessen, das **liebevoll von der Hausfrau bereitet wurde.**

b Als Grundnahrungsmittel dienten lange Zeit Mehlspeisen, die **in Süddeutschland und Österreich beliebt sind.**

c Die Spanier brachten aus der Neuen Welt die Kartoffel mit, die **vielfach als „Erdapfel" bezeichnet wird.**

4.1

Sowohl die Erfindung **des Buchdrucks mithilfe ganzseitig geschnitzter Holzblöcke** (868) als auch der **erste** Einsatz von **beweglichen Schrifttypen** (um 1045) und der Mehrfarbendruck (1107) gehen auf China zurück.

Dennoch war es die **erst um 1440 wohl unabhängig von den Chinesen gemachte** Wiedererfindung **des Buchdrucks mit beweglichen Lettern durch Johannes Gutenberg**, die die kulturelle Landschaft der Welt und namentlich Europas nachhaltig veränderte. Der Buchdruck bewirkte eine „Kommunikationsrevolution". Das **mühsame handschriftliche und oft mit Fehlern behaftete** Kopieren **von Büchern** wurde überflüssig. Die Menge **der verfügbaren Informationen** nahm explosionsartig zu. Bereits um 1500 produzierten Druckerpressen in ganz Europa über 13.000 Buchtitel. Die **überall entstehenden** Druckereien trugen mit dazu bei, das Bildungsmonopol der Universitäten zu brechen und mit den gelehrten Laien eine neue Gesellschaftsschicht hervorzubringen. Der Buchdruck leistete zudem einen **wichtigen** Beitrag **zur Wiederentdeckung und Verbreitung vieler antiker griechischer Originaltexte**, z. B. von Archimedes, und hatte so **maßgebliche** Wirkung **auf die zeitgenössische Wissenschaft**.

4.2 **a** Es war die Wiedererfindung des Buchdrucks mit beweglichen Lettern durch Johannes Gutenberg, **die erst um 1440 wohl unabhängig von den Chinesen gemacht wurde**, die Europa nachhaltig veränderte.

b Das mühsame handschriftliche Kopieren von Büchern, **das oft mit Fehlern behaftet war**, wurde überflüssig.

c Die Menge der Informationen, **die verfügbar waren**, nahm explosionsartig zu.

d Die Druckereien, **die überall entstanden**, trugen mit dazu bei, das Bildungsmonopol der Universitäten zu brechen.

5 **a** Das **erste, 1974 gescannte** Produkt war eine Packung Kaugummi.

b Der **aus 15 Zeichen bestehende** Code hat einen rechten und linken Rand sowie einen Mittelbalken.

c Jedes **in Deutschland über den Buchhandel vertriebene** Buch hat eine ISBN-Nummer.

d Die **ersten drei, immer für die Branche stehenden** Ziffern beginnen mit 978 und 979.

e Die ersten sechs Ziffern bezeichnen das **die Karte ausstellende** Institut.

f **Mit einer 5 am Anfang beginnende** Kartennummern stammen von einer Bank.

g Aber auch **eine nach den Regeln erfundene** Kartennummer funktioniert praktisch nie.

6.1 › Es wird eine Notwendigkeit ausgedrückt (die Arbeit, die geleistet werden muss / die Arbeit, die abgegeben werden muss).

6.2

Die Bundesanstalt für Arbeitsschutz befragte mehrere Arbeitnehmer zu der Arbeitsmenge, die sie **täglich leisten müssen**.	Man benennt eine Deadline auf die Sekunde genau, lässt aber die Arbeit, die **dann abgegeben werden muss**, so weit wie möglich im Ungefähren.

› Es wird in beiden Beispielen eine Notwendigkeit ausgedrückt.

6.3 die Arbeit und Mühe, die man einsparen kann: **die einzusparende Arbeit und Mühe**
die Deadline, die man genau benennen kann: **die genau zu benennende Deadline**

3.4 Attributsätze und Appositionen

1.1

Um den Erfinder des Telefons streiten sich die Gelehrten

Es war ein Kopf-an-Kopf-Rennen: Im Endspurt um das Patent für das Telefon hatte der Wahl-Amerikaner Alexander Graham Bell am 14. Februar 1876 um zwei Stunden die Nase vor Elisha Gray, **dem Mitbegründer des größten Herstellers für telegrafische Geräte, der Western Electric Manufactoring Company**. Dieser dramatische Konkurrenzkampf hatte noch einen zweiten Verlierer, **den hessischen Schulmeister Johann Philipp Reis, der bereits am 26. Oktober 1861 einem Kreis Frankfurter Wissenschaftler und Honoratioren ein ähnliches Gerät vorgeführt hatte**. „Über Fortpflanzung musikalischer Töne auf beliebiger Entfernung durch Vermittlung des galvanischen Stromes" lautete der Titel des Vortrags vor dem Physikalischen Verein im Senckenberg-Museum. Er war unvorsichtig gewählt. Denn die Tatsache, **dass Reis aus dem 300 Fuß entfernten Bürgerhospital ein so zu sagen live gespieltes Lied übertragen ließ**, kostete ihn vermutlich den unbestrittenen Platz als Erfinder des Telefons in der Technik-Geschichte. Denn die Frage, **ob der damals 27-jährige Pädagoge aus Friedrichsdorf tatsächlich den Fern-Sprecher, oder vielleicht eher eine Art Radio im Sinn hatte**, entzweit seitdem die Gelehrten. Während die „Encyclopedia Britannica" nur Bell nennt und den Deutschen mit keinem Wort erwähnt, hält Meyers Konversations-Lexikon im Jahr 1909 den US-Erfinder lediglich für den Urheber eines weiteren Telefons, **das ohne Batterie auskam**. Arthur Fürst schrieb 1923 in seinem Mammutwerk „Weltreich der Technik": „Es ist Tatsache, dass ein einfacher Lehrer in einem stillen deutschen Dörfchen zuerst einen solchen Apparat erdachte."

1.2 Die Tatsache, dass Reis ein live gespieltes Lied übertragen ließ, kostete ihn den Platz als Erfinder des Telefons.
→ **Der *dass*-Satz bezieht sich nur auf das Nomen.**

Es ist Tatsache, dass ein einfacher Lehrer einen solchen Apparat erdachte. → **Der *dass*-Satz bildet das Subjekt.**

1.3 Bell hatte die Nase vor Elisha Gray, **dem** Mitbegründer der Western Electric Manufactoring Company.

Elisha Gray ist Mitbegründer des größten Herstellers für telegrafische Geräte, **der** WEM Company.

Der Kampf hatte einen zweiten Verlierer, **den** hessischen Schulmeister Johann Philipp Reis.

1.4 > Relativsätze müssen ein Verb enthalten.

> Das Relativpronomen entspricht zwar in Numerus und Genus, nicht aber im Kasus der nominalen Gruppe, auf die es sich bezieht. Der Kasus ist bestimmt durch die Rolle des Relativpronomens im Relativsatz.

2

Der lange Kampf um die Gleichberechtigung

Am 1. Juli 1958 tritt endlich das Gleichberechtigungsgesetz in Kraft. Fortan dürfen Frauen nach der Heirat ihren Mädchennamen als Zusatz behalten. Die Ehegatten werden gegenseitig zum Unterhalt verpflichtet. Die Frau darf den Haushalt in alleiniger Verantwortung führen und hat nun das Recht, erwerbstätig zu sein. 1977 wird Ehebruch als Straftatbestand abgeschafft. Es gilt das Zerrüttungsprinzip – eine alte Forderung von Elisabeth Selbert. Für Selbert gab es keine Lorbeeren. Im Gegenteil. Weder ihr Wunsch, **in den Bundestag zu kommen**, noch der Traum, **eine der ersten Richterinnen am Bundesverfassungsgericht zu werden**, gehen in Erfüllung. Die SPD verweigert ihr die Unterstützung. „Sie war als streitbare Frau verschrien, die anderen so lange mit ihren Forderungen auf die Nerven gehen konnte, bis sie sich durchgesetzt hatte", sagte die frühere Verfassungsrichterin Jutta Limbach einmal in einem Interview.

3 Den Wunsch von Herrn Werner, im Alter auf einem Kreuzfahrtschiff zu leben, halte ich für ...

Prinzipiell finde ich Herrn Werners Vorstellung, **statt im Altersheim auf einem Kreuzfahrtschiff zu wohnen**, ...

Die Ansicht von Herrn Friedman, dass sich zivilisierte Völker nebeneinander dulden können, ...

Im Grunde halte ich Herrn Friedmans Feststellung, zivilisierte Kulturen ließen die Möglichkeit zu, sich im Irrtum zu befinden, für ...

Die Prognose von Herrn Friedman, demnächst mit einem Arbeitskräftemangel konfrontiert zu sein, finde ich ...

Das von Herrn Friedman angesprochene Problem, ob wir genügend Einwanderungswillige finden, halte ich ...

3.5 Nominalisierung und Genitivattribute

1.1

In Deutschland ist Kaffee das Volksgetränk Nr. 1. Kein anderes Getränk wird mehr getrunken als Kaffee. Die meisten Menschen bevorzugen ihn mit Zucker oder Milch oder mit beidem.

Wem Kaffee ohne Zucker oder Milch zu bitter ist, dem könnte vielleicht bald ein Kaffee mit einem Bitter-Blocker angeboten werden. Seit längerem suchen Biotech-Unternehmen nämlich nach Substanzen, die bitteren Geschmack unterdrücken. Ein erstes Patent auf einen solchen Bitter-Blocker hat das US-amerikanische Biotech-Unternehmen „Linguagen" angemeldet. Das weiße Pulver mit der Bezeichnung AMP soll bei der Herstellung **von Chips, Cola, Fertigsuppen und anderen Lebensmitteln** eingesetzt werden.

Präpositionalattribut

Wolfgang Meyerhof, Professor am Deutschen Institut für Ernährungsforschung (DIFE) in Potsdam erwartet, dass sich derartige Substanzen eines Tages für die Hersteller auch finanziell lohnen werden. Doch er sieht diese Entwicklung mit gemischten Gefühlen, denn die Bitter-Blocker gefährden die ursprüngliche Funktion des Bitter-Geschmacks: Die Bitterwahrnehmung dient dem Menschen eigentlich dazu, giftige Substanzen möglichst schnell zu erkennen. Sollten Nahrungsmittel verdorben sein oder ungenießbare Substanzen enthalten, wird der Körper durch den bitteren Geschmack gewarnt. Der bittere Geschmack signalisiert, dass das Produkt möglicherweise giftig ist. Gerade in der frühen Warnung vor einem Verzehr liegt der besondere Vorteil **der Bitterwahrnehmung**.

Genitivattribut

Neben der Lebensmittelindustrie dürfte auch die Pharmaindustrie an Bitter-Blockern interessiert sein, um den bitteren Geschmack **von Arzneimitteln** zu überdecken. Vor einer Anwendung solcher Substanzen in Lebensmitteln und Medikamenten müssten allerdings erst umfangreiche Tests durchgeführt werden, um die Wirksamkeit und Ungefährlichkeit **dieser Stoffe** zu prüfen.

Präpositionalattribut

Genitivattribut

Der Münsteraner Lebensmittelchemiker Hofmann geht da lieber einen ganz anderen Weg. Er versucht die Herstellungsprozesse in der Industrie so zu optimieren, dass unerwünschte Bitterstoffe gar nicht erst entstehen. Dazu hat er neue physikalisch-chemische Verfahren erprobt, bei denen im Labor die verschiedenen Aroma- und Geschmacksstoffe **eines Lebensmittels** voneinander getrennt werden. Diese verschiedenen Bestandteile prüft er dann mit seinen speziell geschulten Personen, die den Geschmack **der einzelnen Substanzen** testen.

Genitivattribut

Die Industrie zeigt bereits großes Interesse an den Forschungen **des Münsteraner Lebensmittelchemikers**. Die Hersteller **von Babynahrung** zum Beispiel hatten viele Jahre lang das Problem, dass ihr Baby-Karottenbrei immer wieder zu bitter wurde. Die Hersteller luden Hofmann ein, ihre Fabriken zu besuchen und Proben aus den verschiedenen Produktionsabschnitten zu nehmen. In seinem Labor entdeckte er schließlich einen Bitterstoff namens Falcarindiol, der im Zusammenhang mit Karotten bislang nicht erwähnt worden war.

Präpositionalattribut

Zunächst ging Hofmann davon aus, dass die Bittersubstanz während der Sterilisation **des Karottenbreis** entstanden war. Dann aber stellte sich heraus, dass einige Karottensorten diesen Bitterstoff bereits in sich trugen, als sie in die Fabrik kamen. Das Ergebnis **der Studie** war: Die Hersteller müssen die Rohware, die sie einkaufen, genau auf den Bitterstoff hin kontrollieren.

Genitivattribut

Hofmann meint, dass man die Geschmacksqualität verbessern kann, wenn man hochwertige Ausgangsstoffe nimmt und die Verarbeitung **der Lebensmittel** optimiert. Seiner Ansicht nach sind dann viele Zusatzstoffe überflüssig.

1.2

die Herstellung von Lebensmitteln	die Herstellung eines Lebensmittels
die Hersteller von Babynahrung	–
der Geschmack von Arzneimitteln	der Geschmack eines Arzneimittels
die Wirksamkeit von Stoffen	**die Wirksamkeit eines Stoffes**
der Geschmack einzelner Substanzen der Geschmack **von einzelnen Substanzen**	der Geschmack einer einzelnen Substanz

› Das Präpositionalattribut mit „von" wird verwendet, wenn die nominale Gruppe kein Artikelwort enthält (Nullartikel).

1.3

den Geschmack testen	der Test des Geschmacks
den Geschmack beeinflussen	die Beeinflussung des Geschmacks
chemische Substanzen anwenden	die Anwendung chemischer Substanzen
nach einer giftigen Substanz suchen	die Suche nach einer giftigen Substanz
die Kosten reduzieren	**die Reduzierung der Kosten**
die Qualität beeinflussen	**die Beeinflussung der Qualität**
vor giftigen Stoffen warnen	**die Warnung vor giftigen Stoffen**
verschiedene Bestandteile prüfen	**die Prüfung verschiedener Bestandteile**
Aroma verlieren	**der Verlust von Aroma**
Es wird Kaffee mit Bitterblockern angeboten.	das Angebot von Kaffee mit Bitterblockern
Der Körper wird gewarnt.	**die Warnung des Körpers**
Umfangreiche Tests werden durchgeführt.	**die Durchführung umfangreicher Tests**
Die Wirksamkeit der Stoffe wird geprüft.	**die Prüfung der Wirksamkeit der Stoffe**
Der Herstellungsprozess wird optimiert.	**die Optimierung des Herstellungsprozesses**
Proben werden entnommen.	**die Entnahme von Proben**
Der Bitterstoff Falcaridiol wurde entdeckt.	**die Entdeckung des Bitterstoffes Falcaridiol**
Die Geschmacksqualität wird verbessert.	**die Verbesserung der Geschmacksqualität**

1.4 **b** Biotech-Unternehmen suchen nach Substanzen, um bitteren Geschmack zu unterdrücken.
      Biotech-Unternehmen suchen nach Substanzen **zur Unterdrückung des bitteren Geschmacks.**
    c Die Bitterwahrnehmung dient dazu, giftige Substanzen möglichst schnell zu erkennen.
      Die Bitterwahrnehmung dient **der schnellen Erkennung giftiger Substanzen.**
    d Es werden Tests durchgeführt, um die Ungefährlichkeit der Stoffe zu prüfen.
      Es werden Tests **zur Prüfung der Ungefährlichkeit der Stoffe durchgeführt.**
    e Die Geschmacksqualität kann verbessert werden, indem man hochwertige Rohstoffe auswählt.
      Die Geschmacksqualität kann durch die Auswahl hochwertiger Rohstoffe verbessert werden.
    f Die Hersteller luden Hoffmann ein, ihre Fabriken zu besuchen.
      Die Hersteller luden Hoffmann zu einem Besuch ihrer Fabriken ein.
    g Der Einsatz von Bitter-Blocker führt dazu, dass die ursprüngliche Funktion des Bitter-Geschmacks gefährdet wird.
      Der Einsatz von Bitter-Blocker führt zur Gefährdung der ursprünglichen Funktion des Bittergeschmacks.

1.5   Wozu dienen umfangreiche Tests von Bitter-Blockern?

    Sie dienen zur Prüfung der Wirksamkeit und Ungefährlichkeit dieser Stoffe.
    Man will durch die Tests prüfen, ob Bitter-Blocker wirksam und ungefährlich sind.
    führen zu + Dativ (Folge)
    Welche Auswirkungen kann die Sterilisation von Lebensmitteln haben?
    Die Sterilisation von Lebensmitteln **kann zur Entstehung von Bitterstoffen führen.**
    Die Sterilisation von Lebensmitteln **kann dazu führen, dass Bitterstoffe entstehen.**
    liegen in + Dativ
    Worin liegt der Vorteil der Bitterwahrnehmung?
    Der Vorteil liegt **in einer frühzeitigen Warnung vor dem Verzehr giftiger Substanzen.**
    Der Vorteil liegt **darin, vor dem Verzehr giftiger Substanzen frühzeitig zu warnen.**

4   Wörter, Wortbildung und Wortverbindungen

4.1   Präpositionen

4.1.1   Kasus der Präpositionen

1

Was ausländische Studierende **von** Deutschland erwarten

von Amory Burchard

Studierende **aus** dem Ausland kommen gerne **nach** Deutschland, sind aber nicht durchweg zufrieden **mit** den Studienbedingungen. So lautet die Grundaussage **gemäß** dem International Student Barometer **für** Deutschland, das der Deutsche Akademische Austauschdienst (DAAD) und die Hochschulrektorenkonferenz (HRK) veröffentlicht haben. Befragt wurden 2009 rund 12.000 Studierende **an** 45 deutschen Hochschulen und 2010 rund 17.000 junge Ausländer **an** 46 Standorten.

Den Angaben der Befragten **nach** sei der Studienort Deutschland **für** 83 Prozent die erste Wahl gewesen. Gut zwei Drittel hatten sich auch nicht anderswo beworben. **Für** 97 Prozent war der gute Ruf der deutschen Hochschulausbildung ausschlaggebend, 86 Prozent gaben aber auch an, sich **wegen** der guten Sicherheitslage **für** Deutschland entschieden zu haben. Die niedrigen Kosten **für** ein Studium spielten eine ebenso große Rolle.

Doch die hohen Erwartungen der jungen Leute **an** ein deutsches Hochschulstudium wurden teilweise enttäuscht. Während **nach** der internationalen Umfrage des Student Barometer 83 Prozent der insgesamt 160.000 befragten internatonalen Studierenden **mit** den Lernbedingungen zufrieden sind, waren es **in** Deutschland nur 73 Prozent. Die Sprachförderung bekam nur **von** 65 Prozent gute Noten (international 82 Prozent). **Laut** DAAD wünschen sich ausländische Studierende **bei** der Betreuung **durch** Lehrende transparente Anforderungen und eine kontinuierliche Unterstützung. **Hinsichtlich** der Betreuung äußerten viele der Befragten den Anspruch, auch Hilfen **beim** Berufseinstieg zu bekommen, was die Hochschulen überraschte.

Infolge der Ergebnisse sollen Universitäten und Fachhochschulen Hinweise erhalten, wie sie **an** der Verbesserung ihres internationalen Marketings arbeiten können, damit auch in Zukunft viele ausländische Studierende **in** die deutschen Studienorte **an** die verschiedenen Hochschulen kommen möchten.

Kasus der wichtigsten Präpositionen

Präpositionen mit Akkusativ	Präpositionen mit Dativ	Präpositionen, die Dativ oder Akkusativ fordern („Wechselpräpositionen")	Präpositionen mit Genitiv
bis, **durch**, **für**, gegen, ohne, um, entlang (nachgestellt)	ab, **aus**, außer, **bei**, gegenüber, **gemäß**, laut*, mit, **nach**, seit, (an)statt*, trotz*, während*, wegen*, zu	**an**, auf, hinter, **in**, neben, über, unter, vor, zwischen	aufgrund, außerhalb, entlang (vorgestellt), **hinsichtlich**, **infolge**, innerhalb, laut*, mithilfe, (an)statt*, trotz*, während*, wegen*

* Präposition kann sowohl mit Dativ als auch mit Genitiv vorkommen.

Verb + Präposition	Nomen + Präposition	Adjektiv + Präposition
erwarten von, **arbeiten an**	Kosten für, **Hilfe bei**, **Erwartungen an**	**zufrieden mit**

2.1 an + dem: *am*
an + das: **aus**
bei + dem: **beim**
in + dem: **im**

in + das: **ins**
von + dem: **vom**
zu + dem: **zum**
zu + der: **zur**

2.2 auf + *das* : aufs
durch + das : durchs
für + das : fürs
hinter + dem : hinterm
hnter + das : hinters
über + dem : überm

über + das : übers
um + das : ums
unter + dem : unterm
unter + das : unters
vor + dem : vorm
vor + das : vors

4.1.2 Lokale Präpositionen

1.1

Berlin in drei Stunden

Tourverlauf: Hauptbahnhof - Regierungsviertel - Brandenburger Tor - Holocaust-Mahnmal - Potsdamer Platz
Dauer: ca. 3 Stunden

Zum Start der Tour verlässt man den Berliner Hauptbahnhof, der übrigens der größte Kreuzungsbahnhof Europas ist, **durch** den Südausgang Richtung Spree. Dort angekommen kann man rechts das Kanzleramt und links den Reichstag sehen. Eine kleine Fußgängerbrücke führt geradewegs **ins** Regierungsviertel.
Geht man nun rechts, kommt man direkt **zum** Bundeskanzleramt, das 2001 fertig gestellt wurde. **In** ihm befinden sich Büros und Arbeitsräume der Bundeskanzlerin. Was man **von** außen nicht ahnt: **Innerhalb** des Areals befindet sich ein Hubschrauberlandeplatz und mit dem Kanzlerpark ist das Grundstück 7000 Quadratmeter groß.
Dem Kanzleramt schräg **gegenüber** steht **in** einiger Entfernung der Reichstag, die gläserne Kuppel ist gut zu erkennen.

Nachdem man den Reichstag einmal umrundet hat – der Spree **entlang** gibt es hier einiges zu sehen – führt die Tour weiter **zum** Brandenburger Tor. Das Brandenburger Tor, von 1788 bis 1791 erbaut, ist Symbol für die deutsche Einheit und die bekannteste Sehenswürdigkeit der Stadt. **Auf** dem Brandenburger Tor thront die Skulptur eines vierspännigen Wagens, einer sogenannten Quadriga.

Vom Brandenburger Tor verläuft die Tour weiter Richtung Süden **zum** Potsdamer Platz, wobei man die Botschaft der Vereinigten Staaten von Amerika und das Holocaust-Mahnmal passiert. **Auf** der anderen Straßenseite des Holocaust-Museums beginnt der Tiergarten, der große Park **im** Zentrum der Stadt. **Vom** Holocaust-Mahnmal läuft man nun **auf** die Skyline des neuen Potsdamer Platz zu. Das Ende der Tour ist zugleich einer der Höhepunkte: Das Sony Center mit seinem zeltartigen Dach, das die Skyline des Potsdamer Platzes prägt.

2 **b** Fahren Sie bis – / **nach** Hamburg? – Nein, ich fahre nur bis **zur** nächsten Station.
  c Warte, ich begleite dich bis **zum** Bahnhof.
  d Ich bin den Bahnsteig von vorn bis hinten abgelaufen, aber ich hab dich nicht gesehen.
  e Er geht bis **zum / an den** Rand des Bahnsteigs.

3 **b** > Er streut Käse auf die Pizza
  c > In dem Zimmer gibt es keinen Schrank.
  d > Lisa steckt das Handy in ihre Tasche.
  e > Die Schule liegt mitten in der Hauptstadt.
  f > An der Decke hängt eine helle Lampe.
  g > Man kommt über die Brücke ins Zentrum.
  h > Das Kind hält einen Stift zwischen seinen Fingern.

4

> **Auf dem** Gemälde „Badestelle in Asnières" von Georges Suerat kann man Personen beim Baden
> **an/in einem** Fluss sehen. **In der** Mitte des Bildes sitzt ein Junge **auf einem** Handtuch, seine
> Beine hängen **ins** Wasser und er schaut **nach** vorne. **Am** Ufer des Flusses gibt es weitere
> Personen: **Neben dem** Jungen liegt eine Frau mit ihrem Hündchen **auf der** Wiese und **hinter** ihm
> sitzt ein Mann mit Hut. **Im** Fluss sind zwei Personen. Weiter hinten im Bild fahren kleine Boote
> **auf dem** Wasser. **Von der** Badestelle **bis zur** Stadt ist es nicht weit, denn sie ist **im** Hintergrund
> zu erkennen. Der Fluss fließt **unter einer** Brücke durch, die die Stadthälften verbindet.

5 Beispiellösung:

> Ich laufe vom Hauptbahnhof zum Schlossplatz. Zuerst gehe ich durch die Unterführung, dann die
> Königsstraße entlang. Ich biege nach einigen Metern nach links in den Schlossgarten. Ich komme an
> einem Teich vorbei und stehe vor dem Schloss. Ich gehe nach rechts und stehe auf dem Schlossplatz. ...

4.1.3 Temporale Präpositionen

1 Janosch wurde am 11. März 1931 als Horst Ecker geboren. **Mit** 13 Jahren begann er eine Schmiede-
und Schlosserlehre. 1946, **nach** Kriegsende, flüchtete die Familie in den Westen. In der Gegend von
Oldenburg arbeitete Janosch in einer Textilfabrik. **Für** kurze Zeit ging er auf die Textilschule in
Krefeld. **Ab** 1953 lebte er in München, wo er an der Akademie der Bildenden Künste sein
Kunststudium wegen „mangelnder Begabung" **nach** einigen Probesemestern abbrechen musste.
Danach war er freier Künstler. **Im** Jahr 1960 erschien sein erstes Kinderbuch. **Innerhalb** der
nächsten 10 Jahre folgten zahlreiche Kinderbücher und er erhielt verschiedene Literaturpreise.
Janosch ist heute einer der bekanntesten deutschen Künstler und Kinderbuchautoren und lebt **seit**
1980 auf Teneriffa.

2 Die Orientierungstage für internationale Studierende finden **vom** 10. **bis** 14. Oktober statt. **Am**
Montag ist Anreisetag! **Zwischen** 8 und 16 Uhr bietet das Akademische Auslandsamt einen
Abholservice, sowie gratis Kaffee im Internationalen Zentrum auf dem Campus an. (Achtung:
Außerhalb der angegebenen Zeit ist das IZ leider geschlossen!) **Am** Abend beginnt **um** 20 Uhr
die offizielle Begrüßung durch den Rektor der Universität. **Am** nächsten Tag kommen Sie bitte
pünktlich **um** 9 Uhr zum Studierendensekretariat, damit Sie sich einschreiben und sofort Ihren
Studentenausweis mitnehmen können. Da wir eine große Gruppe sind, kann dieser Termin
bis 11 Uhr dauern. Danach erhalten Sie eine Campusführung **für** eineinhalb Stunden, die wir
Ihnen sehr empfehlen. **Nach der** Führung würden wir gerne mit Ihnen zusammen in der Mensa
Mittagessen. **Während des** Essens beantworten wir gerne Ihre Fragen.

3 Beispiellösung:

> Ich bin am 30. März 1988 in Oldenburg geboren. Als ich fünf Jahre alt war, ist meine Familie nach
> Hamburg gezogen. Nach der Grundschule bin ich aufs Gymnasium gegangen. Im Sommer 2007 habe ich
> Abitur gemacht und im Herbst eine Ausbildung zum Mediengestalter angefangen. Während der
> Ausbildung habe ich gemerkt, dass ich mich sehr für Gestaltung interessiere. Seit fast drei Jahren
> studiere ich nun Mediendesign in Berlin. ....

4.1.4 Kausale und finale Präpositionen

1 Kausale und finale Präpositionen

- Kausale Präpositionen geben den Grund für etwas an und antworten auf die Frage: Warum?
  anlässlich, **aufgrund**, **aus**, **dank**, **durch**, **mangels**, um … willen, **wegen**, zuliebe
- Finale Präpositionen geben das Ziel oder den Zweck einer Sache an und antworten auf die Frage: Wozu?
  für, **zu**, zwecks

2 Beispiellösung:

> Aufgrund der zahlreichen Beispiele ist die Arbeit leicht nachvollziehbar.
> Mangels klarer Argumente ist die Hausarbeit jedoch nicht überzeugend.
> Die Hausarbeit ist dank des ausführlichen Anhangs gut illustriert.
> Wegen gravierenden formalen Mängeln wird die Arbeit als ungenügend bewertet.
> Durch die Wahl des schwierigen Titels bleibt das Thema unklar.
> Mangels korrekter Zitatangaben wird die Gesamtnote gemindert.
> Dank der präzisen Fragestellung ist die Arbeit sehr gut verständlich.

3 **b** > Der Mann wurde zwecks / zur Feststellung seiner Identität auf die Polizeiwache gebracht.
   c > Unser Kind fährt nicht mehr mit uns in den Urlaub, weil / da es sich nicht dafür interessiert.
   d > Da / Weil es keine finanzierbaren Wohnungen gibt, ziehen wir doch nicht um.
   e > Aufgrund / Wegen eingeschränkter Sicht kommt es auf der A3 Richtung Frankfurt zu Behinderungen.
   f > Damit ich meinen Besuch unterbringen kann, kaufe ich ein Schlafsofa. / Um meinen Besuch unterzubringen, kaufe ich ein Schlafsofa.

4.1.5 Modale, konzessive und adversative Präpositionen

1.1

Gestern Nacht ereigneten sich zahlreiche Unfälle infolge* winterglatter Straßen. Auf der B31 kam es zu einer Vollsperrung, nachdem ein LKW, der ins Schleudern geriet, **mitsamt** seiner Ladung umgekippt ist. Den Beamten **zufolge** war eine unangepasste Geschwindigkeit Ursache für den Unfall. Für die komplizierte Bergung **mittels** eines Spezialkrans benötigte die Feuerwehr mehrere Stunden. Auch im gesamten Bodenseekreis kam es **ent- sprechend** den Witterungsverhältnissen zu vielen Auffahrunfällen. In den meisten Fällen blieb es bei Blechschäden. **Laut** den örtlichen Polizeistellen wird der Sachschaden auf ca. 1,5 Mio. geschätzt.

*> Die Präposition „infolge" hat hier eher kausale Bedeutung.

Mit Gentechnik hergestellte Lebensmittel müssen **gemäß** EU-weiten Vorschriften gekennzeichnet werden. **Trotz** dieser Regelung kann Gentechnik zum Einsatz gekommen sein, ohne dass Verbraucher davon erfahren. **Ungeachtet** der mehrheitlichen Ablehnung innerhalb der Bevölkerung wird Gentechnik zum Beispiel bei Futtermitteln für Nutztiere eingesetzt. Auch Zusatzstoffe, Enzyme, Vitamine und Aromen können **mithilfe** gentechnisch veränderter Mikroorganismen hergestellt werden, ohne dass ein entsprechendes Kennzeichen angebracht werden muss. Für mehr Transparenz sorgte die Bundesregierung 2009: **Anhand** des Siegels „ohne Gentechnik" können die Verbraucher seither sicher erkennen, bei welchen Lebensmitteln keine Gentechnik eingesetzt wurde.

Hallo du, wie geht's? Ich bin gestern umgezogen und einiges ging schief! Die Autovermietung hat mir **anstelle** des reservierten Transporters nur ein kleineres Modell geben können. Und es hat **entgegen** der Wettervorhersage den ganzen Tag pausenlos geregnet! Aber **mithilfe** meiner Freunde habe ich den Umzug trotzdem an einem Tag geschafft. Jetzt sitze ich inmitten von Kisten und teste die Internetverbindung – der Anzeige **nach** funktioniert sie hervorragend. Wir sehen uns spätestens bei meiner Einweihungsparty!

LG Anne

1.2   Bedeutung

Präposition	Umschreibung	Präposition	Umschreibung
trotz ungeachtet	ohne Berücksichtigung (von)	entgegen	**im Gegensatz zu**
mitsamt	**zusammen mit**	mithilfe mittels anhand	**unter Verwendung von / mit Unterstützung (von)**
gemäß zufolge entsprechend laut nach	**in Übereinstimmung mit / so wie … sagt**	anstelle	**statt**

1.3

  b Eine Kritik an der Reform ist, dass sie **ungeachtet der Meinung der Studierenden** durchgeführt wurde.

  c Außerdem wird oft bemängelt, dass **entgegen der Zielsetzung der Reform** heute weniger Studenten als früher im Ausland studieren.

  d **Mithilfe einer Studie** konnte die Uni Hamburg feststellen, dass Bachelorstudenten im Durchschnitt nur 26 Stunden pro Woche für ihr Studium aufwenden.

2

Sehr geehrter Herr Moser,
Anlässlich Ihres Schreibens vom 29.06.2011 haben wir Ihren Fall noch einmal geprüft. **Aufgrund** Ihres leichten Alkoholkonsums haben Sie auf jeden Fall eine Mitschuld an dem Unfall. Und **entgegen** Ihrer Aussagen, dass Ihr Mitfahrer angeschnallt war, müssen wir das Gegenteil annehmen. Unserem Gutachter **zufolge** können solche starken Verletzungen nicht entstehen, wenn man angeschnallt ist. Außerdem hatten Sie **anstelle** des erlaubten Tempos von 100 km/h mindestens eine Geschwindigkeit von 120 Stunden-kilometern. Und **gemäß** der Verkehrsordnung sind außerhalb einer geschlossenen Ortschaft lediglich 100 km/h erlaubt. Wegen des hohen Tempos kam es zu dem Unfall. Wir stellten aber fest, dass auch der andere Fahrer nicht schuldlos ist. **Zwecks** weiterer Untersuchungen bezüglich Ihres Falls brauchen unsere Gutachter noch etwas Zeit und wir bitten Sie weiterhin um Geduld.

Mit freundlichen Grüßen
Ihre Versicherungsgesellschaft AG

Hallo Mama,
ich möchte meiner neuen Mitbewohnerin Eva zu ihrer Begrüßung etwas kochen, denn dank ihr haben wir jetzt viele neue Küchengeräte. Finde aber leider das Rezept für die Lasagne nicht mehr. Meiner Meinung nach liegt es im roten Ordner in der Küche. Kannst du mir kurz die Zutaten schicken, bin gleich im Supermarkt!
LG Petra

Hallo Petra,
schön, dass du mal wieder etwas kochst! Habe das Rezept leider nicht gefunden, aber schau mal im Supermarkt nach einem Fix-Produkt für Lasagne. Mithilfe der Anleitung von dort müsste es dir auch gelingen.
LG Mama

3

Noah Gordon: Der Medicus

Im Alter **von** neun Jahren verliert Rob Jeremy Cole seine Eltern. Seine Mutter stirbt **bei** der Geburt des jüngsten Bruders, der Vater erliegt einer Krankheit. Die Londoner Zimmermannszunft bringt Robs Geschwister **in** verschiedenen Familien unter. Nur Rob bleibt übrig, bis ein Bader **vor** seiner Tür steht, der ihn als Lehrling aufnimmt. **Auf** den gemeinsamen Reisen **durch** das Land lernt Rob das Jonglieren, das Zaubern und erlangt erste medizinische Grundkenntnisse. Als der Bader **nach** einigen Jahren stirbt, zieht der junge Mann allein weiter. Seine Reise führt ihn weit **über** die Grenzen Europas hinaus, was ihn nicht selten **vor** Schwierigkeiten stellt. **In** der Zweckgemeinschaft einer Karawane stößt er **auf** die unterschiedlichsten Leute …

Noah Gordon

Noah Gordon wurde 1926 **in** Worcester, Massachusetts, geboren. **Nach** seinem Studium wandte er sich dem Journalismus zu. **Während** seiner Tätigkeit als wissenschaftlicher Redakteur **beim** Bostoner Herald veröffentlichte er eine Reihe **von** Artikeln und Erzählungen **in** verschiedenen Zeitschriften. Sein erster Roman „Der Rabbi" verhalf ihm zu einem spontanen Durchbruch. Besonders erfolgreich sind seine Romane um die Mediziner-Familie Cole.

Noah Gordon hat drei Kinder und lebt **mit** seiner Frau Lorraine **auf** einer Farm in den Berkshire Hills **im** westlichen Massachusetts.

4 b > Zur Vermeidung von Stress sollte man in seiner Freizeit einen Ausgleich zum Studium haben.

c > Sie schwitzte während der Prüfung vor Nervosität.

d > Durch das regelmäßige Sprechen mit Muttersprachlern lerne ich eine Sprache am besten.

e > Beim letzten Treffen mit meinem Professor hatte er meine Hausarbeit noch nicht korrigiert.

f > Wegen der Demonstration gegen Studiengebühren fahren zwischen 14 und 16 Uhr im Zentrum keine Busse.

g > Ohne Abitur kann man in Deutschland nicht studieren.

h > Laut Rektor fallen ab nächstem Semester die Studiengebühren weg.

4.2 Adverbien

1

	Adverb	Adjektiv
besseres		X
am besten	X	
schnell	X	
freudig	X	
gemeinsamen		X
gemeinsam	X	
stark	X	

2

In regelmäßigen Abständen werden in Leipzig **geschickt** Banken <u>überfallen</u>. Die Geldräuber konnte man noch nicht ergreifen. Werden die Täter von einem Maulwurf bei der Polizei **rechtzeitig** über den Stand der Ermittlungen <u>informiert</u>? Die Dresdner Kommissare Ehrlicher und Kain sollen den potenziellen Verräter **umgehend** <u>finden</u>. Da gerät eine Streifenpolizistin in einen der Banküberfälle und stirbt im Kugelhagel. Die nächste Zielscheibe: Ehrlicher selbst! Eine junge Frau namens Jenny, die von Beamten als Augenzeugin des letzten Bankraubes registriert wurde, <u>führt</u> die beiden ermittelnden Polizisten **schnell** auf die Spur zweier Verdächtiger, die **unerklärlich** zu erstaunlichem Reichtum <u>gelangt sind</u>. Bob und Frieder Vodenka leben mit Jenny in einer luxuriösen Villa. Jenny <u>verschwindet</u> **plötzlich** und hinterlässt ein hochinteressantes Notizbuch. Eine Leiche, eine Verschwundene und zu wenig Beweise, um die Brüder hinter Schloss und Riegel zu bringen – die Kommissare stehen unter Druck. Doch Ehrlicher lässt sich nicht aus der Ruhe bringen und <u>vertraut</u> **unbeirrt** der Psychologie: „Gefühle tragen weiter".

3 Adverbien

Zeit	Ort	Art und Weise
montags	**dort**	beispielsweise
wieder (2x)	**drinnen**	**möglichst**
ständig		unbeholfen
manchmal		amüsiert
schließlich		eigentlich

4 a immer am Nachmittag: **nachmittags**
   b in zwei Tagen: **übermorgen**
   c mehr als einmal: **mehrmals / mehrfach / wiederholt**
   d jede Stunde: **stündlich**
   e in früheren Zeiten: **damals / früher**
   f nicht regelmäßig: **unregelmäßig / ab und zu /
      manchmal / gelegentlich**

   g für viele Tage: **tagelang**
   h nicht oft: **selten / kaum / manchmal**
   i als Erstes: **zuerst**
   j in diesem Moment: **jetzt / gerade / nun / momentan**
   k ein zweites Mal: **wieder / noch einmal**
   l im Anschluss: **anschließend**
   m in sehr kurzer Zeit: **gleich / sofort**

5 a draußen – **drinnnen**
   b links oben – rechts unten
   c überall – **nirgends / nirgendwo**
   d hinten – **vorn(e)**
   e hierher – **dorthin**
   f aufwärts – **abwärts**
   g rückwärts – **vorwärts**
   h irgendwohin – **nirgendwohin**
   i von außen – **nach innen / von innen**

6 b **Gern** komme ich am Wochenende zu Besuch. Ich freue mich darauf.
   c **Hauptsächlich** junge Menschen kommen zu der Veranstaltung. Ältere interessieren sich nicht so
      sehr für dieses Thema.
   d **Glücklicherweise** habe ich die Prüfung bestanden. Ich freue mich.
   e **Am besten** rufst du mich abends an. Da bin ich meistens zu Hause.
   f **Wahrscheinlich** können wir am Samstag grillen. Es soll schön werden.
   g **Anscheinend** hat sie wieder schlecht geschlafen. Sie sieht müde aus.
   h **Eventuell** kauft er sich ein neues Auto. Das hängt aber von den Kosten ab.
   i **Bestimmt** wird dieses Buch ein Bestseller. Der Autor ist so talentiert.
   j **Hoffentlich** sehe ich ihn morgen wieder. Er ist so toll!

7 Beispiellösung:

   Liebes Tagebuch,
   heute Morgen hat um acht Uhr der Wecker geklingelt und nach der üblichen Morgenroutine bin ich wieder
   zu spät zur U-Bahn gegangen. Ein bisschen spät (um 9.30 Uhr) bin ich beim Sender angekommen.
   Hoffentlich hat es mein Chef nicht bemerkt. Dort haben wir weiter an den Aufzeichnungen zu „Bauer sucht
   Frau" gearbeitet, die schon vorgestern begonnen haben. Leider waren zwei Bauern heute krank. Halb eins
   war endlich Mittagspause. Weil so schönes Wetter war, saßen wir draußen in der Sonne. Anschließend bin
   ich mit dem Kamerateam noch ein Eis essen gegangen. Glücklicherweise funktionierte die Kaffeemaschine
   wieder, denn ich brauche nachmittags immer eine große Tasse. Ich war der letzte im Studio!
   Wahrscheinlich mussten meine Kollegen alle irgendwohin. Von 19 bis 20 Uhr bin ich eine Stunde zum
   Schwimmen gegangen. Da habe ich meine Freunde getroffen. Danach waren wir zusammen Abendessen.
   Abends bin ich meistens müde und fühle mich nirgends so wohl wie zu Hause. Es ist 23 Uhr und bestimmt
   fallen mir gleich die Augen zu. Gute Nacht – bis morgen!

4.3 Gradpartikeln

1.1 Frauke: **negativ**
    Rebecca: **eher positiv**
    Claudia: **positiv**
    Conni: **positiv**
    Torsten: **negativ**

1.2

Das Leben in Berlin ist …	Verstärkung	Abschwächung
… furchtbar laut.	+	
… **gar** nicht gemütlich.	+	
… **besonders** toll.	+	
… **total** bunt.	+	
… **ziemlich** verrückt.		+
… **ausgesprochen** spannend.	+	
… **relativ** gesetzlos.		+
… **enorm** vielfältig.	+	
… **verhältnismäßig** aufregend.		+

2

Verstärkung	Abschwächung	Verstärkung einer Negation	über dem Normalmaß
sehr	**vergleichsweise**	**überhaupt**	**zu**
extrem	**recht**	**gar**	**viel zu**
schrecklich	**halbwegs**		
entsetzlich	**einigermaßen**		
ungewöhnlich			
unglaublich			

3 a **Meine Heimatstadt ist recht klein.**
  b **Meine Jugend war deshalb sehr unspektakulär.**
  c **Es passierten keine besonders überraschenden Dinge.**
  d **Ich finde meinen jetzigen Wohnort unglaublich toll.**

4.2

LENA BADKE: Guten Tag. Ich möchte mich gerne vorstellen, ich bin Ihre neue Nachbarin. Mein Name ist Lena Badke.

INGRID SCHÄFERMANN: Oh, das ist ja ganz **(+)** entzückend. Mein Name ist Ingrid Schäfermann. Es ist selten, dass man seine Nachbarn kennenlernt in diesem großen Haus!

LENA BADKE: Das ist aber schade. Ich lege ganz **(+)** viel Wert auf gute Nachbarschaft.

INGRID SCHÄFERMANN: Sind Sie also neu in der Stadt? Wie gefällt es Ihnen bisher?

LENA BADKE: Ach ja, ganz **(-)** gut. Ich verfahre mich oft, weil ich das U-Bahn-Netz noch nicht so kenne und die Stadt ist einfach riesig.

INGRID SCHÄFERMANN: Ach, da machen Sie sich mal keine Sorgen, das U-Bahn-Netz werden Sie ganz **(+)**schnell kennenlernen, das brauchen Sie jeden Tag hier. Haben Sie denn schon eine Stadtführung mitgemacht, das kann am Anfang doch ganz **(-)** gut helfen.

LENA BADKE: Ja, ich fand die Führung ganz **(-)** interessant, aber einen Überblick habe ich noch nicht.

INGRID SCHÄFERMANN: Also wenn Sie möchten, kann ich Ihnen am Wochenende gerne einmal unser Stadtviertel zeigen. Das Wetter soll ja ganz **(-)** schön werden.

LENA BADKE: Das würde mich freuen! Vielen Dank …

4.4 Gradpartikel *ganz*

- Die Gradpartikel *ganz* kann sowohl verstärkend als auch abschwächend wirken.
  Die Bedeutung hängt von der Betonung ab:
  Ist *ganz* **betont**, dann verstärkt es die Bedeutung.
  Ist *ganz* **unbetont**, dann schwächt es die Bedeutung ab.

5 Beispiellösung:

INGRID SCHÄFERMANN: Hallihallo, Frau Badke!!! Wie geht es Ihnen? Haben Sie sich inzwischen ein bisschen eingelebt?

LENA BADKE: Ach, Frau Schäfermann guten Morgen! Danke, mir geht es **ganz** gut.

INGRID SCHÄFERMANN: Das hört sich aber nicht überzeugend an. Was ist denn los?

LENA BADKE: Ach, ich will mich nicht beschweren. Aber es ist **ziemlich** laut in meiner Wohnung. Es liegt an dem Nachbarn aus dem 4. Stock. Ein **sehr** unfreundlicher Mensch!

INGRID SCHÄFERMANN: Oje, das tut mir aber **furchtbar** Leid. Ich kenne den Herrn aus dem 4 gar nicht. Um was für Lärm handelt es sich denn?

LENA BADKE: Nun, es beginnt morgens um 5 Uhr mit dem **entsetzlich** langen Klingeln seines Weckers Am Abend folgt dann laute und **relativ** geschmacklose Musik.

INGRID SCHÄFERMANN: Haben Sie denn schon versucht, mit ihm darüber zu sprechen?

LENA BADKE: Ja, ich habe ihn natürlich sofort auf das Problem angesprochen, aber es interessierte ihn **recht** wenig. Er meinte, er habe einen **ungewöhnlich** tiefen Schlaf und würde nicht so schnell aufwachen, deshalb das lange Klingeln des Weckers. Die Musik sei Geschmackssache, aber er würde die Lautstärke runterfahren. Leider ist sie bisher noch **überhaupt** nicht leiser geworden. Jetzt muss ich eben mit Ohrstöpseln schlafen.

INGRID SCHÄFERMANN: Aber Frau Badke sie geben **viel zu** schnell nach.

4.4 Wortbildung

4.4.1 Nomen: Zusammensetzung (Komposition)

1.1

Das unterschätzte Tier

Rockender Regent des Gartens

Der **Zaunkönig** mutet herrschaftlich an, zumindest dem Namen nach. Doch sein braunes Gewand suggeriert eher „angepasstes **Fußvolk**" als „strahlender **Thronerbe**". Zudem misst der deutsche **Gartenbewohner** gerade mal neun Zentimeter und bringt trotz kleinem und pummeligem Körper maximal 11 Gramm auf die Waage. Das macht ihn zu einem der kleinsten Vögel Europas. So wird der **Zaunkönig** wegen seiner **Unscheinbarkeit** nicht nur gern übersehen, sondern auch unterschätzt. Dabei ist das **Hochverrat**! Denn der **Sperlingsvogel** ist ein **Gesangtalent**, ein talentierter **Häuserbauer** und ein **Flattermann** mit **Familiensinn**. Doch der Regent unter den **Gartenvögeln** kann noch mehr. Der **Zaunkönig** nämlich ist eine echte **Rockröhre**. Mit bis zu 90 Dezibel – das entspricht dem Dröhnen eines **Lastwagens** – schallt sein Gesang durch das **Unterholz**. Die **Lautstärke** braucht es, um sich zur **Paarungszeit** bemerkbar zu machen. Von morgens bis abends schmettert er deshalb **Liebeslieder**. Zwar klingt sein Gesang eher nach naturel- ler Klassik als nach ultimativen **Rock-Love-Songs**, doch die Frauen fliegen drauf. Wohl auch, weil so ein **Zaunkönig** nicht nur ein **Performancekünstler** sondern zudem ein echter Macker ist. Tatsächlich bieten die Männchen ihren **Artgenossinnen** sogar bis zu zwölf runde **Rohbauten** als **Nistplatz** an und bauen erst nach deren kritischer **Auswahl** ein Nest fertig. Auch bei der **Inneneinrichtung** mit kolonialen Federn, Moos im **Landhausstil** oder heimatlicher Wolle lässt er seiner Königin freie Hand. Was für ein Kerl!

1.2/1.3

Wortart Bestimmungswort	Bestimmungswort	Grundwort	Kompositum
Nomen	der Zaun	der König	der Zaunkönig
	der Fuß	**das Volk**	das Fußvolk
	der Thron	der Erbe	**der Thronerbe**
Adjektiv	**laut**	**die Stärke**	die Lautstärke
	roh	die Bauten	**die Rohbauten**
Verb	**flattern**	**der Mann**	der Flattermann
	nisten	der Platz	**der Nistplatz**
Präposition	**unten**	**das Holz**	das Unterholz
	aus	die Wahl	**die Auswahl**
Adverb	**innen**	**die Einrichtung**	die Inneneinrichtung

1.4 Zusammengesetzte Nomen

- **Das Grundwort bestimmt das Genus des gesamten Nomens.**
- **Das Bestimmungswort spezifiziert das Grundwort.**
- **Die Betonung des zusammengesetzten Nomens liegt auf dem Bestimmungswort.**
- Bei Verben als Bestimmungswort fallen die Endungen *-n* (Flattermann) und *-en* (Nistplatz) des Infinitivs weg.

2.1 Kaufhaus = **ein Haus, in dem man etwas kaufen kann**

Überstunden = *die Stunden, die man über die Arbeitszeit hinaus arbeitet*

Großstadt = **eine Stadt, die groß ist**

Prüfungsordnung = **die Ordnung für die Prüfung**

Abendvorstellung = **eine Vorstellung am Abend**

Trinkwasser = **Wasser zum Trinken**

Fahrschule = **eine Schule, wo man fahren lernen kann**

Gegenargument = **ein Argument gegen ein anderes Argument**

Übungsgrammatik = **eine Grammatik zum Üben**

Fingernagel = **der Nagel eines Fingers**

3.1

Kompositum	Bestimmungswort	Fugenzeichen	Grundwort
der Versuchsaffe	der Versuch	-s-	der Affe
die Meeresschnecke	das Meer	**-es-**	**die Schnecke**
der Wüstenfuchs	**die Wüste**	-n-	**der Fuchs**
das Bärenfell	**der Bär**	**-en-**	**das Fell**
die Rinderherde	**das Rind**	**-er-**	**die Herde**

3.2 Beispiellösung:

die Besuchszeit, der Besucherservice, die Fütterungszeit, der Fütterungsplatz, das Streichelgehege, das Streicheltier, das Elefantenbaden, das Elefantengehege, der Waschbär, der Waschplatz, die Eiszeit, der Eisbär, der Eisverkauf, das Jungtier, der Lieblingsplatz, das Lieblingstier

4.4.2 Nomen: Ableitung (Derivation)

1.1

Tiere

Das **Transportieren** von Tieren ist bei einem **Flug** möglich. Abhängig von Gewicht und Größe werden die Tiere entweder in der Passagierkabine oder in tiergerechten Containern in einem klimatisierten **Abschnitt** des Frachtraums des Flugzeugs transportiert. Bitte beachten Sie, dass die gültigen Bestimmungen des **Tierschutzes** sowie die **Ein**- und Aus**fuhr**bestimmungen der betroffenen Länder eingehalten werden müssen. Zur Anmeldung ihres Tieres ist ein **Anruf** bis spätestens 24 Stunden vor dem **Abflug** nötig.

Gerne können Sie Ihr Tier auch persönlich anmelden. Erlauben Sie uns den Hinweis darauf, dass aufgrund begrenzter Kapazitäten und abweichender **Verbote** bei Partnerairlines Ihre Buchung die Akzeptanz von Sondergepäck und Tieren nicht in jedem Falle garantiert werden kann. Das **Mitnehmen** kleiner Hunde und Katzen in der Passagierkabine ist erlaubt, wenn ein Gewicht von 8 kg (inkl. Transportbehälter) nicht überschritten wird.

…

schützen: **der (Tier-)Schutz**

einführen: **die Einfuhr**

verbieten: **das Verbot**

anrufen: **der Anruf**

abschneiden: **der Abschnitt**

abfliegen: **der Abflug**

mitnehmen: **das Mitnehmen**

fliegen: **der Flug**

transportieren: **das Transportieren**

1.2　**Ableitung ohne Affixe: Nominalisierung aus Verben**

- Nomen können von Verben abgeleitet werden: Entweder aus dem Infinitiv (*das Schützen*) oder aus dem Verbstamm (*der Schutz*).
- Die Nominalisierung des **Infinitivs** hat immer den Artikel *das*.
- Die Nominalisierung des **Verbstamms** hat oft den Artikel *der*.

1.3　der Versuch, das Versuchen; der Lauf, das Laufen; der Ritt, das Reiten; der Sprung, das Springen; der Vergleich, das Vergleichen; der Schluss, das Schließen; der Verkauf, das Verkaufen; der Schnitt, das Schneiden; der Unterschied, das Unterscheiden; der Unterricht, das Unterrichten; der Riss, das Reißen

Beispiellösung:

Ich mache seit ein paar Monaten einen Reitkurs. Es gibt große Unterschiede zwischen den Pferden, auf Nathan macht mir der Unterricht am meisten Spaß. Am besten gefällt mir das Springreiten, wir springen natürlich nur über ganz niedrige Hindernisse. Vor dem ersten Sprung hatte ich ein bisschen Angst, aber nach ein paar Versuchen hat es mir schon riesigen Spaß gemacht …

2.1

VORSITZENDER DES TIERSCHUTZVEREINS:

Liebe Mitglieder des Vereins, liebe Anwesende. Ich begrüße Sie herzlich zu der heutigen Versammlung. Im vergangenen Jahr ist sowohl **Gutes** als auch **Unerfreuliches** in unserer Region passiert. Ich möchte das **Wichtigste** kurz in Erinnerung rufen und Ihnen einen Ausblick auf unsere kommenden Projekte liefern.

Im Sommer hatten wir große Probleme mit **Jugendlichen**, die im Seewald Partys veranstaltet haben. Die **Betrunkenen** hinterließen neben Müll auch viele Scherben, die gefährlich für alle Waldbewohner sind. Durch Kontrollgänge von Vereinsmitgliedern an den Wochenenden bekamen wir dieses Problem aber in den Griff. Vielen Dank nochmal an alle **Freiwilligen**! Ebenfalls im Seewald passierte das wohl **Schlimmste** in diesem Jahr: Wilddiebe haben Fallen aufgestellt. Ein Jogger trat in eine Falle. Der **Verletzte** musste sofort ins Krankenhaus gebracht werden. Bis heute konnten die **Kriminellen** leider nicht gefasst werden.

Doch nun möchte ich auch noch **Positives** mitteilen: Unser Verein zählt 8 neue Mitglieder, die uns tatkräftig unterstützen. Im kommenden Jahr beginnen zwei tolle neue Projekte: In Zusammenarbeit mit der Agentur für Arbeit werden uns **Arbeitsuchende** bei verschiedenen Aktionen über das Jahr hinweg unterstützen. Zum Abschluss noch ein aktueller Tipp: Jetzt im Herbst sorgen sich Tierfreunde vermehrt um Igel, die in Gärten und der freien Natur noch auf Futtersuche sind. Die Tiere brauchen aber nur in Ausnahmefällen menschliche Hilfe. Für **Ratsuchende** hilft der Verein gerne unter der bekannten Infonummer.

Ich danke Ihnen für Ihre Aufmerksamkeit und wünsche allen noch einen schönen Abend.

2.2

Adjektiv	Partizip I	Partizip II
gut > Gutes	anwesend > Anwesende	betrunken > die Betrunkenen
unerfreulich › Unerfreuliches	arbeitsuchend › Arbeitsuchende	verletzt › der Verletzte
wichtig › das Wichtigste	ratsuchend › Ratsuchende	
jugendlich › Jugendlichen		
freiwillig › alle Freiwilligen		
schlimm › das Schlimmste		
kriminell › die Kriminellen		
positiv › Positives		

2.3

Journalist: Anna, wird das kommende Jahr aufregend für dich?
Anna: Ja, sehr. Das **Aufregende / Aufregendste** sind all die neuen Dinge, die ich lernen werde.
Journalist: Aber viel **Neues** gibt es für dich doch gar nicht zu lernen, da du schon seit drei Jahren Mitglied im Tierheim bist. Du kennst bestimmt schon viele Kollegen?
Anna: Ja, das stimmt. Im Tierheim habe ich viele **Bekannte**, aber es gibt noch viele, die ich nicht kenne, auch viele Vereinsmitglieder. Es ist schön, dass hier Menschen jeden Alters aktiv sind. Die **Erwachsenen** und die **Jugendlichen** arbeiten oft eng zusammen, da sie die Liebe zu den Tieren verbindet.

Journalist: Tierliebe ist wohl das **Wiichtigste**, was man mitbringen muss, wenn man im Tierheim angestellt ist?
Anna: Als **Angestellte** in einem Tierheim ist die Tierliebe natürlich Grundvoraussetzung. Aber auch Liebe und Verständnis für die Mitmenschen ist wichtig. Manche machen aus Unwissenheit Fehler bei der Tierpflege und brauchen unseren Rat. Das **Schönste** an meinem zukünftigen Beruf ist doch, wenn es den Tieren und den Menschen gut geht.
Journalist: Vielen Dank für das Gespräch und alles **Gute** für deine Zukunft!

3.1

SÜSSE WELPEN ZU VERKAUFEN

Biete vier schwarz-weiße Labrador Mischlinge mit wenig weißer **Färbung**. Ihr Vater ist der perfekte Familienhund und zeichnet sich durch **Freundlichkeit**, Geduld und **Zurückhaltung** aus. Beide Elterntiere haben alle erforderlichen Gesundheitstests und sind frei von erblichen **Krankheiten**. Die Welpen wachsen in einer Großfamilie mit vielen verschiedenen Tieren auf. Sie kennen also das Verhalten und **Geschrei** kleiner Kinder.

EIN TRAUMTYP FÜR DIE GANZE FAMILIE

Der Schimmelhengst hat eine **Größe** von 160 cm und ist sehr gut gebaut, auch für einen größeren **Reiter** geeignet. Seine Stärken sind **Lebhaftigkeit**, aber auch Ausgeglichenheit. Seine Eltern können auch besichtigt werden, sie stehen in der **Nachbarschaft** (ca. 6 km entfernt).

3.2

backen: Bäcker, **Bäckerei**	wecken: **Wecker**
Verständnis: **Missverständnis**	sprechen: **Sprache, Sprecher**
Brot: **Brötchen**	Glück: **Unglück**

3.3

Basiswortart: Verb	Präfix / Suffix	abgeleitete Nomen
reden **schreien**	Ge- (und –e)	das Gerede **das Geschrei**
färben **zurückhalten**	-ung	die Färbung **die Zurückhaltung**
abgeben **sprechen**	-e	die Abgabe **die Sprache**
drucken **backen**	-ei	die Druckerei **die Bäckerei**
gewinnen **reiten** **backen** bohren **wecken**	-er	der Gewinner **der Reiter** **der Bäcker** der Bohrer **der Wecker**
prüfen **mischen**	-ling	der Prüfling **der Mischling**

Basiswortart: Adjektiv	Präfix / Suffix	abgeleitete Nomen
schön **krank**	-heit	die Schönheit **die Krankheit**
abhängig **freundlich**	-keit	die Abhängigkeit **die Freundlichkeit**
süß **lebhaft**	-igkeit	die Süßigkeit **die Lebhaftigkeit**
stark **groß**	-e	die Stärke **die Größe**

Basiswortart: Nomen	Präfix / Suffix	abgeleitete Nomen
Ruhe **Glück**	Un-	die Unruhe **das Unglück**
Achtung **Verständnis**	Miss-	die Missachtung **das Missverständnis**
Partner **Nachbar**	-schaft	die Partnerschaft **die Nachbarschaft**
Kopf **Brot**	-chen	das Köpfchen **das Brötchen**

4

	abgeleitet von	Nomen	Verb	Adjektiv	Suffix	maskulin	feminin	neutrum
Sauberkeit	sauber			x	-keit		x	
Forscher	**forschen**		x		-er	x		
Absolventin	**Absolvent**	x			-in		x	
Gemeinheit	**gemein**			x	-heit		x	
Bedrohung	**bedrohen**		x		-ung		x	
Druckerei	**drucken**		x		-ei		x	
Verwandtschaft	**verwandt**			x	-schaft		x	
Beinchen	**Bein**	x			-chen			x
Lehrling	**lehren**		x		-ling	x		

Ableitung mit Suffixen

Die Suffixe können Informationen über den Artikel des Nomens geben:
- Nomen mit den Suffixen -er, -ling sind immer **maskulin**.
- Nomen mit den Suffixen -keit, -heit, -igkeit, -in, -e, -ung, -ei, -schaft sind immer **feminin**.
- Nomen mit den Suffixen -chen, -lein sind immer **neutrum**.

5

Adjektiv	Nomen	Gegenteil
klug	Klugheit	Dummheit
ehrlich	**Ehrlichkeit**	**Unehrlichkeit / Verlogenheit**
nah	**Nähe**	**Ferne / Weite**
ähnlich	**Ähnlichkeit**	**Verschiedenheit / Unähnlichkeit**
schnell	**Schnelligkeit**	**Langsamkeit**
offen	**Offenheit**	**Verschlossenheit**

6.1

Ruhe im Büro

Büroarbeit heute ist geprägt von einer Mischung aus **Konzentration** und **Kommunikation**. Sowohl Besprechungen im Team als auch konzentrierte Denkarbeit im Wechsel mit Telefonaten gehören zur **Normalität** eines Büroalltags. Die erfolgreiche **Kombination** dieser gegensätzlichen Anforderungen in den modernen Großraumbüros ist eine machbare Herausforderung.

Lärm gilt als Störfaktor Nummer 1 im Büro. Er senkt die Fähigkeit der geistigen **Produktion** und macht auf Dauer krank. Als besonders störend wird Gesprächslärm empfunden. Jede **Information**, die beim unfreiwilligen Mithören ankommen, beeinträchtigt die **Stabilität** der Konzentrationsfähigkeit enorm.

Weitere Krachmacher im Büro sind alle Arten von elektronischen Geräten wie zum Beispiel EDV-Geräte. Darum ist es wichtig, bei einer Neuanschaffung eines Gerätes nicht nur auf **Funktionalität**, sondern auch besonders auf den Geräuschpegel zu achten.

Der Zusammenhang zwischen Bürolärm, Leistungsfähigkeit und Gesundheit ist nachgewiesen: Je informationshaltiger und intensiver der Lärm ist, desto mehr steigen die Fehlerquoten. Alle Arbeitgeber, die die **Ambition** haben, einen perfekten Arbeitsplatz zu bieten, sollten Möglichkeiten schaffen, die **Flexibilität** der Arb eitnehmer und die **Individualität** ihrer Arbeitsweise zu wahren.

6.2

Adjektiv / Partizip	Suffix	Nomen
konzentriert	-ation	Konzentration
kommuniziert		Kommunikation
kombiniert		Kombination
informiert		Information
produziert	-tion	Produktion
ambitioniert		Ambition

Adjektiv / Partizip	Suffix	Nomen
normal	-alität	Normalität
funktional		Funktionalität
individual / individuell		Individualität
stabil	-ilität	Stabilität
flexibel		Flexibilität

6.3 Fremde Suffixe: *-ation, -tion, -alität, -ilität*

- Nomen mit dem Suffix *-ation* und dem Suffix *-tion* werden von Partizipien / Adjektiven, die auf *-iert* enden, abgeleitet:
  konzentriert → *Konzentration, kommuniziert* → **Kommunikation**, *kombiniert* → **Kombination**, *informiert* → **Information**, *produziert* → **Produktion**, *ambitioniert* → **Ambition**
- Nomen mit dem Suffix *-alität* werden von Adjektiven auf *-al* und auf *-ell* abgeleitet:
  normal → Normalität, funktional → Funktionalität, individual / individuell → Individualität
- Nomen mit dem Suffix *-ilität* werden von Adjektiven auf *-il* und *-el* abgeleitet:
  stabil → Stabilität, flexibel → Flexibilität
- Alle Nomen mit diesen Suffixen sind feminin.

7.1

Artikel	Nomen	Suffix(e)
die	Strateg**ie**, Bürokrat**ie** Harmon**ie** , Demokrat**ie**	-ie
	Krit**ik**, Lyr**ik** Takt**ik**, Opt**ik**, Log**ik**	-ik
	Toler**anz**, Eleg**anz** Domin**anz**, Tend**enz** Konsequ**enz**, Differ**enz**	-anz / -enz

Artikel	Nomen	Suffix
das	Argu**ment**, Parla**ment** Instru**ment** , Ele**ment** Doku**ment**	-ment
	Vokabul**ar** , Gloss**ar**, Invent**ar**	-ar
der	Egois**mus** , Kapitalis**mus** Feminis**mus**, Kommunis**mus**	-mus

8

-ar / -är	-ant / -and	-ent	-urg	-ist	-e	-or	-ör / -eur
Millionär	Doktorand	Konsument	Chirurg	Journalist	Astrologe	Autor	Frisör
Notar	Migrant	Absolvent	Dramaturg	Polizist	Geologe	Professor	Redakteur
Bibliothekar	Praktikant	Student		Pessimist	Biologe	Direktor	Jongleur
Visionär	Demonstrant	Agent		Artist		Doktor	Regisseur
		Präsident		Zivilist		Lektor	Ingenieur

4.4.3 Trennbare und nicht trennbare Verben

1.2

	trennbar	nicht trennbar		trennbar	nicht trennbar
entscheiden		X	beruhigen		x
beraten		x	erweitern		x
einzahlen	x		erarbeiten		x
genießen		x	ausgeben	x	
verreisen		x	zurücklegen	x	
besuchen		x	vorhaben	x	
zerbrechen		x			

Nicht trennbare Verben

- Die Präfixe *miss-*, **be-**, **ge-**, **ver-**, **zer-**, **er-**, **ent-** sind nicht trennbar.

1.3 > Die Kaffeelandschaft in Berlin **enttäuschte** 1994 die Amerikanerin Cynthia Barcomi. Sie **eröffnete** deshalb ihr erstes Café „Barcomi's Kaffeerösterei" in Kreuzberg. Ihr Konzept **beinhaltete** verschiedene Kaffeesorten, selbstgemachte Kuchen und Gebäck. Sie **verbesserte** damit den schlechten Ruf der amerikanischen Esskultur. Sie **veröffentlichte** schon mehrere Backbücher und **erweiterte** ihr Geschäft um eine weitere Filiale in Berlin-Mitte.

1.4 > Die Geschäftsidee fiel dem damals Arbeitslosen René Frauenkron im Imbiss **ein**. Er **dachte** über die Frage **nach**: „Was passiert mit dem verbrauchten Frittierfett?" Seine Firma stellt Öl **her**, bietet dieses zum Verkauf **an** und holt dafür das Altöl von seinen Kunden **ab**, welches seine Partnerunternehmen **wiederverwerten**. Sein Geschäftsmodell hält modernen Ansprüchen von Recycling **stand**. Deshalb schlug man ihn für den Umweltpreis 2007 **vor**.

4.4.4 Bedeutungen der nicht trennbaren Präfixe

1.1 a Der Rock **endet** knapp oberhalb ihres Kries.
  b Unser Garten **endet** hier an diesem Baum.
  c Warum hast du den Streit mit deinem Freund nicht **beendet**?
  d Der Lehrer **beendet** den Unterricht heute 10 Minuten früher.
  e Der Unterricht **endet** heute 10 Minuten früher.
  f Ich muss mir einen neuen Anbieter suchen, denn bald **endet** mein Handyvertrag.

1.2 b > **Du musst auch die anderen Fahrer auf der Straße beachten.**
  c > **Ich möchte deinen Kaffee bezahlen.**
  d > **Können Sie bitte meine Frage beantworten?**
  e > **Viele Leute bezweifeln die geplante Schulreform der neuen Regierung.**

1.3 Präfix *be-*

Das Präfix *be-* kann auch mit anderen Wortarten zu einem Verb kombiniert werden:
be- + *Adjektiv*: *be-* + *frei*: befreien („frei machen"), *be-* + *unruhig*: **beunruhigen** („**unruhig** machen"),
be- + *richtig*: **berichtigen** („**richtig** machen")
be- + *Nomen*: *be-* + *Titel*: betiteln („einen Titel hinzufügen"), *be-* + *Schaden*: **beschädigen**
(„**Schaden** hinzufügen"), *be-* + *Strafe*: **bestrafen** („**Strafe** hinzufügen")

ent-

1.1 a ent-**decken**: eine Pflanzenart, einen Fehler, ein Talent
b ent-**werten**: die Fahrkarte, das Geld, das Kinoticket
c ent-**sorgen**: die alte Zeitung, das kaputte Handy, den Biomüll
d ent-**lassen**: einen Gefangenen, einen Arbeiter, einen Patienten
e ent-**führen**: die Braut, das Flugzeug, das Publikum

1.2

	Trennung / Wegnehmen	Anfang
Beim Fasten wird der Körper entgiftet.	x	
Viele Ökonomen fordern, die Banken zu entmachten.	x	
Durch die Diskussion ist ein heftiger Streit entbrannt.		x
Die Donau entspringt im Schwarzwald.		x
Beim Fensterputzen ist mir heute Morgen mein Vogel entflogen.	x	
Die Zusammenarbeit ist zufällig entstanden.		x
Die Milliardärstochter wurde schon zweimal entführt.	x	
Die neue Methode wurde von einer Gruppe Studierender entwickelt.		x

1.3 Präfix *ent-*

- Das Präfix *ent-* kann eine Trennung oder ein Wegnehmen bedeuten: Etwas wird entfernt, etwas
  wird weggenommen oder etwas / jemand wird befreit oder gefunden:
  entdecken, **entgiften, entmachten, entfliegen, entführen**
- Das Präfix *ent-* kann auch den Anfang einer Handlung / Sache bedeuten:
  entbrennen, **entspringen, entstehen, entwickeln**

er-

1.1

Verb	abgeleitet von	Verb	Adjektiv
erblinden	blind		x
erleichtern	**leicht**		x
ermüden	**müde**		x
erklären	**klar**		x
erfrieren	**frieren**	x	
errechnen	**rechnen**	x	

1.2 errechnen:
[x] **etwas zu Ende rechnen**
[ ] sehr schnell rechnen
[ ] nicht rechnen

erfrieren:
[ ] nicht frieren
[x] **sich tot frieren**
[ ] sehr stark frieren

1.3 **b** Sie bekommt eine befriedigende Antwort, indem sie viel fragt. – **erfragen**

　　　c Sie gewinnt das Geld, indem sie erfolgreich spielt. – **erspielen**

　　　d Sie kann sich ein Haus kaufen, indem sie ange gespart hat. – **ersparen**

1.1 **a** Sie hat jetzt lange Haare: Sie hat ihre Haare **verlängern** lassen.

　　　b Die Fenster haben jetzt Gitter: Die Fenster wurden **vergittert**.

　　　c Er hat die falsche Hausnummer aufgeschrieben: Er sich bei der Hausnummer **verschrieben**.

　　　d Die Kinder haben die Vögel gefüttert, jetzt sind keine Körner mehr da:

　　　　Die Kinder haben alle Körner **verfüttert**.

1.2 **Das Präfix *ver-***

Mit dem Präfix *ver-* werden Verben von Adjektiven, Nomen oder Verben abgeleitet.
Dabei verleiht es verschiedene Bedeutungen:
- Ableitung von Verben:
  - Handlung ist falsch oder unerwünscht, z. B. falsch schreiben: **verschreiben**
  - Ende eines Vorgangs, z. B. zu Ende füttern: **verfüttern**
  - Gegenteil des Ausgangsverbs, z. B. kaufen – **verkaufen**
- Ableitung von Adjektiven:
  - Zustandsveränderung, etwas oder jemand wird in den Zustand, den das Adjektiv ausdrückt, gebracht, z. B. *länger machen*: **verlängern**
- Ableitung von Nomen:
  - etwas mit dem ausstatten, das vom Nomen ausgedrückt wird, z. B. *mit Gittern ausstatten*: **vergittern**
- etwas zu dem machen oder zu dem werden, was vom Nomen ausgedrückt wird,
  z. B. *zu Dampf werden* – verdampfen

1.3 **Ableitung von Verben**

Handlung ist falsch oder unerwünscht: *verschlucken* , **verhören**, **verfahren**, **versalzen**

Ende eines Vorgangs: *verblühen* , **verhungern**, **verheilen**

Gegenteil des Ausgangsverbs: **vermieten**

Ableitung von Adjektiven

Zustandsveränderung, etwas oder jemand wird in den Zustand, den das Adjektiv ausdrückt, gebracht:

verengen , **verdunkeln**, **verschönern**, **verdoppeln**, **verarmen**, **verhärten**

Ableitung von Nomen

etwas mit dem ausstatten, das vom Nomen ausgedrückt wird: *verhüllen* , **vergolden**, **verglasen**

etwas zu dem machen, was vom Nomen ausgedrückt wird: *verschrotten* , **verfilmen**, **verbeamten**

1 **b** **> Heute ist mein Pechtag, alles missglückt mir!**

　　c **> Ich denke, dass du ihre Reaktion missdeutet hast.**

　　d **> Anna missachtet alle meine Regeln zu Hause.**

　　e **> Du musst den Text genau lesen, man kann ihn leicht missverstehen.**

　　f **> Eva missgönnt ihrem Kommilitonen die gute Note.**

1 **Beispiellösung:**

eine Tasse: *Beim Abwaschen ist eine Tasse zerbrochen.*

eine Wohnung: *Der Rockstar hat eine ganze Wohnung zertrümmmert.*

einen Kuchen: *Sie hat den Kuchen in 12 Stücke zerschnitten.*

ein Bonbon: *Das Kind hat das Bonbon nicht gelutscht, sondern zerbissen.*

das Blumenbeet: *Beim Spielen hat Marius das Blumenbeet völlig zertrampelt.*

Fotos: *Sie hat alle Fotos, auf denen sie sich nicht gefiel, zerrissen.*

4.4.5 Trennbare und nicht trennbare Erstglieder

1.2 Erstglied betont: *durchsetzen*, **wiedergeben, umdrehen, durchlesen**
Verbstamm betont: *wiederholen*, **überholen, unterschätzen, unterhalten, widersprechen,
durchschauen, umarmen, übersehen, unterschreiben**

Trennbare und nicht trennbare Erstglieder

- Liegt die Betonung auf dem **Erstglied**, so ist das Verb trennbar.
- Liegt die Betonung auf dem **Verbstamm**, so ist das Verb untrennbar.

1.3 b > **Der Journalist hat letzte Nacht den Artikel umgeschrieben.**
c > **Das Kind hat den Ball wiedergeholt.**
d > **Die Bürger haben dem Bürgermeister Korruption unterstellt.**
e > **Ich habe den Text ins Deutsche übersetzt.**
f > **Die Lehrerin hat das unbekannte Wort umschrieben.**
g > **Er hat seine Pflanzen während der Reise bei einem Freund untergestellt.**
h > **Wir sind mit der Fähre auf die Insel übergesetzt.**

4.4.6 Adjektive

1.2

Silvester in New York!

Happy New York! Verbringen Sie einen **unvergesslichen**
Jahreswechsel in der Stadt, die niemals schläft.
Auszüge aus dem Programm:
1. Tag: Hello New York!
2. Tag: New-York Update
Auf zur Rundfahrt durch Manhattan! Wir streifen durch
die verschiedenen Viertel und grüßen vom Battery Park
hinüber zur Freiheitsstatue. Gleich um die Ecke: die
Wall Street und Ground Zero, der **ehemalige** Standort
der Türme des World Trade Centers. Nach der Tour gibt
es genug für unser Dinner am **heutigen** Abend.
3. Tag: Manhattan-Skyline
Zu Fuß flanieren wir zwischen Wolkenkratzern über
den Broadway. Im Museum of Modern Art (MoMA)
präsentiert sich uns eine **fantastische** Sammlung
moderner und **zeitgenössischer** Kunst. Hoch hinauf
geht es im Rockefeller Center: Von ganz oben genießen
wir den **dortigen** Blick über die Stadt! Und danach?
Auf ein spätes Frühstück mit Audrey Hepburn zu
„Tiffany's"?
4. Tag: …

Grand Beach, Hurghada ****

Lage
Das „Grand Beach" verfügt über einen **unvergleichlich**
langen Privatstrand. Das Hotel eignet sich besonders für
Familien. Der Ortskern von Hurghada liegt ca. 8 km entfernt
und ist durch einen **täglichen kostenlosen** Bustransfer
problemlos erreichbar.
Ausstattung
Die Hotelanlage verfügt über insgesamt 550 Zimmer. In
dem Haupthaus mit großräumiger Empfangshalle befinden
sich die Rezeption sowie die zwei Hauptrestaurants. Am
Pool finden Sie ein Grillrestaurant, verschiedene Bars und
Geschäfte. In der angrenzenden „Siva Mall" befinden sich
viele weitere Geschäfte. Besondere Highlights sind das
orientalische Kaffeehaus und ein **libanesisches** Restaurant.
Zimmer
Die Deluxe- und Bungalow-Zimmer (Gartenblick) befinden
sich in 20 kleinen Nebengebäuden. Ihre komfortable
Einrichtung ist **auffällig** schön.
Sport & Unterhaltung
Es gibt ein wechselndes Animationsprogramm mit
sportlichen Aktivitäten, Spielen und Wettbewerben.
Abendliche Unterhaltung bei Live-Musik und Billard.

Suffix	Adjektiv	Grundwort	Wortart des Grundwortes
-isch	fantastisch	Fantasie	Nomen
	zeitgenössisch	**Zeitgenosse**	**Nomen**
	orientalisch	**Orient**	**Nomen**
	libanesisch	**Libanese**	**Nomen**
-lich	unvergesslich	vergessen	Verb
	täglich	**Tag**	**Nomen**
	sportlich	**Sport**	**Nomen**
	unvergleichlich	**vergleichen**	**Verb**
	abendlich	**Abend**	**Nomen**
-ig	ehemalig	ehemals	Adverb
	heutig	**heute**	**Adverb**
	dortig	**dort**	**Adverb**
	auffällig	**auffallen**	**Verb**
-bar	erreichbar	erreichen	Verb
-los	kostenlos	Koster	Nomen
	problemlos	**Problem**	**Nomen**

1.3   b > **Es ist bezahlbar**.
        c > **Preislich ist das Angebot konkurrenzlos**.
        d > **Leider ist es nach der Buchung nicht mehr kündbar**.
        e > **Das Kleingedruckte ist nicht lesbar**.
        f > **Aber ich denke, dieses Problem ist lösbar**.
        g > **Ich frage, ob das Angebot auch in einem größeren Format lieferbar ist**.
        h > **Dann kann ich die Bilder von grenzenlosen Stränden und wolkenlosem Himmel betrachten.**

2.1   unklar – klar   instabil – **stabil**   asozial – **sozial**
       zufrieden – **unzufrieden**   klein – **groß**   unproblematisch – **problematisch**
       legal – **illegal**   tolerant – **intolerant**   real – **irreal**

Adjektivbildung mit Präfixen

- Das Präfix **miss** kann die Bedeutung von „schlecht" oder „falsch" haben.

2.2   a Der Satz kann nicht verstanden werden. Er ist **unverständlich**.
       b Der Satz kann falsch verstanden werden. Er ist **missverständlich**.
       c Der Baum ist sehr schräg, er ist nicht richtig gewachsen. Er ist **misswachsen**.
       d Die Tat kann man nicht verzeihen. Sie ist **unverzeihlich**.
       e Sie konnte nicht schlafen und ist schlecht gelaunt. Sie ist **missgelaunt**.
       f Auf meinen Bruder kann ich mich leider nicht verlassen. Er ist **unzuverlässig**.

3

Liebe Sandra,
wir wünschen dir ein gutes neues Jahr! Dieses Jahr sind wir mit unserer Kleinen zum ersten Mal verreist und es ist traumhaft! Ich liege gerade am schneeweißen Strand mit Blick auf das türkisblaue Meer, mit einem **alkoholfreien**, aber zuckersüßen Cocktail in der Hand bei einer Temperatur von 39 Grad. Unsere Reise ist ein Volltreffer! Wir haben ein riesengroßes, vollklimatisiertes Familienzimmer in einem wunderschönen Hotel. Es ist sehr **geschmackvoll** eingerichtet. Das Essen ist superlecker und es gibt **zahlreiche** Restaurants. Das Personal ist freundlich und immer **hilfsbereit** und mindestens dreisprachig. Besonders gut gefällt mir aber, dass alles so **kinderfreundlich** ausgerichtet ist. Du kennst mich, ich bin übervorsichtig bei Marie, aber hier ist alles sehr **liebevoll** und ungefährlich.
Ich und meine Familie schicken sonnige Grüße ins kalte Berlin,
Heike

Happy New Year!!!
Wir sind gerade in New York und alles ist sehr **eindrucksvoll**, v.a. die himmelhohen Wolkenkratzer. Dagegen scheint unser Berlin so klein. Das Programm ist sehr **abwechslungsreich** gestaltet und mein **bildungshungriger** Mann kriegt genug geboten. Das amerikanische Essen ist lecker, aber alles andere als **fettarm**. Aber was soll's, im Urlaub darf man ja schlemmen! Mir gefällt besonders gut, dass es hier so viele **rauchfreie** Orte gibt. Die extremen Staus während der Rush Hour sind allerdings **gewöhnungsbedürftig**. Aber wir benutzen sowieso nur die öffentlichen Verkehrsmittel, weil das viel **preisgünstiger** ist. Unsere Gruppe geht weiter, ich muss Schluss machen.
LG auch von Thomas
Marie

Adjektiv	Bestimmungswort + Grundwort	Bedeutung
alkoholfrei	Alkohol + frei	frei von Alkohol / ohne Alkohol
geschmackvoll	Geschmack + voll	mit gutem Geschmack
zahlreich	**Zahl + reich**	„reich an Zahlen", viele
hilfsbereit	Hilfe + bereit	**bereit, anderen zu helfen**
kinderfreundlich	**Kinder + freundlich**	Kindern gegenüber positiv eingestellt
liebevoll	**Liebe + voll**	mit viel Liebe
eindrucksvoll	**Eindruck + voll**	viel Eindruck hinterlassen
abwechslungsreich	**Abwechslung + reich**	**reich an Abwechslung / mit viel Abwechslung**
bildungshungrig	**Bildung + hungrig**	hungrig nach Bildung / streben nach Bildung
fettarm	**Fett + arm**	**arm an Fett / mit wenig Fett**
rauchfrei	**Rauch + frei**	**frei von Rauch**
gewöhnungsbedürftig	**Gewöhnung + bedürftig**	man muss sich erst an etwas gewöhnen
preisgünstiger	**Preis + günstig**	einen günstigen / guten Preis haben

4   Vergleich   Verstärkung

eiskalt: kalt wie Eis   todmüde: sehr müde

schneeweiß: weiß wie Schnee   **vollklimatisiert: komplett klimatisiert**

zuckersüß: süß wie Zucker   **übervorsichtig: extrem vorsichtig / zu vorsichtig**

riesengroß: groß wie ein Riese   **superlecker: sehr lecker**

wunderschön: schön wie ein Wunder

4.5 Wortverbindungen

4.5.1 Kollokationen

1.1　Geld(summe): **anlegen, abheben, einzahlen**
　　ein Konto: **eröffnen**　　ein Formular: **ausfüllen**　　ein Risiko: **eingehen**

2

Starker Raucher beendet jahrelange
Sucht mit Nicotinpflastern!

Wer hier nicht aufpasst,
wird es **bitter bereuen**!

Ehe schließen und eine Woche
danach scheiden lasser ?!

Die **Hoffnung** nie **aufgeben!**

Junge Künstler ernteten **stürmischen Applaus**

Überleben auf dem Meer: **Durst löschen** *mit Salzwasser*

Warum muss eigentlich immer eine
Prise Salz in den Kuchenteig?

Kollokationen

- Es gibt Kombinationen aus verschiedenen Wortarten:
  Nomen-Verb-Kollokationen: **Hoffnung aufgeben, Ehe schließen, Durst löschen**
  Adjektiv-Nomen-Kollokationen: **starke Raucher, stürmischer Applaus**
  Adverb-Verb-Kollokationen: **bitter bereuen**
  Nomen-Nomen-Kollokationen: **Prise Salz**

3　Konto:
　eröffnen, überziehen, ~~übertreten~~, sperren, ~~lösen~~, auflösen, ausgleichen, ~~gehen~~, belasten, entlasten
　anlegen:
　einen Park, ein Beet, ~~Gemüse~~, eine Kartei, eine Uniform, Geld, jemandem einen Verband, ~~ein Fenster~~
　aufgeben:
　ein Paket, eine Anzeige, ~~einen Hinweis~~, Hausaufgaben, ~~ein Auto~~, das Rauchen, das Geschäft, den Widerstand
　eröffnen:
　ein Konto, ~~eine Dose~~, einen Laden, ein Lokal, ~~sein Herz~~, eine Autobahn, neue Perspektiven, ein Testament
　ausgeben:
　Geld, Essen, eine Runde Bier, ~~ein Wort~~, einen Befehl
　stark:
　ein Charakter, ein Glaube, Nerven, eine Brille, ~~eine Jacke~~, ~~ein Regal~~, ~~Apfelsine~~, Zigaretten, Kaffee, Verkehr

4.1
Streit um Frauenquote

Zurzeit sind in Deutschland **mehr als** 90 % der Führungskräfte Männer. Um dem entgegenzuwirken, fordern viele die **sogenannte** Frauenquote. **Das heißt,** dass es bei der Besetzung von Führungspositionen eine Quotenregelung geben soll. Dieses Thema wird jedoch kontrovers diskutiert. Ein zentraler Streitpunkt der Gegner ist „Qualifikation statt Quote": Sie sind der Meinung, dass die Qualifikation eines Bewerbers zugunsten des Geschlechtes in den Hintergrund rücken würde. Ein Argument der Befürworter ist **unter anderem** dass weibliche Führungskräfte **in der Regel** häufiger als Männer auf Führungseigenschaften wie **zum Beispiel** „Inspiration" und „partizipative Entscheidungsfindung" zurückgreifen. Es gibt viele Argumente **sowohl** für, **als auch** gegen die Quote. Für mehr Chancengleichheit – ohne ein Geschlecht zu bevorzugen – könnte die anonyme Bewerbung sorgen. Dabei erhalten die Personalverantwortlichen **erst mal** nur Informationen über die beruflichen Qualifikationen der Bewerber. Bei der Entscheidung über die Einladung zum Bewerbungsgespräch kennen sie **weder** das Geschlecht, **noch** das Alter, den Namen, die Herkunft oder den Familienstand des Bewerbers. Eine Studie ergab, dass **vor allem** jüngere Frauen von diesem Bewerbungsverfahren profitieren würden.

4.2

z.B. = **zum Beispiel**	sog. = **sogenannt**	i.d.R. = **in der Regel**
v.a. = **vor allem**	usw. = **und so weiter**	z.T. = **zum Teil**
d.h. = **das heißt**	u.a. = **unter anderem**	z.Zt. / zz = **zurzeit**

4.5.2 Funktionsverbgefüge

1.1   Thema: **Führung**   Studienfach: **Wirtschaftswissenschaft, BWL**

1.2

> Um was soll es ganz grob gehen? Es wird um den Begriff der Führung gehen. Innerhalb dieses Führungskonzepts werden wir einige klassische Überlegungen **anstellen**. Gibt es Merkmale, die einen Unterschied zwischen dem Führenden und dem nicht Führenden **machen**? Wenn man nämlich erkannt hat, welche Persönlichkeitsmerkmale bei einem Führenden eine große Rolle **spielen**, dann kann man sie gezielt bei Leuten suchen. Und da sind wir jetzt bei den Methoden für die Auswahl von Führungskräften oder von Führungsnachwuchskräften. In vier Jahren werden viele von Ihnen vermutlich in eine eignungsdiagnostische Situation geraten. Man macht Assessment Center mit Ihnen, man testet Sie, man **führt** Gespräche mit Ihnen, man **setzt** sie unter Druck, um zu sehen, ob Sie in der Lage **sind**, einmal eine Führungskraft zu werden. Diese Methoden werden wir zur Diskussion **stellen**.

1.3

Funktionsverbgefüge	Entsprechung
eine Überlegung anstellen	überlegen
einen Unterschied machen zwischen	unterscheiden zwischen
eine Rolle spielen	wichtig sein
Gespräch(e) führen	sprechen
unter Druck setzen	bedrängen
in der Lage sein	fähig sein
zur Diskussion stellen	diskutieren

2   **c**   Auf einer Lehrertagung bringen die Lehrer unterschiedliche Herausforderungen des neuen Systems zur Sprache. > **Anfang und Verursacher (die Lehrer) der Handlung**

     d   Auch das Problem des Lehrermangels kommt zur Sprache. > **Anfang der Handlung**

     e   Die Schülersprecher kommen auch miteinander ins Gespräch.> **Anfang der Handlung**

     f   Den Schülern steht seit dieser Reform nicht genug Freizeit zur Verfügung. > **Dauer der Handlung**

     g   Die Eltern stellen dieses Problem noch einmal grundsätzlich zur Debatte.
        > **Anfang und Verursacher (die Eltern) der Handlung**

     h   Zur Debatte steht seit Jahren auch die Zahl der verfügbaren Studienplätze. > **Dauer der Handlung**

     i   Die Behörden haben das alte Schul- und Gymnasiumgesetz außer Kraft gesetzt.
        > **Ende und Verursacher (die Behörden) der Handlung**

     k   Das neue Gesetz tritt in Kraft. > **Anfang der Handlung**

     l   Das neue Gesetz ist seit ein paar Monaten in Kraft. > **Dauer der Handlung**

3.1   **a**   Unterstützung erfahren: **unterstützt werden**   **f**   Bestätigung erfahren: **bestätigt werden**

     b   Verständnis finden: **verstanden werden**   **g**   den Entschluss fassen: **entschließen**

     c   zum Ausdruck bringen: **ausdrücken**   **h**   das Versprechen geben: **versprechen**

     d   in Gefahr bringen: **gefährden**   **i**   aufs Spiel setzen: **riskieren**

     e   sich zur Wehr setzen: **wehren**   **k**   Verantwortung tragen für: **verantworten**

3.2   > *Viele Menschen drücken ihre Abneigung gegen Rauchen in der Öffentlichkeit deutlich aus. Sie sagen, dass das passive Rauchen ihre Gesundheit gefährdet. Diese Klage ist durch wissenschaftliche Untersuchungen schon lange bestätigt worden. Schon seit den 70ern versuchen sich Nicht-Raucher durch organisierte Initiativen zu wehren. Die Nicht-Raucher werden seit einigen Jahren durch verschieden starke Rauchverbote in den Bundesländern von den Landesregierungen unterstützt. Außerdem entschloss man sich 2007, das Rauchen in Taxis, öffentlichen Gebäuden und in Zügen bundesweit zu verbieten. Seitdem verantworten die Raucher ordnungswidriges Handeln. Denn wer sich an diesen Orten trotzdem eine Zigarette anzündet, der riskiert eine Geldstrafe bis zu 1000 EUR. Das wird von vielen nicht verstanden, da sie das Rauchverbot als Eingriff in ihre Privatsphäre empfinden. Doch das Bundesgesundheitsministerium hat versprochen, die Gesundheit aller Bürger zu schützen.*

5 Zeiträume

5.1 Gegenwart

1.1 Lesen Sie den Text und markieren sie die Verben im Präsens.

Wölfe in Deutschland

Es **gibt** sie wieder: Seit Ende der neunziger Jahre **leben** wieder Wölfe in Deutschland. Inzwischen **sind** nach Angaben von Naturschützern rund 60 Wölfe bei uns heimisch. Die zwölf nachgewiesenen Rudel **leben** vor allem in Sachsen, Sachsen-Anhalt und Brandenburg.[1]

Die graubraunen europäischen Wölfe (Canis lupus) **leben** im Familienverband, dem Rudel. Auf ihren Streifzügen **legen** Wölfe oft 40 Kilometer oder mehr in einer Nacht **zurück**. Sie **beanspruchen** große Reviere, wo sie vor allem Hirsche, Rehe und Wildschweine **jagen**.[2]

Wölfe **sehen** den Schäferhunden ähnlich, sind aber kräftiger, **haben** längere Beine und einen kürzeren Hals. Wölfe **sind** 110 bis 140 Zentimeter lang, der buschige Schwanz **misst** zusätzlich 30 bis 40 Zentimeter. Sie werden 65 bis 80 Zentimeter hoch und **wiegen** zwischen 25 und 50 Kilogramm. Das Fell der europäischen Wölfe **ist** dunkelgrau bis dunkelbraun und mit einigen gelblich-blonden Haaren durchsetzt.[3]

Wölfe **sind** sehr vorsichtig und **meiden** Menschen gewöhnlich. Selbst Wissenschaftler, Förster und Jäger **bekommen** sie nur selten zu Gesicht. Wolfsforscher **müssen** daher sehr gute Spurenleser **sein**, um Hinweise auf Wölfe zu bekommen.[4]

Die Rückkehr des Wolfes 150 Jahre nach seiner Ausrottung **ist** ein erster Erfolg für den Artenschutz, denn seit die Wölfe nicht mehr geschossen werden dürfen, **leben** sie wieder bei uns. Vielerorts **haben** die Menschen noch Vorurteile gegenüber Wölfen – Rotkäppchen **lässt grüßen**. Den Märchen und Legenden **begegnet** der NABU mit vielen sachlichen Informationen über das seltene Säugetier.[5]

1.2 In Abschnitt 1 und 5 wird von aktuellen Entwicklungen und Zuständen berichtet, in den Abschnitten 2, 3 und 4 bekommt der/die Leser/in allgemeingültige Informationen über den europäischen Wolf.

1.3 Beispiellösung:

Mein Lieblingstier ist der Gepard. Geparden sind Raubtiere, sie haben einen schlanken Körper und lange Beine. Sie messen vom Kopf bis zur Schwanzspitze bis zu 220 Zentimeter, die Schulterhöhe beträgt bis zu 80 Zentimeter. Trotz dieser Größe wiegen sie nur ca. 60 Kilogramm. Ihr Fell ist bräunlich-gelb gefärbt und besitzt schwarze Flecken. Typisch sind auch die schwarzen Streifen im Gesicht, die sogenannten Tränenstreifen.

Geparden gelten als die schnellsten Landsäugetiere, sie erreichen Geschwindigkeiten von über 100 Kilometer pro Stunde. Allerdings können Geparden dieses Tempo nicht lange halten, schon nach etwa 600 Metern werden sie langsamer.

Aufgrund ihres schlanken Körperbaus haben Geparden keine großen Reserven, sie brauchen deshalb viel regelmäßiger Nahrung als andere Raubtiere und müssen häufiger jagen.

Früher waren Geparden in ganz Afrika sowie in Vorderasien und Teilen Zentralasiens verbreitet, heute leben sie nur noch südlich der Sahara in freier Wildbahn.

Die Raubkatzen werden zwar oft in Zoos gehalten, pflanzen sich dort aber kaum fort.

2.2

PETRA: Was ist mit dir los, seit 20 Minuten blätterst du in diesem langweiligen Möbelhauskatalog?

JULIA: **Ich bin gerade dabei**, mein Wohnzimmer **zu renovieren**. Die Dielen habe ich schon abgeschliffen, jetzt sind die Wände dran!

PETRA: Ich **bin auch am Überlegen**, ob ich meine Wände neu streichen sollte … Hast du schon Ideen?

JULIA: Meine alten Möbel möchte ich behalten, ich wollte erst mal Jens fragen, ob er Lust auf rote Wände hat, aber der ist zurzeit nicht ansprechbar.

PETRA: Was ist denn los mit Jens?

JULIA: Dauernd **ist er beim Joggen**, in zwei Monaten ist der Berlin-Marathon, da will er fit sein.

PETRA: Also ich finde, dass rote Wände gut zu deinen dunklen Möbeln passen, das ist auf jeden Fall mal was anderes als dieses Sonnenblumengelb überall …

3 Beispiellösung:

Person 1: Hallo Doreen, ich bin's, Gabi. Schön, dass ich dich erreiche.

Person 2: Hallo Gabi!

Person 1: Annika hat doch gerade Abitur gemacht und sie war richtig gut.

Person 2: Du, Gabi …

Person 1: Ach was sag ich, sie hat ein hervorragendes Abitur gemacht, als Klassenbeste! Jetzt ist sie dabei, sich an den besten Unis zu bewerben.

Person 2: Wow, das freut mich sehr. Bei mir ist es gerade etwas schlecht, ich warte dringend auf den Rückruf des Klempners. Unsere Toilette funktioniert nicht – Jonathan, hör auf, deine Schwester zu ärgern!

Person 1: Ach so, das ist doch kein Problem, ich bin nur so stolz auf Annika, sie macht uns soviel Freude und …

Person 2: Außerdem bin ich am Kochen, du, kann ich dich später zurückrufen?

Person 1: Ja, das geht auch. Ich bin nur heute Abend nicht da, wir gehen mit der ganzen Familie essen, weil …

Person 2: Alles klar, also dann bis später! Tschüss.

4 Bei allen sechs Beispielen werden mit dem Perfekt abgeschlossene Vorgänge ausgedrückt, jedoch sind nicht diese Vorgänge für die Kommunikation wichtig, sondern der Zustand zum Sprechzeitpunkt:

A: Manuela ist jetzt im Büro.

B: Das Abiturzeugnis von Lukas ist besser als gedacht.

C: Christoph ist schuld, dass wir jetzt keine Schokolade naschen können.

D: Die Kleine schläft jetzt endlich.

E: Peter ist jetzt mindestens 5 cm größer als vor einem Jahr.

F: Jetzt liegt Schnee und man kann Schlitten fahren.

5.1

Sehr geehrter Herr Dr. Bäumer,

zu meinem Geburtstag habe ich von meiner Frau ein Rennrad geschenkt bekommen und ich will damit **im** nächsten Sommer die Alpen überqueren. Leider steht das Fahrrad **seit** Monaten im Keller und langweilt sich. Nur **ab und zu** drehe ich mit dem Rad eine Feierabendrunde, **fast nie** mache ich eine längere Tour. **Meistens** setze ich mich **nach** der Arbeit zu Hause aufs Sofa, trinke ein Bier und esse Kartoffelchips. **Jedes Mal** wenn ich meine Sportkleidung anziehe, klingelt das Telefon, oder es fängt an zu regnen. **Bislang** war ich also sehr faul, aber **ab** März will ich endlich ernsthaft trainieren: Ich brauche **jetzt** einen Trainingsplan und Tipps zur richtigen Ernährung. Können Sie mir helfen?

5.2   Beispiellösung:

Lieber Herr Berg,

zuerst sollten Sie aufhören, faule Ausreden zu erfinden: Ein Telefonat kann man
verschieben und auch bei Regen kann man trainieren. Gehen Sie dann zu einem Arzt und
lassen Sie sich gründlich durchchecken! Danach müssen Sie sich ein realistisches Ziel
setzen: Eine Alpenüberquerung im Alleingang würde ich eher durchtrainierten
Leistungssportlern empfehlen. Es gibt mittlerweile jedoch auch begleitete Routen ohne
extrem steile Abfahrten und Steigungen. Melden Sie sich jetzt für eine Tour an, dann
haben Sie einen guten Ansporn zu trainieren und können nicht so leicht kneifen.
Wenn Sie gesund sind, sollten Sie jetzt Ihre Grundausdauer trainieren: Dazu fahren Sie
fünfmal in der Woche drei bis vier Stunden Rad (Pulsbelastung: 60-70 Prozent,
Trittfrequenz: 85-100). Wählen Sie anfangs nur flache bis leicht wellige Strecken.
Montags und donnerstags sollten Sie jeweils einen Ruhetag einlegen.
Nach einem Monat können Sie die Pulsbelastung etwas erhöhen und einmal wöchentlich
eine längere Tour von bis zu sechs Stunden in den Trainingsplan einbauen. Außerdem
sollten Sie beginnen, Strecken mit Steigungen zu fahren.
Achten Sie auch auf Ihre Ernährung: Mittags können Sie Kohlenhydrate zu sich nehmen,
abends sollten Sie darauf verzichten, besonders in der Kombination mit Fett, also Finger
weg von den Kartoffelchips!

Viel Erfolg bei Ihrem Training und alles Gute!
Ihr
Dr. Bäumer

5.2   Vergangenheit

5.2.1   Vergangenheit in der geschriebenen Sprache

1.1   **1. Zeitschriftenartikel**   **3. Mahnung**   **5. Zeitungsmeldung**
       2. literarische Erzählung   **4. E-Mail**

1.2

❶

Sie **hatte** auf der Couch eines Psychotherapeuten **gelegen**, spiritistische Sitzungen **besucht** und in Kirchen **gebetet**. Als ihre Depression trotzdem nicht besser **wurde**, **suchte** Sabine Wolter im Internet nach Hilfe und **stieß** auf eine Website, die Besserung **versprach** – innerhalb weniger Monate, ohne Therapeuten, mit einem automatischen Behandlungsprogramm.

„Ich **war** so verzweifelt, dass ich mich darauf **eingelassen habe**. Obwohl ich mir überhaupt nicht **vorstellen konnte**, wie das funktionieren **sollte**", erinnert sie sich. „Jahrelang **habe** ich tagsüber als Verkäuferin und abends als Köchin in einem Restaurant **gearbeitet**. Irgendwann **bin** ich einfach **zusammengebrochen**."

Zwar **verschrieb** ihr ein Psychiater Tabletten, doch wirklich besser **wurde** es erst mit dem Online-Selbsthilfeprogramm. „Ich **habe** sehr schnell Vertrauen zu dem Programm **gefasst** und **gedacht**: Das könnte funktionieren!", sagt sie.

Sabine Wolter **hat** die Hilfe aus dem Internet **überzeugt**, allerdings stoßen Onlinebehandlungen besonders bei Problemen mit tief liegenden Ursachen schnell an ihre Grenzen. Eine viel versprechende Variante könnte die Kombination von klassischer und Internet-Therapie darstellen.

❷

Das Rad an meines Vaters Mühle **brauste** und **rauschte** schon wieder recht lustig, der Schnee **tröpfelte** emsig vom Dache, die Sperlinge **zwitscherten** und **tummelten** sich dazwischen; ich **saß** auf der Türschwelle und **wischte** mir den Schlaf aus den Augen; mir **war** so recht wohl in dem warmen Sonnenscheine. Da **trat** der Vater aus dem Hause; er **hatte** schon seit Tagesanbruch in der Mühle **rumort** und die Schlafmütze schief auf dem Kopfe, der **sagte** zu mir: „Du Taugenichts! [...]

❸

Sehr geehrte Damen und Herren,

ich **habe** am 13.7.2010 bei Ihnen die Lieferung und Montage der Einbauküche ‚Akkurat' zu einem Kaufpreis von 5999 Euro **bestellt** und 3000 Euro Anzahlung **geleistet**. Mit dem Schreiben vom 15.8.2010 **habe** ich Ihnen eine Lieferfrist bis zum 1.9.2010 **eingeräumt**. Diese Lieferfrist **ist** inzwischen ungenutzt **verstrichen**. Damit befinden Sie sich im Lieferverzug. Ich setze Ihnen daher eine Nachfrist von 3 Wochen bis zum 21.9.2010 und erkläre bereits jetzt für den Fall, dass auch dieser Termin ungenützt verstreichen sollte, meinen Rücktritt vom Kaufvertrag.

Mit freundlichen Grüßen

Renate Musterfrau

❹ `_ □ X`

Hi!
Bin grad aus Tallinn **zurückgekommen**. Wir **sind** morgens mit der Fähre von Helsinki nach Tallinn **gefahren** – **war** eine tolle Fahrt durch das Eis (Minna **hat** auch Fotos **gemacht**…). In Tallinn **hat's geschneit, war** aber nicht so kalt wie in Helsinki („nur" minus 10 Grad) – die Finnen **sind** alle mit ihren großen Einkaufstaschen in den Supermärkten **verschwunden**, wir **haben** dann die schöne Altstadt **besichtigt** und in tollen Keramikläden hübsche Geschenke **gekauft**. Leckere Schokolade **gab's** auch – wenn du mich morgen Abend vom Flughafen abholst, bekommst du vielleicht ein Stückchen davon ab…
Na dann bis morgen, Küsschen, Maus

❺

1. Mai: Weniger Gewalt

In Berlin-Kreuzberg **ist** die Nacht nach den 1.-Mai-Demonstrationen ohne größere Zwischenfälle **verlaufen**. Nach Mitternacht **räumte** die Polizei das Gelände um das Kottbusser Tor. Dort **waren** aus einer Gruppe von etwa 1500 Menschen immer wieder Polizisten **angegriffen worden**. Es **gab** nach Angaben der Polizei auch mehrere Festnahmen. Bei einer Demonstration **flogen** Flaschen, Steine und Feuerwerkskörper gegen Polizisten. Polizeifahrzeuge **wurden attackiert** und Geschäfte **angegriffen**. Brennende Barrikaden wie in der Vergangenheit **gab** es jedoch nicht. Mehrere Polizisten **wurden** in einem Wartehäuschen **eingekesselt**. Die Polizei **sprach** von 9000 Teilnehmern, die Veranstalter **gaben** 13.000 Demonstranten **an**.

1.3   Ausdruck von Vergangenheit in schriftlichen Texten

- *Haben* und *sein* als Vollverben sowie die Modalverben werden meistens im **Präteritum** verwendet.
- In schriftlichen Texten ist das **Präteritum** das übliche Tempus, in dem eher distanziert und entspannt erzählt wird. Typische Textsorten sind Märchen, Romane, Erzählungen sowie Berichte und Nachrichten.
- In Texten mit argumentativem Charakter, Mahn- und Beschwerdebriefe sowie in wissenschaftlichen Texten benutzt man häufig das **Perfekt**. Es signalisiert: Bei dem Thema handelt es sich um etwas Kontroverses bzw. Wichtiges, das uns alle angeht!
- Am Anfang von journalistischen Texten wird der Inhalt häufig schon kurz im **Perfekt** zusammengefasst, um die Aktualität und Relevanz des Ereignisses zu signalisieren. Die Einzelheiten werden dann meist in ihrem Ablauf im Präteritum geschildert.
- Das **Perfekt** wird auch bei der schriftlichen Wiedergabe der gesprochenen Sprache und in schriftlichen Texten mit informellem Charakter verwendet, vor allem dann, wenn das geschilderte Ereignis wichtig für die Gegenwart ist. Typisch sind E-Mails, Blogs, Notizen und informelle private Briefe.
- Das **Plusquamperfekt** drückt die Vorzeitigkeit eines Vorgangs, Zustands oder einer Handlung gegenüber einem anderen Ereignis in der Vergangenheit aus, liegt zeitlich also vor der Aussage im Präteritum oder Perfekt. Signalwörter für die Verwendung des **Plusquamperfekt** sind Konnektoren wie *nachdem, sobald, vorher, zuvor*.

1.4   a: Das Plusquamperfekt drückt Vorzeitigkeit der Ereignisse aus: Noch bevor Sabine Wolter Hilfe im Internet fand, hatte sie auf der Couch eines Psychotherapeuten gelegen, spiritistische Sitzungen besucht und in Kirchen gebetet.

b: Im Präteritum wird im eher neutralen Zeitschriftenartikel von Frau Wolters Handlungen berichtet.

c: Die gesprochene Sprache von Frau Wolter wird im Perfekt wiedergegeben („eingelassen habe"). ‚Sein' und die Modalverben werden jedoch auch in der gesprochen Sprache, die hier wiedergeben wird, meist im Präteritum verwendet.

d: Präteritum wird als typisches Erzähltempus in der literarischen Erzählung verwendet.

e: Die Verwendung des Perfekts zu Beginn des formellen Briefs (Mahnung) zeigt an, dass es um etwas Wichtiges geht.

f: Das Perfekt ist typisch für informelle, private Texte wie die E-Mail, es ist von aktueller Relevanz, dass „Maus" wieder da ist.

g: Der erste Satz im Zeitungsartikel fasst den Inhalt kurz zusammen. Perfekt signalisiert die Aktualität und Relevanz des Ereignisses.

2.1   Baron Münchhausen ist als „Lügenbaron" berühmt geworden, am bekanntesten ist seine Erzählung vom Ritt auf einer Kanonenkugel.

2.2

Doktor Münchhausen

von Simone Utler

Mit einem zusammengesponnenen Lebenslauf und gefälschten Zeugnissen <u>schlich</u> sich der Banker Christian E. als Assistenzarzt in eine chirurgische Klinik ein. Erst nach 14 Monaten <u>brach</u> sein Lügengebäude zusammen – ein anonymer Tipp <u>ließ</u> den falschen Doktor <u>auffliegen</u>.

Der weiße Kittel <u>war</u> für Christian E. ein Symbol – für Erfolg, für Anerkennung und für die Möglichkeit, Kranken den Klinikalltag zu erleichtern. Er <u>versorgte</u> Patienten, <u>war</u> bei fast 190 Operationen dabei und <u>schulte</u> sogar OP-Kräfte. Doch jedes Mal, wenn er den Kittel <u>überzog</u>, <u>spürte</u> er auch seine Angst. Die Angst, einen Fehler zu machen, einen Patienten zu verletzen – und aufzufliegen.

In seinem ersten und echten Leben <u>arbeitet</u> Christian E. als Banker. Nach seinem Realschulabschluss 1995 <u>macht</u> der hochgewachsene Dunkelhaarige eine Ausbildung zum Bankkaufmann – so wie seine beiden älteren Geschwister. Er <u>wird</u> Wertpapierhändler, <u>verdient</u> gut und <u>kauft</u> sich eine Eigentumswohnung.

Doch dann <u>lernt</u> E. eine andere Welt kennen, jenseits von Ölkontrakten und Wetten auf Getreidepreise. Während seines Zivildienstes bei den Maltesern <u>arbeitet</u> er mit Menschen mit Idealen und Engagement. Das <u>imponiert</u> ihm dermaßen, dass er sich auch ehrenamtlich <u>engagiert</u>, in der Altenbetreuung, als Sanitäter bei Festen, später im Rettungsdienst. E. <u>gilt</u> als begabt und <u>wird</u> gefördert. Irgendwann <u>will</u> er Arzt <u>werden</u> [...]

› Der Wechsel ins Präsens steigert die Spannung.

4

Ich **bin** wegen der Liebe nach Deutschland **gekommen**. Meine Freundin und ich **haben** uns auf einer Party von Freunden in Berlin **kennengelernt**. Ein Jahr lang hatten wir eine Fernbeziehung – wir **haben** uns nur an langen Wochenenden und in den Ferien **gesehen**. Dann **bin** ich im April zu ihr nach Hamburg **gezogen** Die ersten Wochen **haben** mir Angst **gemacht**: Es **hat** fast ununterbrochen **gestürmt** und **geregnet**, die Leute auf der Straße **sind** schnell aneinander **vorbeigelaufen** und **haben** fast nie miteinander **geredet**. Dann kam endlich der Sommer und ich **bin** jeden Tag spazieren **gegangen** und **habe** die Altstadt, St. Pauli und natürlich den Hafen **entdeckt** – ich liebe den Fischmarkt am Sonntagmorgen! Als meine Freundin Urlaub hatte, **sind** wir an die nordfriesische Küste **gereist**. Wir hatten viel Spaß, **sind** viel mit dem Fahrrad **gefahren**, **haben** Sandburgen **gebaut** und **sind** jeden Tag im eiskalten Wasser **geschwommen**. Vor 2 Tagen **sind** wir nach Hamburg **zurückgekehrt**, jetzt beginnt für uns wieder der Ernst des Lebens: Meine Freundin sitzt schon in ihrem Büro und ich muss mich ernsthaft auf mein Studium an der Hamburger Universität vorbereiten …

5

Betreff: Reise nach Olbia   Buchungsnummer: 334455

Anspruchsanmeldung

Sehr geehrte Damen und Herren,

wir haben bei Ihnen **am** 4. April 2011 eine Reise für 2 Personen nach Olbia auf Sardinien gebucht. In dem gebuchten Hotel wurden **während** unseres Aufenthaltes **von** 7 Uhr **bis** 22 Uhr Bauarbeiten innerhalb und außerhalb des Hauses durchgeführt, die eine Erholung stark eingeschränkt haben. Auch **in** der Nacht sind wir wegen der lauten Tanzmusik kaum zur Ruhe gekommen. **Am** 2. Juli 2011 **um** 11 Uhr haben wir Ihrer örtlichen Reiseleitung, Frau Sauer, den Mangel angezeigt. Wir haben ihr eine Frist gesetzt, den Mangel **innerhalb** von 24 Stunden zu beseitigen. Sie konnte uns aber **bis zum** Ende unseres Urlaubs kein Zimmer in einem anderen Hotel zuweisen.

Die Reise war durch den ständigen Lärm erheblich beeinträchtigt. Wir verlangen daher eine Minderung des Reisepreises um mindestens 30 %. Wir setzen Ihnen zur Erledigung der Angelegenheit eine Frist **bis zum** 1.10.2011.

Mit freundlichen Grüßen
Käthe und Heinrich Lehmann

6   Beispiellösung:

> Hallo Jona,
>
> danke für deinen Tipp mit dem Gardasee! Ich bin seit gestern wieder da und hab mich
> fantastisch erholt. Eigentlich wollte ich ja vor allem Mountainbike fahren, aber tagsüber
> war es selbst mir zu heiß. Also lag ich bis zum späten Nachmittag am See und bin
> geschwommen. Abends habe ich mich dann immer aufs Rad geschwungen.
> Wie war denn dein Urlaub? Wollen wir morgen zusammen ne Radtour machen? Ich
> könnte am Samstag den ganzen Tag, am Sonntag bis 18 Uhr.
> Liebe Grüße
> Micha

7   Beispiellösung:

> Sehr geehrte Damen und Herren,
>
> am 17.09. habe ich mich in Ihrer Sprachschule angemeldet und besuche seit gestern den
> A2-Sprachkurs. Eigentlich wurde mir bei der Anmeldung letzte Woche mitgeteilt, dass
> ich noch einen Einstufungstest absolvieren soll, aber dazu ist es nie gekommen.
> Ich bitte dringend um die Versetzung in einen Kurs mit höherem Sprachniveau! Gern bin
> ich bereit, noch diese Woche einen Einstufungstest abzulegen. Spätestens ab nächster
> Woche möchte ich dann einen passenden Kurs besuchen.
>
> Mit freundlichen Grüßen
> Tatjana Flath

5.2.2   Vergangenheit in der gesprochenen Sprache

1.2

> REPORTER: Haben Sie mit dem Mauerbau gerechnet?
> PETRA MÜLLER: Nein, überhaupt nicht. Das **war** für mich vollkommen überraschend.
> Ich **hab** drüben im Osten gelebt und zufälligerweise **war** ich am 12.
> August, einen Tag vor dem Mauerbau, am Gesund-brunnen im Kino.
> Ich **hab** in Pankow **gewohnt** und da **war** der Bahnhof Gesundbrunnen
> der Anlaufspunkt für uns, wenn man einkaufen **wollte**. Als ich nach
> dem Kino auf den Bahnhof Gesundbrunnen **kam**, **hab** ich mich
> **gewundert**, dass da 'ne Menschenmauer **stand**. Alle **standen** da und es
> **kam** kein Zug. Es **kam** auch keine Ansage, keine Durchsage und wir
> **haben** uns **gefragt**, warum nichts **passierte**. Ich **hab's** mir eine halbe
> Stunde **angehört** und dann dachte ich: „Läufst du zu Fuß nach Hause
> – das dauert auch nur 'ne halbe Stunde." Gesagt, getan. Ich **komme** in
> der Wollankstraße an die Grenze und **seh** da schon
> Maschendrahtzäune. Ich **bin** dann nach Hause **gegangen** und da **hab**
> ich im Radio **gehört**, dass der Grenzverkehr nicht mehr möglich ist …

1.3   Sie verwendet Präteritum bei „sein", Modalverben (wollte) und häufig verwendeten Verben (kam, stand, standen), ansonsten verwendet sie in ihrer Erzählung das Perfekt.

1.4   Ausdruck von Vergangenheit in der gesprochenen Sprache

- Bei den meisten Verben wird das **Perfekt** benutzt, um von Vergangenem zu berichten.
  Ich hab drüben im Osten gelebt ...
  Ich hab in Pankow gewohnt ...
  ... hab ich mich gewundert, ...
  ... wir haben uns gefragt, ...
  Ich hab's mir eine halbe Stunde angehört ...
  Ich bin dann nach Hause gegangen und da hab ich im Radio gehört, ...
- Die Verben *haben* und *sein* stehen häufig im **Präteritum**.
  Das war für mich vollkommen überraschend
  ... zufälligerweise war ich am 12. August ... im Kino.
  ... da war der Bahnhof Gesundbrunnen der Anlaufspunkt für uns
- Die Modalverben stehen ebenfalls im **Präteritum**.
  ... wenn man einkaufen wollte.
- Auch einige häufig verwendete Verben wie *denken, geben (es gab), gehen, heißen, kennen, kommen, laufen, meinen, sitzen, stehen, wissen* stehen meistens im **Präteritum**
  Als ich nach dem Kino auf den Banhof Gesundbrunnen kam, ...
  ... dass da 'ne Menschenmauer stand.
  Alle standen da und es kam kein Zug. Es kam auch keine Ansage ...
- Auch das **Präsens** wird zur Schilderung von vergangenen Ereignissen verwendet, damit soll die Darstellung besonders ‚lebendig' gestaltet werden.
  Ich komme in der Wolkanstraße an die Grenze und seh da schon Maschendrahtzäune.

2

Wir **hatten** es nicht leicht nach dem Krieg, ich **dachte** oft, wir schaffen das nicht. Opa **war** in der Kriegsgefangenschaft in Amerika und ich **musste** zwei kleine Kinder durch den Winter bringen. Schon im Dezember **waren** die Kartoffeln alle und dann **gab** es nur noch Steckrüben: Montags Steckrübeneintopf, dienstags Steckrübeneintopf, mittwochs Steckrübeneintopf, ich **konnte** das Zeug nicht mehr sehen! Gott sei Dank **brachte** Tante Marie vom Landweg uns jede Woche eine Kanne Milch und einen Laib Brot. Das Brot **war** köstlich, die **backten** das da noch selbst. Ja, ja, die Zeiten **waren** hart, aber man **half** sich doch, wo man **konnte** ...

3   Beispiellösung:

Erinnerungen an den Mauerfall

Petra Müller lebte 1989 mit ihrem Mann in Magdeburg. Sie konnte es damals, wie wahrscheinlich die meisten, kaum glauben, dass man nun endlich „rüber" durfte. Ihr Sohn, der zu der Zeit in Berlin studierte, rief mitten in der Nacht an und erzählte überglücklich, dass er im „Westen" war.
Sie ging am nächsten Tag nach der Arbeit gleich zur Polizei und holte sich ihren Visa-Stempel im Personalausweis. Am folgenden Tag ging es los: Die ganze Familie wartete schon mit dem Trabant und die Fahrt zu den Verwandten in West-Berlin begann. An der Grenze mussten sie stundenlang warten, aber in Berlin wurden sie mit Freudentränen empfangen und es wurde ein sehr langer Abend mit der ganzen Familie.

5.3 Zukunft

1.2

JANINE: Hi Mike, ich sitz' grad' in der U-Bahn, was **machst** du morgen Abend so?

MIKE: Ich **geh** morgen auf Kevins Party, da **spielt** die Band von meiner Schwester.

JANINE: Ohje … Wer **kommt** denn sonst noch so?

MIKE: Uwe, Silke, Murat, Melanie, Kati – die ganze Bande **erwartet** dich.

JANINE: Und Andy?

MIKE: Der **kommt** auch und zwar mit seiner neuen Flamme*!

JANINE: Wer ist denn die Glückliche?

MIKE: Manuela aus der 10b**.

JANINE: Ich dachte, Andy hätte ein bisschen mehr Geschmack … Na gut, ich **werde** mal kurz **vorbeischauen**.

MIKE: Okay, dann sehen wir uns da. Tschüss!

JANINE: Bis dann.

› Es wird überwiegend Präsens zum Ausdruck der Zukunft verwendet.

2

Zukunftsprognosen: Die Welt in 520 Wochen

Wie sieht die Welt im Jahr 2021 aus? Ein Zukunftsforscher, ein Historiker und ein Wissenschaftsjournalist spekulieren über die Schweiz von morgen.

Die Prognosen des Zukunftsforscher Lars Thomsen klingen utopisch: «Schon in wenigen Jahren **werden** Elektroautos billiger als Benzinfahrzeuge **sein** und mit erneuerbaren Energien wie Wasser-, Wind- und Sonnenkraft **angetrieben werden**.» Zudem **wird** sich das Problem mit Erdöl und Atomkraft in 10 Jahren von selbst **lösen**, meint der optimistische Wissenschaftler.

Für Daniele Ganser ist es unrealistisch zu glauben, dass die Solarenergie das Erdöl in zehn Jahren **ersetzen wird**. Die Energiewende **wird** erst **eintreten**, wenn der Erdölpreis drastisch steigt. Um die Umwelt und das Klima zu schützen, sieht der Historiker den Weg des sparsamen Umgangs mit Energie als realistisch an: «Wir sollten leichtere Autos fahren, die weniger Benzin benötigen und alte Häu-ser sollten neu isoliert und mit Wärme-pumpen, Sonnenenergie oder Fernhei-zungen geheizt werden. Dadurch könnte fast die Hälfte des Schweizer Erdölbe-darfs eingespart werden».

Der Wissenschaftsjournalist Marcel Hänggi stimmt dem zu und ergänzt, dass wir die Klimakatastrophe nur mit einer Änderung des Lebensstils in den indust-rialisierten Ländern abwenden können. Lars Thomsen widerspricht dem und sieht die Lösung in der Intelligenz der Technik. «Ich vertraue der Technologie mehr als dem Bewusstseinswandel der Menschen.» Die meisten Geräte des All-tags **werden** miteinander **vernetzt sein** und sich durchs Internet gegenseitig op-timieren und Energie einsparen. «Es be-ginnt die Zeit der schlauen Maschinen und in den nächsten 30 Jahren **werden** wir kaum mehr CO_2-Emissionen **produ-zieren**.» Große Windparks und Solar-Kraftwerke **werden** dafür erneuerbare Energien **liefern**.

3 Ausdruck von Zukunft

- Zum Ausdruck von zukünftigen Vorgängen und Zuständen wird im Deutschen häufig das **Präsens** benutzt. Um deutlich zu machen, dass von der Zukunft die Rede ist, werden Temporalangaben eingefügt (z. B. *morgen, nächste Woche, nach der Arbeit*). Oft erschließt sich der Zukunftsbezug auch durch den Kontext: Beim Telefongespräch in Aufgabe 1 ist klar, dass vom folgenden Abend gesprochen wird.
  ... was machst du morgen abend so?
  Ich geh morgen auf Kevins Party, da spielt die Band von meiner Schwester.
  Der kommt auch ...

- Auch das **Futur I** kann verwendet werden, um zukünftige Sachverhalte auszudrücken. Hier wird jedoch der Zukunftsbedeutung häufig eine modale Komponente hinzugefügt. Der Sprecher möchte deutlich machen, dass es sich bei der Aussage um ein Versprechen, eine Absicht oder um einen Vorsatz handelt und Versprechen, Absichten und Vorsätze sind immer mit einer gewissen Unsicherheit behaftet.
  Na gut, ich werde mal kurz vorbeischauen.

- Das **Futur I** wird auch für Vorhersagen und Prognosen benutzt, auch hier will der Sprecher keine Garantie dafür übernehmen, dass die Vorhersagen genau so eintreffen, wie prophezeit.
  Schon in wenigen Jahren werden Elektroautos billiger als Benzinfahrzeuge sein und mit erneuerbaren Energien ... angetrieben werden.
  Die Energiewende wird erst eintreten, wenn der Erdölpreis drastisch steigt.
  Große Windparks und Solar-Kraftwerke werden dafür erneuerbare Energien liefern.

- Das **Futur II** bezeichnet Vorgänge, die in der Zukunft abgeschlossen sein werden. Diese Form ist vor allem in der gesprochenen Sprache sehr selten und wird meist durch das Perfekt ersetzt.
  Die meisten Geräte des Alltags werden miteinander vernetzt sein.

4

KLARA: **Samstagabend** ist endlich **wieder** die Lange Nacht der Museen, hast du **schon** einen Plan für uns gemacht?

TOBIAS: Na klar, **zuerst** kaufen wir uns das Ticket und **dann / danach** geht's direkt ins Bode-Museum.

KLARA: Hat das denn **schon / wieder / noch** geöffnet, ich dachte, da wird **noch / schon / wieder** renoviert?

TOBIAS: Klar ist das auf und **bis** zum 31. Oktober kann man da dieses berühmte Bild von Leonardo da Vinci sehen, wie hieß das noch einmal?

KLARA: Du meinst die Frau mit dem Tier? Da müssen wir hin!

TOBIAS: Und **hinterher / danach** möchte ich in die Humboldt-Uni, **ab** 21 Uhr hält Professor Meierbusch einen Vortrag über die Kultur der Renaissance in Italien.

KLARA: Ach der alte Langweiler – gehen wir lieber **gleich** ins Neue Museum – die Cafeteria da ist toll!

TOBIAS: Dein Wunsch ist mir Befehl, aber **nach** dem Imbiss fahren wir in die Neue Nationalgalerie ...

KLARA: ... und **dann / danach / hinterher** lädst du mich **noch** in dieses tolle Restaurant am Landwehrkanal ein, ich freu' mich **schon** drauf!

5 Beispiellösung:

Person 1: Hallo Daniel, hier ist David.

Person 2: Hey David, lang nix mehr gehört. Was gibt's?

Person 1: Am Freitag geht das Melt! los, ich wollte eigentlich mit Tim hinfahren, aber der ist krank und hat keine Lust auf Zelten. Jetzt hab ich ein Ticket übrig.

Person 2: Melt!, was ist das noch mal für ein Festival? Rock?

Person 1: Hm, auch Rock, aber vor allem Elektro. Warst du nicht 2009 mit dabei, als Oasis gespielt hat? Ich war seit dem auch nicht mehr und werd diese Jahr auf jeden Fall hinfahren.

Person 2: Nee, da musste ich doch in die Abi-Nachprüfung …

Person 1: Ach ja, stimmt. Aber es war super damals.

Person 2: Wer spielt denn dieses Jahr?

Person 1: Gossip und Bloc Party, das sind die Bekanntesten. Und kennst du Caribou, den find ich richtig gut!

Person 2: Nee, aber ich werd ihn kennenlernen: Ich bin dabei! Bloc Party gefällt mir nämlich richtig gut.

Person 1: Genial, dann hol ich dich Freitagmittag ab. Hast du nen Schalfsack? Sonst bring ich Tims Zeug mit …

6 Beispiellösung:

Ich komme aus Brasilien und dort wird sich der Klimawandel schon sehr bald deutlich bemerkbar machen. Das Klima wird völlig außer Kontrolle geraten. Die Durchschnittstemperaturen im Sommer werden immer weiter ansteigen, dafür könnten die Winter extrem kalt werden. Das Land wird anfällig sein für Überschwemmungen und Dürren. Großteile des Landes werden entweder überschwemmt oder völlig austrocknen.

Die Caatinga wird sich in eine Wüste verwandeln, der Amazonas-Regenwald wird schrumpfen und schwere Dürren durchmachen. Es wird zu einem weitreichenden Artensterben kommen. Außerdem wird es immer häufiger Orkane geben, die Menschen und Landschaften noch viel stärker als heute bedroht werden.

6   Perspektiven
6.1   Handlung, Betroffene und Handelnde

1.1   › Man erfährt vom Mord an Karin Rosenherz im Jahr 1966, aber nichts über den Täter.
Außerdem gibt es eine Verbindung zu einem aktuellen Verbrechen, dem Raubüberfall, bei dem Tereza
verletzt wird (Opfer), über den Täter erfährt man auch im aktuellen Fall nichts.

1.2

	Handelnde(r)	Betroffene(r)	sprachliche Struktur
In einer heißen Augustnacht des Jahres 1966 wird in Frankfurt Karin Rosenherz ermordet.	?	**Karin Rosenherz**	wird … ermordet Passiv
An einem nebligen Morgen kommt es im Frankfurter Stadtwald bei einem Kunsttransport zum Raubüberfall.	?	?	Es kommt zu einem Überfall. Nomen-Verb-Verbindung
Hauptkommissar Marthalers Freundin Tereza wird dabei schwer verletzt.	?	**Tereza**	**wird … verletzt** **Passiv**
Robert Marthaler wird von den Ermittlungen ausgeschlossen.	?	**Robert Marthaler**	**wird … ausgeschlossen** **Passiv**
Marthaler sieht sich gezwungen, mit der jungen Journalistin Anna Buchwald zusammenzuarbeiten.	?	**Robert Marthaler**	sich sehen + Partizip II (Reflexivkonstruktion)

1.4   › Der Krimi soll für den Leser spannend sein, deshalb werden die Täter / Handelnden in der Buchvorstellung
nicht genannt. Damit man trotzdem eine Vorstellung von der Handlung bekommt, werden aber die
Betroffenen genannt.

2   **Warum wird der Handelnde nicht genannt?**

Es gibt verschiedene Gründe, warum der Handelnde in einem Text nicht genannt wird:
- Der Handelnde / Täter ist **nicht** bekannt.
- Der Handelnde / Täter ist **unwichtig** oder selbstverständlich.
- Der Handelnde / Täter wird **bewusst** nicht genannt. Man will nicht sagen, wer verantwortlich ist.
- Der Handelnde ist allgemein **bekannt** oder bereits aus dem Kontext bekannt.
- Es geht um allgemein **gültige** Sachverhalte und Wissensbestände.

3.1

Perspektive Empfänger	Perspektive Handelnder
Robert Marthaler *erhält* von einem Informanten den entscheidenden Tipp.	–
Robert Marthaler **wird von einem Informanten der entscheidende Tipp gegeben.**	Ein Informant gibt **Robert Marthaler den entscheidenden Tipp.**

3.2   [x] **bekommen** (gehoben)   [x] **kriegen** (umgangssprachlich)

85

4.1 Weitere sprachliche Mittel, mit denen man das Geschehen in den Mittelpunkt stellen kann, ohne den Handelnden zu nennen:
- sogenannte Passiversatzformen oder Passivalternativen, u.a. Konstruktionen mit *lassen* + *sich* + Infinitiv; *sein* + *zu* + Infinitiv; *sein* + Adjektive auf *-bar, -aber, -lich, -fähig*; *geben* + *zu* + Infinitiv; *es gilt* …
- Nomen-Verb-Konstruktionen in passivischer Bedeutung
- das unpersönliche Pronomen *man*
- das Pronomen *jemand*
- Passiv mit *sein* (Zustandspassiv) um das Resultat einer Handlung auszudrücken

4.2

○ Sie sind im Büro und Ihr Kollege informiert
○ Sie sofort:
○
○ Du wirst am Telefon verlangt
○ ~~Du wirst angerufen.~~
○ Es will dich jemand am Telefon sprechen.
○ Da will dich jemand sprechen.
○

○ Sie kommen erst am Nachmittag ins
○ Büro:
○
○ Heute hat jemand für dich angerufen.
○ ~~Du bist heute angerufen worden.~~

4.3 › Die Aussagen zum richtigen Verhalten bei Arbeitslosigkeit gelten nicht für eine spezielle Person, sondern für alle. Um dies auszudrücken, wir vor allem das unpersönliche Pronomen man verwendet.
Der / Die Arbeitslose muss sich sofort arbeitslos melden. Das Arbeitsamt kürzt sonst das Arbeitslosengeld und straft damit Arbeitslose.

4.4

erfahren + Nomen	*werden*-Passiv
Der Arbeitslose erfährt eine Kürzung des Arbeitslosengeldes.	Dem Arbeitslosen wird das Arbeitslosengeld gekürzt.
Leider erfährt der Beruf der Erzieher gesellschaftlich noch immer nicht ausreichend Anerkennung.	Leider **wird** der Beruf der Erzieher gesellschaftlich noch immer nicht ausreichend **anerkannt**.
Kosten- und Unternehmensplanung - wer sich in Unternehmen mit diesen Fragen auseinander setzen muss, **erfährt mit dem vorliegenden Handbuch eine praxisnahe Unterstützung**.	Kosten- und Unternehmensplanung – wer sich in Unternehmen mit diesen Fragen auseinander setzen muss, wird mit dem vorliegenden Handbuch praxisnah unterstützt.

Infinitivkonstruktion	*dass*-Satz
Es ist wichtig, sich arbeitslos zu melden.	Es ist wichtig, dass man sich arbeitslos meldet.
Man muss aufpassen, **andere mit Worten nicht unnötig zu verletzen**.	Man muss aufpassen, dass man andere mit Worten nicht unnötig verletzt.
Man sollte darauf achten, seine Energie für lohnende Ziele einzusetzen.	Man sollte darauf achten, dass **man seine Energie für lohnende Ziel einsetzt**.

6.2 Passiv mit *werden*

1   Fahrzeugbrief und Fahrzeugschein

Der Fahrzeugbrief ist eine amtliche Urkunde, mit der die allgemeine Zulassung eines Kraftfahrzeuges für den öffentlichen Straßenverkehr **bescheinigt wird**. Wechselt ein Auto den Besitzer, dann erhält der neue Fahrzeughalter auch den Fahrzeugbrief. Ein anderes Dokument ist der Fahrzeugschein, der für die konkrete Zulassung eines Fahrzeuges für jeden Autobesit-zer neu **ausgestellt wird**. **Wird** ein Auto **abgemeldet**, dann **wird** der Fahrzeug-schein **eingezogen**, der Fahrzeugbrief **wird** dagegen nur mit einem Abmeldevermerk **versehen**. Der Fahrzeugbrief bleibt also immer beim Auto und man kann dem Fahrzeugbrief entnehmen, wie viele Besitzer ein Auto schon gehabt hat.

› Im Mittelpunkt des Textes stehen die Dokumente Fahrzeugbrief und Fahrzeugschein. Es wird gesagt, was mit diesen Dokumenten getan wird. Welche Person das genau tut, ist nicht wichtig.

2.1

Adjektiv + Vollverb *werden*	vierstellig werden, **wertlos werden**
Adjektiv im Komparativ + Vollverb *werden*	blasser werden, **billiger werden**
werden + Partizip II = Passiv	erzählt werden, **geändert werden, demoliert werden, geprügelt werden, tiefergelegt werden, zerkratzt werden, umlackiert werden**

2.2 Beispiellösung:

Besitzerwechsel	verschenkt werden (an), verloren gehen, gefunden werden (von), **verkauft werden (an), vererbt werden (an), verliehen werden (an)** ...
Veränderung	wertlos werden, wertvoller werden, kaputt gehen, verbeult werden, Kratzer bekommen, **repariert werden, neu gestrichen werden,** ...

2.3 Beispiellösung:

Früher wurden manchmal Kleiderschränke oder Truhen zur Hochzeit an das Brautpaar verschenkt.
Möbel waren sehr wertvoll und wurden an die nächsten Generationen vererbt.
Natürlich bekamen die Schränke oder Truhen mit der Zeit Kratzer und andere kleine Schäden. Deshalb
wurden sie abgeschliffen oder neu gestrichen.
Da alte Möbel meist im Laufe der Zeit immer wertvoller werden, lohnt es sich, etwas Geld zu investieren
und sie entweder selbst zu restaurieren oder sie von Fachleuten aufarbeiten zu lassen.

3

Die Entscheidung liegt bei dir

Oder nehmen Sie Howard Schultz, der sich beim Überprüfen seiner Verkaufslisten
darüber wunderte, dass eine kleine, gerade mal vier Läden umfassende Firma große
Mengen Kaffeemaschinen bei seinem Unternehmen bestellte und offenbar zusammen
mit dem Kaffee verkaufte, der daraufhin nach Seattle flog, sich um einen Job bei der
Firma bewarb, **abgelehnt wurde**, sich 14 Monate lang einmal wöchentlich telefonisch in
Erinnerung brachte, **endlich den Job bekam**, sofort weitere Expresso-Bars in Kaufhäusern
eröffnete, **von den Inhabern deswegen gefeuert wurde**, bei 242 Kapitalgebern vorsprach,
von 217 abgewimmelt wurde, mit dem Geld der restlichen eine eigene Coffee-Shop-Kette
eröffnete, den Wettbewerb gewann und schließlich das Unternehmen kaufte.

› Es wird v. a. *werden*-Passiv und das Verb *bekommen* verwendet.

4.1

David Wagner: Vier Äpfel

L. und ich besuchten einmal ein Museum, in dem neben
anderen kuriosen Dingen auch alte Konservendosen
ausgestellt wurden. Die Exponate **durften angefasst werden**,
weshalb uns auffiel, dass das Haltbarkeitsdatum des
Mexikanischen Feuerzaubers, die Dose sah noch gut aus,
im Frühjahr 1988 **überschritten worden war** und das
Serbische Reisfleisch bis Ende 1985 **hätte verzehrt werden
sollen**, statt dessen war es, wir fanden das komisch, in
einem Museum gelandet, vor den Jugoslawienkriegen ist
es einmal ein populäres Gericht gewesen. Am besten gefiel
uns die Indonesische Reistafel, die aus zwölf kleinen
Konservendosen bestand, die im Wasserbad zu erhitzen
waren. Die Vorstellung, zwölf Dosen öffnen zu müssen,
hat allerdings etwas Abschreckendes, aber wahrscheinlich
gab es deshalb – meine Großmutter hatte einen –
elektrische Dosenöffner.

› In einem Museum ist es normalerweise nicht erlaubt, die Ausstellungsstücke anzufassen. Die
Erlaubnis wird deshalb mithilfe des Modalverbs *dürfen* explizit ausgedrückt. Das
Mindesthaltbarkeitsdatum auf den Dosen drückt eine Empfehlung aus, diese wird durch dem
Modalverb *sollen* zum Ausdruck gebracht.

Da nicht gesagt wird, wer die Erlaubnis und die Empfehlung ausgesprochen hat, die handelnden
Personen also nicht genannt werden, wird Passiv verwendet.
So kommt es im Text zu der Verwendung von Passiv mit Modalverben

4.2

Passiv im Nebensatz	
Präteritum	In einem Museum wurden alte Konservendosen ausgestellt.
	Wir besuchten ein Museum, in dem **alte Konservendosen ausgestellt wurden**.
Perfekt	Das Haltbarkeitsdatum ist im Frühjahr 1988 überschritten worden.
	Uns fällt auf, dass das Haltbarkeitsdatum **im Frühjahr 1988 überschritten worden ist**.
Plusquamperfekt	Das Haltbarkeitsdatum war im Frühjahr 1988 überschritten worden.
	Uns fiel auf, dass das Haltbarkeitsdatum **im Frühjahr 1988 überschritten worden war**.

Passiv im Nebensatz mit Modalverben	
Präteritum	Die Exponate durften angefasst werden.
	Es überraschte uns, dass die Exponate **angefasst werden durften**.
	Das Serbische Reisfleisch sollte bis Ende 1985 verzehrt werden.
	Es fiel uns auf, dass das Serbische Reisfleisch **bis Ende 1985 verzehrt werden sollte**.
	Bei der Indischen Reistafel mussten zwölf Dosen geöffnet werden.
	Es schreckte uns ab, dass bei der Indischen Reistafel **zwölf Dosen geöffnet werden mussten**.
	Die Indische Reistafel sollte im Wasserbad erhitzt werden.
	Am besten gefiel uns die Indische Reistafel, die **im Wasserbad erhitzt werden sollte**.
Perfekt (mit Konjunktiv)	Das Serbische Reisfleisch hätte bis Ende 1985 verzehrt werden sollen.
	Es fiel uns auf, dass das Serbische Reisfleisch **bis Ende 1985 hätte verzehrt werden sollen**.

4.3   *sein* + *zu* + Infinitiv

Die Reistafel bestand aus zwölf Dosen, die im Wasserbad zu erhitzen waren.
Die Reistafel bestand aus zwölf Dosen, die im Wasserbad **erhitzt werden sollten / mussten**.
- Die Konstruktion *sein* + *zu* + Infinitiv entspricht einer Passivkonstruktion mit den Modalverben *können*, **sollen** oder **müssen**.

> Der Satz enthält eine Notwendigkeit.

5   Beispiellösung:

Seit langem werden Placebos eingesetzt, um zu testen, ob ein Medikament wirkt oder nicht. Dabei werden Arzneimittel in Doppelblindstudien geprüft, bei denen weder der Arzt noch der Patient wissen, wer das Arzneimittel und wer das Placebo bekommt. Dadurch sollen falsche Untersuchungsergebnisse verhindert werden. Der Einsatz von Placebos ist seit dem Jahr 2002 in der Deklaration von Helsinki festgelegt. Diese Deklaration besagt, dass in Studien Placebos nur eingesetzt werden dürfen, wenn noch kein bewährtes Mittel auf dem Markt ist. Ansonsten muss das neue Medikament im Vergleich zum bereits vorhandenen Medikament getestet werden.

6.3   Resultate festhalten: Passiv mit *sein*

1 > Um den Ist-Zustand bzw. das Resultat auszudrucken, wird das Passiv mit *sein* (Zustandspassiv) verwendet.

Resultat	... und das wird / wurde gemacht
Die Zutatenliste ist oft sehr klein gedruckt.	Die Zutatenliste wird oft sehr klein gedruckt.
Die Zutaten **sind nach ihren Mengen aufgelistet.**	Die Zutaten **werden nach ihrer Menge aufgelistet.**
Das Produkt ist vielfältig industriell bearbeitet.	Das Produkt wurde vielfältig industriell bearbeitet.
Im offenen Verkauf sind Zutatenlisten nicht direkt am Produkt angebracht.	**Im offenen Verkauf werden Zutatenlisten nicht direkt am Produkt angebracht.**
Das Brötchen ist mit echtem Käse überbacken.	Das Brötchen wurde mit echtem Käse überbacken.

2

Handlung	Resultat
Die Messe wurde am 1. April eröffnet.	Die Messe ist seit dem 1. April eröffnet.
Sie ist vor zwei Tagen operiert worden.	Sie ist seit zwei Tagen operiert.
Das Stadion **wurde** vor fast **zwei Jahren geschlossen**.	Das Stadion ist seit fast zwei Jahren geschlossen.
Das Kind **wurde im Juni bei einer Familie in Köln untergebracht**.	Das Kind ist seit Juni bei einer Familie in Köln untergebracht.
Sie wurde vor 30 Jahren an der Universität angestellt.	Sie **ist seit 30 Jahren an der Universität angestellt**.

3   › Die Formulierung „Das Studium war mit dem akademischen Grad M.A. abgeschlossen." klingt wie eine
allgemeine Information zum Studium. In einem Bewerbungsschreiben geht es aber nicht um allgemeine
Aussagen, sondern darum, welchen Abschluss man selbst wann erworben hat.
(Würde man allgemein über das Studium an der Universität Jena informieren, würde man das Passiv
mit werden verwenden, da das sein-Passiv eine individuelle Situation voraussetzt, bei der nach
einer Handlung ein bestimmtes Resultat eingetreten ist.)

Auch Aussagen zum Thema der eigenen Magisterarbeit sollten nicht im Passiv formuliert werden.
Besonders zusammen mit dem Possessivartikel mein entsteht sonst der merkwürdige Eindruck,
dass bei der Arbeit eine andere handelnde Person eine Rolle gespielt hat. Weitere mögliche
Formulierungen – neben der in Version B – wären:
Meine Magisterarbeit habe ich zum Thema ... geschrieben.
In meiner Magisterarbeit habe ich mich mit ... auseinandergesetzt.

6.4 Unpersönliche Ausdrucksformen: modale Verwendung

1.1   Bei der EU in Brüssel z. B. verwendet man eine automatische Übersetzungsmaschine.

1.2

Modalität: Möglichkeit (können)	
sich lassen + Infinitiv	Lassen sich Menschen durch Computer ersetzen?
Infinitiv + *man* + *können*	Kann man **Menschen durch Computer ersetzen?**
sein + Adjektiv auf *-bar*	Sind Menschen durch Computer **ersetzbar?**
sein + *zu* + Infinitiv	Sind **Menschen durch Computer zu ersetzen?**
Passiv + *können*	Können **Menschen durch Computer ersetzt werden?**
Aktiv + *können*	Können **Computer Menschen ersetzen?**

Modalität: Notwendigkeit (müssen)	
sein + *zu* + Infinitiv	Der Computer liefert Übersetzungen, die nochmals zu überarbeiten sind.
Passiv + *müssen*	Der Computer liefert Übersetzungen, die nochmals **überarbeitet werden müssen**.

Modalität: Möglichkeit (können)	
sein + *zu* + Infinitiv	Diese Fragen sind nicht leicht zu beantworten.
Passiv + *können*	Diese Fragen **können nicht leicht beatwortet werden**.

Modalität: Notwendigkeit (müssen)	
sein + *zu* + Infinitiv	Weitere Forschungsergebnisse sind abzuwarten.
Passiv + *müssen*	Weitere Forschungsergebnisse **müssen abgewartet werden**.

Modalität: Notwendigkeit (sollen)	
sollen + Passiv	Wenn literarische Texte übersetzt werden sollen, kann ein Computer einen menschlichen Dolmetscher noch immer nicht ersetzen.
man + Infinitiv + *wollen*	Wenn man literarische Texte **übersetzen will**, kann ein Computer einen menschlichen Dolmetscher noch immer nicht ersetzen.

2

		sprachliche Mittel
Es gibt viele Grafikprogramme, mit denen man Teile eines Bildes austauschen kann, so dass Manipulationen kaum zu erkennen sind.	Es gibt viele Grafikprogramme, mit denen **sich Teile des Bildes austauschen lassen**, so dass **man Manipulationen kaum erkennen kann**.	sich lassen + Infinitiv *man* + Modalverb
Um zu garantieren, dass digitale Bilder nicht verfälscht werden können, …	Um zu garantieren, dass digitale Bilder nicht **verfälschbar sind**, …	sein + Adjektiv auf *-bar*
Mit Hilfe dieses Programms lassen sich digitalisierte Bilder kennzeichnen.	Mit Hilfe dieses Programms **können digitalisierte Bilder gekennzeichnet werden**.	Passiv + Modalverb
Dazu baut der Computer ein unsichtbares Wasserzeichen in das Bild ein.	Dazu **wird vom Computer ein unsichtbares Wasserzeichen in das Bild eingebaut**.	Passiv
Wird ein so geschütztes Bild nachträglich manipuliert, können die Änderungen sichtbar gemacht werden.	**Manipuliert man ein so geschütztes Bild nachträglich, lassen sich die Änderungen sichtbar machen**.	*sich lassen* + Infinitiv

3.1 Der Textteil **ist** in Kapitel/Abschnitte/Unterabschnitte **einzuteilen**.
> *sein + zu* + Infinitiv
Um die Über- und Unterordnung deutlich zu machen, **soll** nach dem Dezimalsystem **gegliedert werden**.
> *werden*-Passiv und Modalverb *sollen*
Die Nummerierung der Kapitel **beginnt** mit der Einleitung und **endet** mit dem zusammenfassenden Kapitel.
> **Aussagesatz im Präsens Aktiv**
Die Seitennummerierung **sollte** mit der Einleitung beginnen.
> **Modalverb** *sollen* **im Konjunktiv II**
Das Schlusskapitel der Arbeit **bildet** die Zusammenfassung.
> **Aussagesatz im Präsens Aktiv**
Hier **müssen** die Fragestellungen oder Thesen der Einleitung wieder **aufgenommen** und die Ergebnisse der Arbeit knapp und prägnant **formuliert** sowie in einen größeren Zusammenhang **eingeordnet werden**.
> *werden*-Passiv und Modalverb *müssen*
Es **sollten** Schlussfolgerungen **gezogen** und ein Ausblick auf mögliche Konsequenzen bzw. noch zu lösende Probleme **gegeben werden**.
> *werden*-Passiv und Modalverb *sollen* im Konjunktiv II
Hier **ist** auch der Platz für eigene Einschätzungen und Vorschläge für weitere wissenschaftliche Arbeiten.
> **Aussagesatz im Präsens Aktiv**
Im Literaturverzeichnis **müssen** alle für die Arbeit benutzten Quellen in bibliographischer Vollständigkeit **wiedergegeben werden**, …
> *werden*-Passiv und Modalverb *müssen*
… wobei im Textteil auf jede dieser Quellen mindestens einmal **verwiesen sein muss**.
> *sein*-Passiv und Modalverb *müssen*
Quellen, die nur über das WWW verfügbar waren, sind ebenfalls in das Literaturverzeichnis **aufzunehmen**.
> *sein + zu* + Infinitiv
Falls diese im WWW gefundenen Quellen in einer Zeitschrift o.Ä. veröffentlicht wurden, **sollten** diese Quellen (zumindest zusätzlich) **angegeben werden**.
> *werden*-Passiv und Modalverb *sollen* im Konjunktiv II

3.2

sein + zu + Infinitiv Aussagesatz im Präsens Aktiv *werden*-Passiv und Modalverb *müssen*	**verbindliche Anweisung** (die Vorgabe muss unbedingt beachtet werden)
werden-Passiv und Modalverb *sollen*	**sehr dringende Empfehlung**
(*werden*-Passiv und) Modalverb *sollen* im Konjunktiv II	**Empfehlung**

3.3

verbindliche Vorgabe	Der Textteil ist in Kapitel / Abschnitte / Unterabschnitte einzuteilen.
	Im Schlusskapitel **müssen die Fragestellungen oder Thesen der Einleitung wieder aufgenommen werden.**
	Im Textteil muss auf jede Quelle mindestens einmal verwiesen sein.
	Im Literaturverzeichnis müssen alle benutzten Quellen wiedergegeben werden.
	Quellen, die nur über das WWW verfügbar waren, sind ebenfalls in das Literaturverzeichnis aufzunehmen.
sehr dringende Empfehlung	(Es) soll nach dem Dezimalsystem gegliedert werden.
Empfehlung	Die Seitennummerierung sollte **mit der Einleitung beginnen.**
	Es sollten Schlussfolgerungen gezogen werden.
	Es sollte ein Ausblick auf mögliche Konsequenzen bzw. noch zu lösende Probleme gegeben werden.

4 Beispiellösung:

Regeln des Zitierens

Alle Zitate, die in die eigene wissenschaftliche Arbeit übernommen wurden, müssen kenntlich gemacht werden und mit einer Quellenangabe einschließlich der Seitennummer belegt werden.

Zitate sollten sparsam verwendet und geschickt in die eigene Darstellung eingebunden werden. Es sollten v. a. kurze Zitate verwendet werden.

Wörtliche Zitate müssen durch Anführungszeichen gekennzeichnet werden. Bei wörtlichen Zitaten werden Orthographie und Interpunktion genau wiedergeben, auch bei Schreibfehlern und veralteter Schreibweise. Auslassungen, Veränderungen und Eingriffe bei Zitaten sind in jedem Fall zu kennzeichnen.

Zitate sollen aus dem Originaltext verwendet werden, nicht aus „zweiter Hand". Nur in Ausnahmefällen sollte aus „zweiter Hand" zitiert werden. Dann erfolgt die Kennzeichnung durch den Zusatz „zitiert in", gefolgt von der Angabe der Sekundärquelle.

6.5 Unpersönliche Ausdrucksformen: nicht modale Verwendung

1.1   **Das erste Kapitel widmet sich** dem Thema interkulturelle Verständigung in China und Taiwan.

> **Reflexivkonstruktion im Präsens Aktiv: „Das erste Kapitel" wird personifiziert und erscheint als „Handelnder".**

Sodann **werden** Möglichkeiten und Voraussetzungen der Definition „interkulturellen Lernens" **diskutiert** und **problematisiert.**

> *werden*-Passiv

Im zweiten Kapitel **erfolgt eine Auseinandersetzung** mit dem Begriff „Alltagstheorie".

> *erfolgen* + Nomen

Es **folgt die Bestimmung** des hier verwendeten Begriffs von „Alltagstheorie" und eine Begründung der Untersuchungsmethode

> *folgen* + Nomen

Das dritte Kapitel beschäftigt sich mit Forschungsarbeiten zum Thema „Gesicht".

> **Reflexivkonstruktion im Präsens Aktiv: „Das dritte Kapitel" wird personifiziert und erscheint als „Handelnder".**

Die Präsentation und Diskussion der Ergebnisse von einigen neueren Forschungsarbeiten dazu **bilden** den Abschluss des Kapitels.

> *bilden*

Im vierten Kapitel **werden** die wichtigsten Ergebnisse der Untersuchungsgrundlagen **rekapituliert** und die Forschungsfrage der empirischen Untersuchung **präzisiert.**

> *werden*-Passiv

Das fünfte Kapitel stellt die Durchführung der empirischen Untersuchung **vor.**

> **Präsens Aktiv: „Das fünfte Kapitel" wird personifiziert und erscheint als „Handelnder" einer Handlung der Autorin (die Autorin hat eine empirische Untersuchung durchgeführt).**

Eine kurze Einführung in die Lebensbedingungen am Untersuchungsort Taiwan **vermittelt** wichtiges Hintergrundwissen, bevor der Untersuchungsablauf, die eingesetzte Methode und Auswertungsverfahren **geschildert werden.**

> *vermitteln*; *werden*-Passiv

Im sechsten Kapitel **werden** die Ergebnisse der Analyse **dargestellt.** In fallübergreifenden Analysen **wird** sodann der Frage **nachgegangen**, welche Bedeutung chinesische Sprachkenntnisse sowie verschiedene Lernstrategien für das Lernen über „Gesicht" besitzen.

> *werden*-Passiv

Ergebnisse dieser Analysen **sind Gegenstand des achten Kapitels.**

> *ist Gegenstand* + Genitiv

Im neunten Kapitel **erfolgt die Diskussion** der empirischen Befunde.

> *erfolgen* + Nomen

Erkenntnisse und Beschränkungen der vorliegenden Arbeit **werden** kritisch **betrachtet** und mit den Ergebnissen früherer Forschung **in Bezug gesetzt.**

> *werden*-Passiv; Nomen-Verb-Verbindung in passivischer Bedeutung (*in Bezug setzen*)

1.2   a   Das zweite Kapitel setzt sich mit dem **Begriff „Alltagstheorie" auseinander.**

     b   Dann wird der hier verwendete **Begriff „Alltagstheorie" bestimmt.**

     c   Zum Abschluss werden **die Ergebnisse präsentiert und diskutiert.**

     d   Im fünften Kapitel wird **die Durchführung der empirischen Untersuchung vorgestellt.**

     e   Das sechste Kapitel stellt **die Ergebnisse der Analyse dar.**

     f   Im neunten Kapitel werden **die empirischen Befunde diskutiert.**

2   Beispiellösung:

Das zweite Kapitel stellt die theoretischen Grundlagen der Motivation und der Anreize in der Arbeit dar. In diesem Zusammenhang erfolgt die Diskussion kultur- und menschenbildbedingter Gestaltungsmöglichkeiten der Anreize. Im dritten Kapitel werden arbeitsrelevante Werthaltungen der Chinesen theoretisch analysiert und auf dieser Basis Fragestellungen für die empirische Untersuchung entwickelt.

Die empirische Untersuchung ist Gegenstand des vierten Kapitels.

Das fünfte Kapitel präsentiert die Ergebnisse und fasst sie zusammen. Dabei werden relevante Handlungsempfehlungen und Konsequenzen für die Anwendung und die Praxis herausgearbeitet.

Das sechste Kapitel gibt einen Ausblick auf zukünftige Forschungsfragen in diesem Bereich.

3   Marthaler sieht sich gezwungen, mit der Journalistin Anna Buchwald zusammenzuarbeiten, die im Besitz der Akte Rosenherz ist. **C**

Marthaler ist gezwungen, mit der Journalistin Anna Buchwald zusammenzuarbeiten, die im Besitz der Akte Rosenherz ist. **A**

Marthaler wird gezwungen, mit der Journalistin Anna Buchwald zusammenzuarbeiten, die im Besitz der Akte Rosenherz ist. **B**

4.1   Beispiellösung:

Warum kann sich ein Arbeitnehmer in seinem Job überfordert fühlen?

Arbeitnehmer, die in ihrem Arbeitsalltag zu wenig Unterstützung bei schwierigen Aufgaben und Problemen bekommen, fühlen sich schnell im Stich gelassen. Auch wenn Projekte sehr zeitaufwendig sind, aber innerhalb eines engen Zeitplans fertig werden sollen, kommt es zu Schwierigkeiten: Die Arbeitnehmer sehen sich ungerecht behandelt, wenn sie mit Projekten im Verzug sind und fühlen sich ungerecht behandelt.

Oftmals fühlen sie sich auch bei wichtigen Entscheidungen übergangenen und sehen sich um ihren Erfolg gebracht, wenn Projekte an unrealistischen Zeitvorgaben scheitern.

Wenn sich Arbeitnehmer jeden Tag gezwungen sehen, Aufgaben zu erledigen, für die ihnen die Zeit oder die notwendige Unterstützung fehlt, fühlen sie sich schnell überfordert.

6.6   Passiv mit *bekommen*

1.2

bekommen + **Akkusativ**	*bekommen* + **Akkusativ** + **Partizip II**
Er bekam ein Paket.	Er bekam ein Paket zugeschickt.
Anna hat von Peter Blumen bekommen.	Anne **hat von Peter Blumen geschenkt bekommen.**
Bekomme ich das Ergebnis automatisch per Post?	**Bekomme ich das Ergebnis automatisch per Post zugeschickt?**

6.7   Funktionsverbgefüge in passivischer Bedeutung

1.1

Echt künstlich

In seinem neuesten Buch „Echt künstlich" dokumentiert Hans-Ulrich Grimm eindringlich, wie Verbraucher und besonders Kinder durch den sorglosen Umgang mit Zusatzstoffen und das Wegschauen der Politik gefährdet werden. Ein Beispiel ist Zitronensäure (E330). Sie wird immer dann eingesetzt, wenn etwas frisch und fruchtig schmecken soll. Zitronensäure greift die Zähne stark an und fördert die Aufnahme von Metallen wie Blei und Cadmium ins Blut. Die aggressive Säure **kommt** auch als Entkalker für Kaffeemaschinen oder als WC-Reiniger **zum Einsatz**. Dann sind allerdings Warnhinweise vorgeschrieben. Der politi-sche Skandal ist, dass bis heute nicht untersucht worden ist, wie viel Zitronensäure beispielsweise Kinder tatsächlich zu sich nehmen. Es wird einfach nicht erfasst, wie viel Gummibärchen und Softdrinks die Kleinen verzehren.

In der Europäischen Union (EU) sind über 300 Zusatzstoffe zugelassen. Ein großer Teil von ihnen **steht im Verdacht**, die Gesundheit zu schädigen. Im umfangreichen Lexikon-Teil von Grimms Buch werden alle zugelassenen Substanzen kurz dargestellt und bewertet. Handlich und verständlich.

1.2

Passiv	Funktionsverbgefüge
Zitronensäure wird als Entkalker für Kaffeemaschinen oder als WC-Reiniger eingesetzt.	Zitronensäure kommt **als Entkalker für Kaffeemaschinen oder als WC-Reiniger zum Einsatz.**
Zitronensäure wird immer dann eingesetzt, wenn etwas frisch und fruchtig schmecken soll.	Zitronensäure **kommt immer dann zum Einsatz, wenn etwas frisch und fruchtig schmecken soll.**
Durch **den sorglosen Umgang mit Zusatzstoffen werden Verbraucher und besonders Kinder gefährdet.**	Der sorglose Umgang mit Zusatzstoffen bringt Verbraucher und besonders Kinder in Gefahr.
Ein großer Teil der Zusatzstoffe wird verdächtigt, die Gesundheit zu schädigen.	Ein großer Teil **der Zusatzstoffe steht im Verdacht, die Gesundheit zu schädigen.**
Im umfangreichen Lexikon-Teil werden alle zugelassenen Substanzen kurz dargestellt und bewertet.	Im umfangreichen Lexikon-Teil erfolgt **eine kurze Darstellung und Bewertung aller zugelassener Substanzen.**

2 a Am Samstag hatten viele Zeitungsabonnenten Grund zum Ärger. In der Nacht **kam** es in der Produktion **zu** technischen **Störungen**, so dass die Zeitung die Leser verspätet erreichte.

   b Der Ausdruck „verboten" **findet** in der Jugendsprache **Verwendung**, wenn die noch nicht volljährigen Sprecher ihrer Begeisterung Ausdruck verleihen wollen.

   c Die Geldkarte, die mit Bargeld aufgeladen wird, **findet** beim Handel keine **Unterstützung**.

   d Gentechnisch veränderte Lebensmittel **unterliegen** genauer Registrierung und **Kontrolle**.

   e Die Firma **steht im Verdacht**, ihren Arbeitnehmern nicht den vorgeschriebenen Mindestlohn zu zahlen.

3 Beispiellösung:

Kunststoff aus dem Hühnerstall

Die fossilen Ressourcen werden immer knapper. Aber gerade diese Rohstoffe sind die Quelle für viele chemischer Produkte. Seit Jahren suchen Wissenschaftler nach Alternativen.

Rapsöl findet als Biokraftstoff Verwendung, aus Holz wird Erdgas gewonnen. Allerdings stößt diese alternative Energiegewinnung auf Kritik. Denn ein mit Rapspflanzen bedecktes Feld kann nicht mehr genutzt werden, um Nahrungsmitteln anzubauen.

Wissenschaftler forschen auf dem Gebiet der Nebenprodukte von Hühnerfarmen. Hühnerfleisch und Hühnereier werden in großen Mengen gebraucht. Die Federn können dagegen nur in Bettdecken und Kopfkissen genutzt werden. Der Großteil der Federn findet keine Verwendung.

7 Modalität

7.1 Notwendigkeiten, Möglichkeiten, Pläne

7.1.1 Modalverben zum Ausdruck von Bedingungen des Handelns

1.1

Die Hausverwaltung ...	Modalverb
äußert einen Wunsch.	wollen $_1$
äußert einen höflichen Wunsch.	**möchten** $_3$
äußert, dass etwas möglich ist.	**können** $_5$
äußert eine Erlaubnis.	**dürfen** $_6$
drückt aus, dass etwas notwendig ist. (2x)	müssen $_4$, **müssen** $_8$
fordert Sie auf, etwas zu machen.	**sollen** $_2$
äußert ein Verbot.	**dürfen nicht** $_7$

1.2

Modalverben zum Ausdruck von Bedingungen des Handelns

- Modalverben modifizieren sehr oft ein **Verb im Infinitiv**. Im Hauptsatz steht dieses dann am **Ende**. Wenn die Bedeutung aus dem **Kontext** hergeleitet werden kann, kann der **Infinitiv** weggelassen werden.

2   **Steinbock 22.12. – 20.01.** Positives Denken ist eine starke Kraft. Unter Pluto **können** Sie Berge versetzen, wenn Sie an sich glauben. Sie **sollten** sich dem Außergewöhnlichen öffnen und mit den Energien gehen. Falls sich dann doch die vorsichtige Steinbock-Frau in Ihnen zu Wort meldet, **dürfen** Sie keine Sekunde zweifeln. Ihre Strategie, alles konzentriert zu erarbeiten, **können** Sie ja trotzdem fortsetzen.

Wassermann 21.01. – 19.02. Sie **möchten** Kompromisse bilden, doch diese sind nicht automatisch weise – sagt Ihnen Ihr Instinkt. Unter spannungsreicher Planeten-Konstellation **müssen** Sie aber Abstriche machen. Nur dagegenzuhalten, auch mit ganz wunderbaren Absichten, **könnte** zu verhärteten Fronten führen. Fragen Sie sich also vor jedem wichtigen Gespräch, was Sie um keinen Preis aufgeben **wollen** und was eine Auseinandersetzung mit anderen eigentlich nicht wert ist.

Fische 20.02. – 20.03. Die Planeten zeigen Ihnen bald einen neuen Kosmos in Liebe und Beziehung – um den erkennen und schätzen zu **können**, **müssen** die Weichen richtig gestellt werden. Wenn Sie sich also gebremst fühlen, betrachten Sie das als verordnete Pause, um Ihren Weg überdenken zu **können**. Sie **dürfen** zuversichtlich in die Zukunft blicken.

3

TEENAGER:	Mann, warum denn nicht? Warum **darf** ich **nicht** zum Festival?	VATER:	Ich **kann** dich leider **nicht** ohne die Erlaubnis deiner Mutter fahren lassen.
MUTTER:	Du bist 15 und deine Oma feiert Geburtstag.	MUTTER:	Das ist ja mal wieder typisch, dass du jetzt nicht entscheiden willst.
TEENAGER:	Mann, ich will, ich will da hin!	TEENAGER:	Papa, du musst nur wollen.
BRUDER:	Sie ist 15! Weißt du, was die da machen?	VATER:	Ok, dann entscheide jetzt ich: Du **darfst** fahren, aber nur unter einer Bedingung, dass Max mitfährt!
TEENAGER:	Max, misch dich da nicht ein. Bitte, Papa! Bitte!		
MUTTER:	Warum **willst** du **nicht** stattdessen zu Omas Geburtstag?	TEENAGER:	Na toll! Dann **will** ich jetzt auch **nicht** mehr fahren!
TEENAGER:	Es ist nicht so, dass ich **nicht mag**. Ich möchte ja gerne mitkommen, aber auf dem Festival spielt die Band „Wir sind Helden". Die möchte ich unbedingt sehen!		

4 b muss dieser einen Maulkorb tragen   **d müssen ihren Hund innert drei Monaten dem Veterinäramt melden**
c muss an der Leine geführt werden   **e welche Maßnahmen ergriffen werden müssen**

5

Probleme beim Angeln

SOHN: Was sagst du? Ich hätte den Köder gar **nicht** an den Haken hängen **dürfen / sollen**?
Warum nicht? Wie beißen denn dann die Fische an?

VATER: Ich hab's dir doch schon erklärt, hier wimmelt es nur so von Fischen, die beißen auch
ohne Köder an. Da hättest Du den Köder **nicht** verschwenden **brauchen / müssen**.

6

VATER: Nun musst du den Wattwurm auch nicht mehr an den Haken hängen.

SOHN: Warum nicht?

VATER: Weil wir für heute genug gefangen haben. Wir gehen lieber schon nach Hause, weil
wir morgen noch einiges erledigen müssen. Morgen früh laufen die Krabbenkutter
ein, da können wir denen frische Krabben abkaufen, die wir dann als Köder für
Plattfische nehmen können. Du musst wissen: Das Angeln von Plattfischen erfordert
Techniken, die von jedem guten Angler beherrscht werden müssen.

7.1.2 Modifizierende Verben (Verben mit Infinitiv)

1.1

Was mache ich, wenn der Hund an der Leine zerrt?

Wenn Sie **spazieren gehen** und der Hund dabei ständig an der Leine zerrt, ist das unangenehm. Es gibt dazu zwei Strategien, die Ihnen den Hund **erziehen helfen**: Bei der ersten Methode **bleiben** Sie jedes Mal **stehen**, wenn der Hund an der Leine zerrt und warten, bis er sich Ihnen zuwendet und Sie neben sich **stehen sieht**. Am Anfang werden Sie alle paar Schritte **stehen blei-** **ben** müssen. Bei der zweiten Methode wechseln Sie jedes Mal die Richtung, wenn der Hund an der Leine zerrt. Auch wenn der Hund nach zwei Monaten mal wieder zerrt, **bleiben** Sie **stehen** oder wechseln Sie die Richtung. **Lassen** Sie den Hund dagegen **zerren**, ist die Unart schnell wieder da und er wird nicht reagieren, auch wenn er Sie **rufen hört**.

1.2

	Präsens	Präteritum	Perfekt: *haben* + Infinitiv + Infinitiv
hören	Er hört Sie rufen.	**Er hörte Sie rufen.**	Er hat Sie rufen hören.
lassen	Sie lässt den Hund zerren.	Sie ließ den Hund zerren.	**Sie hat den Hund zerren lassen.**
	Präsens	Präteritum	Perfekt: *sein* / (*haben*) + Partizip II
bleiben	**Er bleibt stehen.**	**Er blieb stehen.**	Er ist stehen geblieben.
gehen	Sie geht spazieren.	**Sie ging spazieren.**	**Sie ist spazieren gegangen.**

---

2 Funktion von *lassen*

Erlaubnis aussprechen	**Nachdem sie etwas gefressen hat, können Sie sie ruhig wieder frei im Garten laufen lassen.**
Auftrag erteilen	**Das Medikament lassen Sie sich von der Schwester geben.**
etwas zurücklassen	**Ja, lassen Sie den Korb hier und legen Sie die Katze ruhig in Ihren Katzenkäfig.**

lassen wird auch als Vollverb verwendet: *Lassen Sie die Medikamentendosierung beim Alten.*

7.1.3 Imperativ: Empfehlung, Ratschlag, Instruktion

1.1

> ### **Gönne** deinem Körper regelmäßig eine Portion Entspannung
>
> Wahrscheinlich hast du auch schon erlebt, dass du dann, wenn du ärgerlich, ängstlich oder eifersüchtig bist, gleichzeitig auch angespannt bist. Das ganze System funktioniert aber auch umgekehrt und du kannst es dir zunutze machen, um dein Wohlbefinden und deine innere Zufriedenheit zu steigern. **Erinnere** dich nur daran, wie wohlig du dich fühlst, wenn du mal ganz entspannt und locker bist. **Bring** dich regelmäßig wieder in einen Entspannungszustand zurück. Ich möchte eine Methode vorschlagen, die ohne allzu lange Übung und Zeitaufwand funktioniert. Man bezeichnet diese Entspannungsübung auch als Ampelübung: **Stell** dir **vor**, du musst bei Rot an einer Ampel halten und willst schnell etwas Gutes für dich tun, anstatt dich über das Rotlicht zu ärgern. Dann **spann** hierzu einmal kurz für ca. 15 Sekunden alle Muskeln deines Körpers vom Nacken bis zu den Füßen an – allerdings nur so weit, dass du dich nicht verkrampfst. Zähle von 1 bis 15. Danach **lass** die Anspannung wieder **los** und **versuche**, alle Muskeln locker zu lassen. Andere großartige Entspannungsmethoden können Folgende sein: **Nimm** ein warmes Bad! **Mache** einen Spaziergang! **Trink** eine Tasse Schokolade! **Hör** schöne Musik oder **lies** ein gutes Buch. Wichtig ist nur: **Nimm** dir Zeit dafür, denn ein entspannter Körper tut sich einfach leichter, Freude zu empfinden. **Sei** achtsam auf die Signale, die dir dein Körper sendet.

1.2

Du-Form des Imperativs: Bildung	
Nimm die 2. Person Singular und streiche die Endung und das Personalpronomen:	Du nimmst ein warmes Bad. → ~~Du~~ Nimmst ein warmes Bad!
Imperative unregelmäßiger Verben werden ohne **Umlaut** gebildet:	Du lässt die Anspannung los. → ~~Du~~ Lasst die Anspannung los!
Ist der Stammlaut ein **Konsonant**, behält der **Imperativ** das -e:	Du antwortest mir. → ~~Du~~ Antwortest mir!
Verben auf -ern und -eln haben ein -e im Imperativ:	erinnern → Erinnere dich! klingeln → Klingle bitte nur ein Mal! ändern → Ändere nichts an dir!

- In der gesprochenen Sprache wird das -e am **Ende** des Imperativs häufig weggelassen: *Erinner dich doch mal!*

1.3

Gönnen Sie Ihrem Körper regelmäßig eine Portion Entspannung.
Gönnt eurem Körper regelmäßig eine Portion Entspannung.

Erinnern Sie sich nur daran, wie wohlig Sie sich fühlen, wenn Sie mal ganz entspannt und locker sind.
Erinnert euch nur daran, wie wohlig ihr euch fühlt, wenn ihr mal ganz entspannt und locker seid.

Bringen Sie sich regelmäßig wieder in einen Entspannungszustand zurück.
Bringt euch regelmäßig wieder in einen Entspannungszustand zurück.

Stellen Sie sich vor, Sie müssen bei Rot an einer Ampel halten.
Stellt euch vor, ihr müsst bei Rot an einer Ampel halten.

Dann spannen Sie hierzu einmal kurz für ca. 15 Sekunden alle Muskeln Ihres Körpers vom Nacken bis zu den Füßen an.
Dann spannt hierzu einmal kurz für ca. 15 Sekunden alle Muskeln eures Körpers vom Nacken bis zu den Füßen an.

Danach lassen Sie die Anspannung wieder los und versuchen, alle Muskeln locker zu lassen.
Danach lasst die Anspannung wieder los und versucht, alle Muskeln locker zu lassen.

Nehmen Sie ein warmes Bad!
Nehmt ein warmes Bad!

Machen Sie einen Spaziergang!
Macht einen Spaziergang!

Trinken Sie eine Tasse Schokolade!
Trinkt eine Tasse Schokolade!

Hören Sie schöne Musik oder lesen Sie ein gutes Buch!
Hört schöne Musik oder lest ein gutes Buch!

Nehmen Sie sich Zeit dafür, denn ein entspannter Körper tut sich einfach leichter, Freude zu empfinden.
Nehmt euch Zeit dafür, denn ein entspannter Körper tut sich einfach leichter, Freude zu empfinden.

Seien Sie achtsam auf die Signale, die Ihr Körper Ihnen sendet.
Seit achtsam auf die Signale, die euch euer Körper sendet.

2.1   Bitte: _a_   Rat / Empfehlung: **c**   Aufforderung: **b, d , e (gilt für beide Sätze)**

2.2   **b** Gib mir bitte den Kaffee! Modalverb: Kann **ich bitte den Kaffee haben.**

         c Gib mir bitte die Butter! Modalverb: **Kannst du mir bitte die Butter geben**?

         d Nimm besser Wurst statt Käse! Frage: Warum **nimmst du nicht Wurst statt Käse**?

         e Räum den Tisch ab! Konjunktiv + Konditionalsatz: **Es wäre toll, wenn du den Tisch abräumen könntest.**

2.3   Als erstes rührst du das Eigelb mit dem Zucker schaumig. Dann rührst du den Quark mit Zitrusschale und Zimt unter. Als nächstes schlägst du das Eiweiß mit einer Prise Salz cremig und ziehst es mit Milch und Mehl unter die Eigelb-Quak-Masse. Danach schälst du nach Belieben die Äpfel und stichst das Kerngehäuse aus. Du schneidest die Äpfel in etwa 5 mm dicke Scheiben. Danach erhitzt du in einer Pfanne etwas Öl. Leg dann einige Apfelscheiben ein und gib auf die Mitte jeder Scheibe ein bis zwei Esslöffel Teig. Wenn die Küchlein von der Unterseite goldbraun sind, wendest du sie vorsichtig und gibst nach Belieben einen Stich Butter hinzu. Zum Schluss legst du die Küchlein mit Zimt-Zucker bestreut auf ein Backblech und bäckst sie im vorgeheizten Backofen fünf Minuten bei 180 Grad.

7.2 Sicherheit und Unsicherheit äußern

7.2.1 Modalverben zum Ausdruck von Wahrscheinlichkeit (subjektiver Gebrauch)

1.1

„Reiche Kommunen würden vom Ende des Infrastruktur-Fonds nicht profitieren"

Kommunalpolitiker aus industriegeprägten Bundesländern fordern ein Ende des Fonds. Das würde den Gemeinden aber nichts bringen, sagt Bürgermeister Markus Starke im Interview.

Herr Starke, in Nordrhein-Westfalen haben die Bürgermeister mehrerer Ruhrgebietsstädte Alarm geschlagen. Ihre Städte **könnten** _weiter verkommen, während man in anderen Bundesländern teilweise nicht mehr wisse, wohin mit dem Geld aus dem Infrastruktur-Fond. Was sagen Sie dazu?_ **Starke:** Erst mal ist die Ausgangslage eine andere. Die genannten industriestarken Kommunen **dürften** in den derzeit laufenden Infrastruktur-Fond II keinen Cent ein-zahlen. Dieser **müsste** nämlich ausschließlich aus Bundesmitteln finanziert sein. Daraus folgt: Ein vorzeitiges Ende des Fonds **dürfte** diesen Kommunen gar nichts bringen, denn sie stehen mit ihren Zahlungen für Dinge gerade, die schon längst erledigt sind. Die gemachten Schulden **müssten** sie trotzdem weiter ab-bezahlen. _Aber es_ **könnte** _den notleidenden Kommu-nen in den industriellgeprägten Bundes-_ _ländern helfen, wenn der Fond nicht mehr ausschließlich ländlich geprägten Bundes-ländern zugute käme, sondern eben auch strukturschwachen Industriegegenden._ **Starke:** Wenn wir anders fördern würden, dann **könnten** die betreffenden Länder mehr erhalten als bisher. Aus dem einfa-chen Grund, dass das stärkste Agrarland immer noch strukturschwächer ist als das schwächste Industrieland.

1.2 Modalverben zum Ausdruck von Wahrscheinlichkeit

Wahrscheinlichkeit	Modalverben	andere sprachliche Mittel
sehr sicher	muss	Mit Sicherheit …,
fast sicher	**müsste**	**Es ist so gut wie sicher, …**
wahrscheinlich	**dürfte**	**vermutlich**
möglich	**könnte**	**möglicherweise; Es ist denkbar, …; vielleicht**

1.3   **b** Es könnte den Kommunen helfen, wenn der Fond nicht mehr nur ländlich geprägten Bundesländern zugute käme.

           Wenn der Fond nicht mehr nur ländlich geprägten Bundesländern zugute kommt, **kann das den Kommunen vielleicht helfen.**

         c Eine Umstrukturierung des Fonds müsste den Kommunen helfen.

           Es ist so gut wie sicher, dass eine Umstrukturierung des Fonds den Kommunen hilft.

         d Wenn wir anders fördern würden, könnten die betreffenden Länder mehr erhalten als bisher.

           Wenn wir anders fördern, ist es denkbar, dass die betreffenden Länder mehr erhalten als bisher.

         e Ein vorzeitiges Ende des Fonds dürfte den Kommunen gar nichts bringen.

           Ein vorzeitiges Ende des Fonds bringt diesen Kommunen vermutlich gar nichts.

7.2.2 Modalwörter

1.2

Kommentaradverbien drücken eine Bewertung des Sachverhalts aus	Adverbien geben den Grad der Wahrscheinlichkeit an
bedauerlicherweise, bekanntermaßen, **dummerweise**, erfreulicherweise, glücklicherweise, **immerhin**, irrtümlicherweise, jedenfalls, klugerweise, leichtsinnigerweise, **leider**, lobenswerterweise, **natürlich**, **schließlich**, seltsamerweise, überraschenderweise, unerwarteterweise, unnötigerweise, **zugegebenermaßen**	kaum, **möglicherweise**, sicherlich, **vermutlich**, **vielleicht**, zweifellos, **zweifelsohne**

2

Anhand eines Tests soll man herausfinden, ob ein Paar für die gemeinsame Wohnung reif sein **könnte**. Man muss sich entscheiden, ob die eigene Beziehung eher einem „alten Baum", einem „jungen Baum" oder einem „Wasserstrudel" gleicht. Nimmt man den alten Baum, **sollte** man bereit sein zusammenzuziehen. Der Wasserstrudel symbolisiert, dass man für eine gemeinsame Wohnung nicht bereit sein **dürfte**. Der junge Baum ist nicht eindeutig und passt in zwei Kategorien: „Sie **sollten** zusammenziehen!" sowie „Sie **könnten** noch nicht bereit sein."

7.2.3 Modalpartikeln

1.1   Tochter: **Nachfrager(in)**
Peter: **Vorwurfsvolle**
Finanzminister Dr. Knauser: **Besserwisser**
Patientin: **Überraschte**

1.2

Peter: Du hast **ja** nie Zeit, wenn man dich braucht.
   Lena: Ich hab **halt** viel um die Ohren.
Peter: Das haben alle. Dann kannst du **ja** wenigstens mal anrufen, aber das machst du **ja** auch nicht.
   Lena: Ich bin meist im Ausland tätig, da ist das **einfach** nicht so leicht anzurufen.
Peter: Ja und? Heutzutage kann man **doch** von überall kostengünstig anrufen!
   Lena: **Eigentlich** hast du **ja** Recht. Hätte ich **doch bloß** mehr Zeit!

Verteidigungsminister Herr von Donnerschlag: Mein ursprüngliches Ziel war es die Berufs- und Zeitsoldaten auf 163.500 zu verringern.
Finanzminister Dr. Knauser: Sie sollten **doch aber** inzwischen wissen, dass die politisch vereinbarte Zahl von 185.000 festgelegt wurde und es bleibt übrigens bei der geltenden Finanzplanung.
Verteidigungsminister Herr von Donnerschlag: Mit dieser Anzahl sind die Einsparungen **aber** finanziell nicht zu schaffen.

Finanzminister Dr. Knauser: Herr von Donnerschlag, ich kann die Grundrechenarten aber nicht außer Kraft setzen. Veranstalten Sie **bloß** nicht so einen Zirkus hier!
Verteidigungsminister Herr von Donnerschlag: Im Rahmen des Sparpakets muss das Verteidigungsministerium bis Ende 2014 insgesamt 6,3 Milliarden Euro einsparen.
Finanzminister Dr. Knauser: Ich sehe, Sie haben **mal** wieder nicht Ihre Hausaufgaben gemacht. Es sind 8,3 Milliarden Euro, Herr von Donnerschlag.

Ärztin: Ich darf Ihnen gratulieren, Sie sind schwanger!
Patientin: Nein, das gibt es **ja** nicht!
Ärztin: Sie befinden sich bereits im 2. Monat. Herzlichen Glückwunsch!
Patientin: Da wird sich mein Mann **vielleicht** freuen!
Ärztin: Dann feiern Sie schön, aber mit alkoholfreiem Sekt.
Patientin: Das versteht sich **doch** von selbst!

1.3

Fragen	denn, eigentlich, etwa
	Ja, aber warum ist denn der Bundespräsident nicht so bekannt?
	Warum hat denn nicht jedes Land einen Bundespräsidenten?
	Haben eigentlich auch andere Länder einen Bundespräsidenten?
	Hat er etwa nicht so viele Aufgaben?
Aussagesätze	doch, aber, doch aber, eben, eigentlich, einfach, halt, ja, mal
	Angela Merkel ist doch das Staatsoberhaupt der Bundesrepublik Deutschland.
	Nein, denn er muss doch Gesetze unterschreiben.
	Sie sollten doch aber inzwischen wissen, dass ...
	Dann kann er doch eigentlich nicht so wichtig sein.
	Mit dieser Anzahl sind die Einsparungen aber finanziell nicht zu schaffen.
	Herr von Donnerschlag, ich kann die Grundrechenarten aber nicht außer Kraft setzen.
	Nein, ist sie ja eben nicht.
	Eigentlich hast du ja Recht.
	Ich bin meist im Ausland, da ist das einfach nicht so leicht anzurufen.
	Ich hab halt viel um die Ohren.
	Du hast ja nie Zeit, wenn man dich braucht.
	Dann kannst du ja wenigstens mal anrufen, aber das machst du ja auch nicht.
	Ich sehe, Sie haben mal wieder nicht Ihre Hausaufgaben gemacht.
Ausrufesätze	doch, ja, vielleicht
	Das versteht sich doch von selbst!
	Heutzutage kann man doch von überall kostengünstig anrufen!
	Nein, das gibt es ja nicht!
	Da wird sich mein Mann vielleicht freuen!
Aufforderungen	bloß
	Veranstalten Sie bloß nicht so einen Zirkus hier!
Wunschsätze	doch bloß
	Hätte ich doch bloß mehr Zeit!

1.4 Modalpartikeln: Verwendung und Bedeutung

- Im Deutschen kommen Partikeln besonders im Dialog und in spontaner Sprache vor, um eine **Einstellung** des Sprechers zu einer Aussage zu verdeutlichen.
- Partikeln können verschiedene **Bedeutungen** haben: Sie können Überraschung ausdrücken (Das ist aber großartig!), etwas Bekanntes (Das war ja klar!) oder etwas Offensichtliches (Das ist halt das Problem.).
- Modalpartikeln stehen meist im **Mittelfeld** und können auch **mehrteilig** vorkommen:
  Dann kann er doch eigentlich nicht so wichtig sein.

2.1

Frau: Wieso geht der Fernseher *den*$_1$ grade heute kaputt?

Mann: Die bauen die Geräte absichtlich so, dass sie schnell kaputt gehen.

Frau: Ich muss nicht unbedingt Fernsehen.

Mann Ich auch nicht. Nicht nur, weil heute der Apparat kaputt ist, ich meine sowieso, ich sehe sowieso nicht gerne Fernsehen.

Frau: Es ist **ja**$_2$ auch wirklich NICHTS im Fernsehen was man gern sehen möchte.

Mann: Heute brauchen wir, Gott sei dank, überhaupt nicht erst in den blöden Kasten zu gucken.

Frau: Nee, es sieht **aber**$_3$ so aus, als ob du hinguckst.

Mann: Ich?

Frau: Ja.

Mann: Nein, ich sehe nur ganz allgemein in diese Richtung. Aber du guckst hin. Du guckst da immer hin.

Frau: Ich? Ich gucke dahin? Wie kommst du **denn**$_4$ darauf?

Mann: Es sieht so aus.

Frau: Das kann gar nicht so aussehen, ich gucke nämlich vorbei. Ich gucke absichtlich vorbei. Und wenn du ein kleines bisschen mehr auf mich achten würdest, hättest du bemerken können, dass ich absichtlich vorbei gucke. Aber du interessierst dich **ja**$_5$ überhaupt nicht für mich.

Mann: Jajajaja.

Frau: Wir können **doch**$_6$ einfach mal ganz woanders hingucken.

Mann: Woanders? Wohin denn?

Frau: Zur Seite, oder nach hinten.

Mann: Nach hinten? Ich soll nach hinten sehen? Nur weil der Fernseher kaputt ist, soll ich nach hinten sehen? Ich lass mir **doch**$_7$ von einem Fernsehgerät nicht vorschreiben, wo ich hinsehen soll.

Frau: Was wäre **denn**$_8$ heute für ein Programm gewesen?

Mann: Eine Unterhaltungssendung.

Frau: Ach.

Mann: Es ist schon eine Unverschämtheit was einem so Abend für Abend im Fernsehen geboten wird. Ich weiß gar nicht, warum man sich das überhaupt noch ansieht. Lesen könnte man stattdessen, Karten spielen oder ins Kino gehen oder ins Theater. Stattdessen sitzt man da und glotzt auf dieses blöde Fernsehprogramm.

Frau: Heute ist der Apparat **ja**$_9$ nu kaputt.

Mann: Gott sei dank.

Frau: Ja.

Mann: Da kann man sich wenigstens mal unterhalten.

Frau: Oder früh ins Bett gehen.

Mann: Ich gehe nach den Spätnachrichten der Tagesschau ins Bett.

Frau: Aber der Fernseher ist **doch**$_{10}$ kaputt.

Mann: Ich lasse mir von einem kaputten Fernseher nicht vorschreiben, wann ich ins Bett zu gehen habe.‘

2.2

Bedeutung	Beispiel
etwas Bekanntes / Selbstverständliches / Offensichtliches ausdrücken	2, **5**, **9**
Kritik / Erstaunen / Überraschung ausdrücken	3
einen (anderen) Rat geben / eine (andere) Problemlösung vorschlagen	6
eine anschließende Frage aus dem Kontext	1, **4**, **8**
Ausruf (Gegensatz)	**7**, **10**

3 eigentlich: **neue Frage zum Thema**

etwa: **Vorwurf**

denn: **anschließende Frage**

übrigens: **Themenwechsel**

---

4 Modalpartikeln: Bedeutung

- Modalpartikeln werden auch verwendet, um an das Wissen der Leserschaft anzuknüpfen.

Modalpartikeln	Bedeutung
eigentlich	eine neue Wendung wird angezeigt oder ein neuer Aspekt wird ins Thema eingebracht
mal	relativiert eine Aufforderung
doch	zeigt an, dass etwas bekannt sein sollte – aber nicht immer ist
nur	dient als Fokussierung / zur Verstärkung
wohl	ziemlich sichere Vermutung

7.3 Wünsche, Bedingungen, Ratschläge, höfliche Bitten

1.1

Was würden Sie tun, falls Sie mal eine Schreibblockade bekommen?
Ich habe nie Schreibblockaden. **Wenn ich beim Schreiben eine Blockade hätte, dann würde ich ab 9 Uhr morgens dafür sorgen, dass die Muse mich küsst.**
Schreiben Sie gerade an einem neuen Roman?
Zurzeit leider nicht. **Wenn ich an einem neuen Roman arbeiten könnte, dann würde ich einen weiteren historischen Roman schreiben.** Ein wahrhaft brenzliges Thema aus Würzburg im Jahr 1628.

Schade, wenn Sie Geschichte studiert hätten, dann wüssten Sie sofort, worum es dabei geht.
Gibt es Geheimtipps beim Schreiben eines historischen Romans?
Es gibt keine Geheimtipps! **Wenn ich tatsächlich bereits einen neuen historischen Roman geschrieben hätte, dann hätte ich profunde Recherche betrieben.** Weiter nichts!
Zu guter Letzt haben Sie noch das Wort speziell an Ihre Leser:
Bleiben Sie mir treu und vor allem offen für Neues.

1.2

Gegenwart (erfüllbar)	Vergangenheit (unerfüllbar)
Wenn ich beim Schreiben eine Blockade hätte, dann würde ich ab 9 Uhr morgens dafür sorgen, dass die Muse mich küsst.	**Wenn ich beim Schreiben eine Schreibblockade gehabt hätte, dann hätte ich ab 9 Uhr morgens dafür gesorgt, dass die Muse mich küsst.**
Wenn ich an einem neuen Roman arbeiten könnte, dann würde ich einen weiteren historischen Roman schreiben.	**Wenn ich an einem neuen Roman hätte arbeiten können, dann hätte ich einen weiteren historischen Roman geschrieben.**
Wenn Sie Geschichte studieren würden, dann würden Sie sofort wissen, worum es dabei geht.	*Wenn Sie Geschichte studiert hätten, dann hätten Sie sofort gewusst, worum es dabei geht.*
Wenn ich tatsächlich einen neuen historischen Roman schreiben würde, dann würde ich profunde Recherche betreiben.	*Wenn ich tatsächlich einen neuen historischen Roman geschrieben hätte, dann hätte ich profunde Recherche betrieben.*

2.1 **Könnte ich doch nur einen weiteren historischen Roman schreiben!**
Wenn ich doch nur einen weiteren historischen Roman schreiben könnte!
oder
Könnte ich doch nur an einem neuen Roman arbeiten!
Wenn ich doch nur an einem neuen Roman arbeiten könnte!

Hätten Sie doch nur Geschichte studiert!
Wenn Sie doch nur Geschichte studiert hätten.

Würde ich doch bloß tatsächlich einen neuen historischen Roman schreiben!
Wenn ich doch bloß tatsächlich einen neuen historischen Roman schreiben würde!

2.2

- Rudolf, ein Überlebender des Holocaust, der nach dem Krieg mit seiner Frau nach Australien zieht, realisiert im Alter, dass er seiner Familie nie all seine Gefühle gezeigt hat: „**Ich wünschte, ich hätte den Mut gehabt, meine Gefühle auszudrücken.**"
- Katharina hat eine Tochter, zu der sie früher ein enges Verhältnis hatte: „Ich dachte, diese Nähe würde immer bleiben. Aber das Leben und unsere Geschäftigkeit kamen dazwischen."
  Ähnlich ist es mit den Freundschaften der alten Dame, sie sind längst eingeschlafen, die

Freunde von früher sind nicht mehr auffindbar. „**Ich wünschte mir, ich hätte den Kontakt zu meinen Freunden aufrechterhalten.**"
- Hilde hat es zu einer der ersten weiblichen Managerinnen in ihrem Unternehmen gebracht, doch das Scheitern ihrer Ehe verwindet sie nicht: „**Ich wünschte, ich hätte mir erlaubt, glücklicher zu sein.**"

Oft werden tragische Geschichten geschildert. Aber nicht alle bedauern etwas:
- Erwin sagt: „Auch wenn ich die Möglichkeit gehabt hätte etwas zu ändern, so hätte ich doch nichts in meinem Leben geändert."

2.3

erfüllbarer Bedingungssatz:	unerfüllbarer / fiktiver Bedingungssatz:
Gegenwart	Vergangenheit:
Konjunktiv II Präteritum *Wenn ich könnte, drückte ich meine Gefühle aus.*	Konjunktiv II Plusquamperfekt *Wenn ich gekonnt hätte, hätte ich meine Gefühle ausgedrückt.*
würde + Infinitiv Präsens *Wenn ich könnte, würde ich meine Gefühle ausdrücken.*	
Konjunktiv II Präteritum *Wenn ich könnte,* **hielte ich den Kontakt zu meinen Freunden aufrecht.**	Konjunktiv II Plusquamperfekt *Wenn ich gekonnt hätte,* **hätte ich den Kontakt zu meinen Freunden aufrecht erhalten.**
würde + Infinitiv Präsens *Wenn ich könnte,* **würde ich den Kontakt zu meinen Freunden aufrecht erhalten.**	
Konjunktiv II Präteritum **Wenn ich könnte, erlaubte ich mir glücklicher zu sein.**	Konjunktiv II Plusquamperfekt **Wenn ich gekonnt hätte, hätte ich mir erlaubt glücklicher zu sein.**
würde + Infinitiv Präsens **Wenn ich könnte, würde ich mir erlauben glücklicher zu sein.**	

erfüllbarer Wunschsatz:	unerfüllbarer / fiktiver Wunschsatz:
Gegenwart	Vergangenheit:
Konjunktiv II Präteritum *Könnte ich doch meine Gefühle ausdrücken!*	Konjunktiv II Plusquamperfekt *Hätte ich doch meine eigenen Gefühle ausdrücken können!*
Konjunktiv II Präteritum (mit wenn) *Wenn ich doch meine eigenen Gefühle ausdrücken könnte!*	
Konjunktiv II Präteritum **Könnte ich bloß den Kontakt zu meinen Freunden aufrecht erhalten!**	Konjunktiv II Plusquamperfekt *Hätte ich bloß* **den Kontakt zu meinen Freunden aufrecht erhalten können!**
Konjunktiv II Präteritum (mit *wenn*) *Wenn ich bloß* **den Kontakt zu meinen Freunden aufrecht erhalten könnte.**	
Konjunktiv II Präteritum *Könnte ich nur* **glücklicher sein!**	Konjunktiv II Plusquamperfekt **Hätte ich nur glücklicher sein können!**
Konjunktiv II Präteritum (mit *wenn*) **Wenn ich nur glücklicher sein könnte!**	

erfüllbarer Konzessivsatz:	unerfüllbarer / fiktiver Konzessivsatz:
Gegenwart	Vergangenheit:
Konjunktiv II Präteritum *Auch wenn ich könnte, drückte ich meine Gefühle nicht aus.* *würde* + Infinitiv Präsens *Auch wenn ich könnte, würde ich meine Gefühle nicht ausdrücken.*	Konjunktiv II Plusquamperfekt: *Auch wenn ich gekonnt hätte, hätte ich meine Gefühle nicht ausgedrückt.*
Konjunktiv II Präteritum *Auch wenn ich könnte,* **hielte ich den Kontakt zu meinen Freunden nicht aufrecht.** *würde* + Infinitiv Präsens *Auch wenn ich könnte,* **würde ich den Kontakt zu meinen Freunden nicht aufrecht erhalten.**	Konjunktiv II Plusquamperfekt: *Auch wenn ich gekonnt hätte,* **hätte ich den Kontakt zu meinen Freunden nicht aufrecht erhalten.**
Konjunktiv II Präteritum **Auch wenn ich könnte, erlaubte ich mir nicht glücklicher zu sein.** *würde* + Infinitiv Präsens **Auch wenn ich könnte, würde ich mir nicht erlauben glücklicher zu sein.**	Konjunktiv II Plusquamperfekt: **Auch wenn ich gekonnt hätte, hätte ich mir nicht erlaubt glücklicher zu sein.**

3.1

Gibt es noch jemanden, dem Sie viel zu verdanken haben?
Schauspielerin: Ja, einem Lehrer auf der Schauspielschule. Er hat uns bayerische Schauspieler immer bestärkt. Dafür bin ich sehr dankbar, weil das damals keine leichte Zeit war. **Ich hatte zwischendurch immer mal Zweifel, ob ich das mit dem Hochdeutschen hinkriegen würde.** → wirklich
Sie haben ja bisher auch fast nur Mundart-Rollen gespielt.
Schauspielerin: **Ja, und selbst wenn's keine Mundart war, habe ich's so gesprochen, als ob es Mundart gewesen wäre.** → nicht wirklich

Ich hab schon ein paarmal auf → nicht wirklich
Hochdeutsch gespielt, aber ich brauche nicht so zu tun, als wäre das meine Stärke, denn die ist wirklich der Dialekt. Es ist nicht so, dass es den → wirklich
Zuschauer nicht doch stören würde, wenn im Film der falsche Dialekt gesprochen wird.

3.2 Fiktive Vergleichssätze

- Fiktive Vergleichssätze drücken einen Vergleich aus, der möglich, aber nicht **wirklich** ist.
- Die Subjunktionen *als, als ob, als wenn* und *wie wenn* leiten die Vergleichssätze ein und werden häufig für Vergleiche mit einem Verb im Konjunktiv II verwendet:
  Als ob sie die Absicht gehabt hätte … Als hätte er nichts zu tun.

Negierte Folgesätze

- Die Folge in negierten Folgesätzen ist möglich oder wahrscheinlich, wird aber negiert.
- Der negierte Folgesatz wird mit **dass** bzw. *als dass* gebildet.
- Im **Hauptsatz** stehen meist abstufende Ausdrücke wie *nicht so, zu, zu wenig, u.a.*
- Im Gegensatz zum Konditionalsatz steht nur der Nebensatz bzw. Folgesatz im **Konjunktiv II**:
  Es ist nicht so, dass es den Zuschauer nicht doch stören würde, wenn im Film der falsche Dialekt gesprochen wird.

3.3   b   Er spricht so deutlich, **als ob / als wenn** er seine Aussprache ewig vor dem Spiegel geübt hätte.

      c   Ich verstehe die Schauspielerin so schlecht, **als ob / als wenn** sie Kaugummi kauen würde.

      d   Die Großeltern applaudieren nach der Vorstellung so laut, **als ob / als wenn** es das beste
Theaterstück gewesen wäre, was sie jemals gesehen haben.

      e   Das wäre ja so, **als ob / als wenn / wie wenn** ich zu dir sagen würde: „Du sprichst zu undeutlich!"

4.1

HANS:	Hallo?
CLAUDIA:	Hi, hier ist Claudia. Hast du deine Bewerbung für das Praktikum schon abgeschickt?
HANS:	Äh, nein, habe ich noch nicht.
CLAUDIA:	Das **solltest** du besser mal schnell machen.
HANS:	Das **könnte** schon sein, allerdings ist meine Druckerpatrone alle und …
CLAUDIA:	Erzähl keinen Unsinn! Geh in einen Copyshop und druck es auf gutes Papier.
HANS:	Hm, das **müsste** funktionieren. Aber dann ist es zu spät, um es heute noch zur Post zu bringen.
CLAUDIA:	Es gibt doch noch die Post am Bahnhof, die hat bis 22 Uhr geöffnet. Das schaffst selbst du auch noch bis dahin!
HANS:	Das **dürfte** schon klappen.
CLAUDIA:	Und das wird es auch. Mach dich auf die Socken!

4.2

Ratschlag, Vermutung, Möglichkeit	
Funktion	**Beispiel**
vorsichtige Vermutung	Das dürfte schon klappen.
Feststellung einer Möglichkeit	**Das könnte schon sein, allerdings …**
starke Vermutung	**Hm, das müsste funktionieren.**
Ratschlag	**Das solltest du besser mal schnell machen.**

5.1   › Die Berliner Mundart wird zum Teil als unfreundlich empfunden, weil sie sehr direkt wirkt. Bei dieser
Aufgabe gibt es keine richtige oder falsche Lösung, da das Empfinden unterschiedlich ist.
Folgende Zuordnungen wären möglich:

An der Bushaltestelle: **eher unfreundlich**
Mit einem Zeitungsverkäufer: **eher unfreundlich**
Im Restaurant: **neutral**
Beim Bäcker: **neutral**
Im Restaurant: **eher unfreundlich**

5.3

Funktionen	Beispiel
höfliche Aufforderung und Bitte	**Ich hätte gerne drei normale Brötchen.**
höfliche Frage	Könnten Sie mir sagen, ob Sie Richtung Alex fahren? **Ich hätte gern gewusst, in welche Richtung es hier zur Volksbühne geht.**
höfliche Aussage	Ich würde nichts aus der Dose essen.

5.4   Zeitungsverkäufer: **Sie sollten eher den Straßen-Wisser-Mann fragen.**
Wirtin: **Wenn ich Sie wäre, würde ich nichts aus der Dose essen.**
Bäckerin: **Sie sollten Schrippe sagen. An ihrer Stelle würde ich das üben.**

7.4 Zitieren und Berichten: Konjunktiv als Mittel der indirekten Rede

1.1 **①** Beschreibung: **direkte Rede**

② Beschreibung: **indirekte Rede, formell, distanziert**

③ Beschreibung: **indirekte Rede, umgangssprachlich, vertraut**

1.2 Zitieren und Berichten

- In der geschriebenen Sprache wird der **Konjunktiv I** verwendet (Text 2).
  Damit stellen z. B. Nachrichtensprecher oder Journalisten eine gewisse Distanz zu den zitierten Aussagen her und übernehmen keine Garantie für den Wahrheitsgehalt.
  Die Jobchancen der Hochschulabsolventen <u>seien</u> generell sehr gut.
- Der Konjunktiv I ist dem **Konjunktiv II** vorzuziehen, wenn er sich vom Präsens Indikativ unterscheidet:
  Das entspreche nicht der Realität.
- Der Konjunktiv I wird oft bei *sein*, den Singular-Formen der Modalverben
  (*müsse, könne, solle*, etc.) und bei Verben, die in der 3. Person Singular des Konjunktivs I auf -e enden (*komme, liege, sehe*, etc.) in der indirekten Rede verwendet.
- Für die indirekte Rede wird in der gesprochenen Sprache und in der Umgangssprache (Text 3) oft der **Indikativ** oder die ***würde*-Umschreibung** verwendet:
  Er hat mir vorhin erzählt, dass er schon seit 3 Monaten ein Praktikum sucht.
  Er hat mir vorhin erzählt, dass er schon seit 3 Monaten ein Praktikum suchen würde.

1.3 Zukunfts- und Vergangenheitsform des Konjunktiv I

- Der Konjunktiv I hat nur eine Vergangenheitsform (Perfekt), diese wird mit dem Konjunktiv I von *haben* oder *sein* und dem Partizip II gebildet.
- Die Zukunftsform Futur I wird mit dem Konjunktiv von *werden* + Infinitiv gebildet.

	es entspricht / sie entsprechen	es ist / sie sind	es gibt / sie geben
Gegenwart Konjunktiv I			
ich	entspreche*	sei	gebe*
er / sie / es	**entspreche**	**sei**	**gebe**
sie (3. Person Plural)	entsprechen*	**seien**	geben*
Vergangenheit Konjunktiv I			
	es habe / sie haben entsprochen	es sei / sie seien gewesen	es habe / sie haben gegeben
Futur Konjunktiv I			
	es werde / sie werden entsprechen	es werde / sie werden sein	es werde / sie werden geben

1.4 Beispiellösung:

> Die meisten Teilnehmer der Studie hätten einen Magistertitel. Über unterschiedliche Jobchancen von Bachelor- und Magisterabsolventen könne deshalb aus dieser Studie noch nicht allzu viel geschlossen werden. Die Einkommen der Bachelor-Absolventen, die Vollzeit arbeiten, liege etwas unter dem Durchschnitt und sie seien häufiger befristet beschäftigt.
> Für eine "Generation Praktikum" spricht jedoch ein anderes Ergebnis der Studie: 62 Prozent der Absolventen hätten während ihrer Ausbildung ein Praktikum gemacht. Dieses Argument ließe er / sie jedoch nicht gelten, da es bei der Diskussion vor allem um fertige Akademiker gehe.

1.6

Man kommt nicht um ein Praktikum drum herum

Auch Barbara Kasper, Jugendsekretärin, sagt im Gespräch, dass sie sehr wohl viele junge Menschen **kenne**, die ein Praktikum – auch unbezahlt – absolvieren. „Teilweise kommt man nicht darum herum." Sie **spreche** aus eigener Erfahrung, sagt Kasper. Im Bachelorstudium **sei** sie verpflichtet **gewesen** ein Praktikum zu machen, um das Studium abschließen zu können. Viele **würden** es in Kauf **nehmen**, keine Bezahlung zu bekommen, anstatt länger zu warten. Sie **zögen** es vor, das Studium abzuschließen. „Oft genug hat man keine andere Wahl", sagt Kasper.

In ihrer Rolle als Jugendsekretärin versucht Kasper jedenfalls Betroffene, die sich an sie wenden, aufzuklären: welche Rechte und Pflichten man als Praktikant hat und dass von einem Praktikanten nicht dasselbe verlangt werden **könne**, wie von einem regulären Angestellten.

„Wie weltfremd muss man sein"

„Wie weltfremd muss man sein, um dann zu resümieren, dass es keine Generation Praktikum gibt?", rügt Schatz die Studie. Die Studie **habe** zudem **ausgeklammert**, dass viele Praktika bereits während des Studiums erfolgten. Danach **solle** es keine Praktika mehr geben, findet Schatz auch wenn es aber oft nicht der Realität **entspreche**.

„Praktika finden in Übergangsphasen statt"

Anna Schopf hat die Plattform Generation Praktikum gegründet. Schon vor mehreren Jahren **habe** sie **begonnen**, sich mit der Thematik zu beschäftigen und sie **finde** es daher erstaunlich, dass heute überhaupt noch thematisiert wird, ob im Zusammenhang mit der Generation Praktikum von einem Mythos gesprochen werden **könne**. „Praktika finden in Übergangsphasen statt', sagt sie. Gerade im ersten Jahr nach Beendigung des Studiums **seien** Praktika Gang und Gäbe.

2.1

MARKUS: Hallo. Du schaust ja nicht grad glücklich aus. Was ist denn passiert?

HOLGER: Ich warte auf meine Frau. Ihr Zug ist irgendwo in Bitterfeld stehen geblieben, es war wohl wieder mal 'ne Weiche kaputt. **Es kam eine Durchsage, dass der Zug außerplanmäßig in Halle halten würde und die Passagiere die S-Bahn nach Leipzig nehmen sollen. Dort würden sie dann einen anderen Zug nach München bekommen.**[1]

MARKUS: Ja, und?

HOLGER: **Maria meinte, dass sie sich noch schnell ein S-Bahn Ticket kaufen würde.**[2] Aber natürlich war vor dem Ticketautomaten eine ewig lange Schlange, da ist ihr die S-Bahn vor der Nase weggefahren.

MARKUS: Ja, aber da muss es doch irgendwelche anderen Möglichkeiten geben, dass sie von dort weiter nach München kommt!

HOLGER: Sollte man meinen, ist aber wohl kompliziert. **Jetzt hat sie mir gesagt, dass sie von Halle die Regionalbahn nach Naumburg nimmt und dann mit dem ICE nach München weiterfährt.**[3] Aber wie's so kommt: Auch der Zug hatte Verspätung! Jetzt sitzt sie in Naumburg fest, auch nicht viel besser.

MARKUS: Das ist ja blöd. Wann geht denn jetzt der nächste Zug nach München?

HOLGER: **Sie haben ihr gesagt, dass der nächste Zug in einer Stunde abfahren würde.**[4] Du wirst es nicht glauben: **Jetzt hat sie mir gesagt, dass auch dieser Zug Verspätung hat.**[5] Manchmal könnt' man schon verzweifeln!

MARKUS: Unglaublich! Da brauchst du wirklich gute Nerven. Gönn dir doch einen Kaffee! Sorry, ich muss jetzt leider los, mein Zug kommt – der ist nämlich ausnahmsweise nicht verspätet.

HOLGER: Danke danke, gute Nerven kann ich brauchen. Mach's gut und bis bald!

MARKUS: Ciao!

2.2 **2** Maria: „**Ich kaufe mir noch schnell ein S-Bahn-Ticket**".

3 Maria: „**Ich nehme die Regionalbahn von Halle nach Naumburg und fahre dann mit dem ICE nach München weiter.**"

4 Bahndurchsage: „**Der nächste Zug fährt in einer Stunde ab.**"

5 Maria: „**Auch dieser Zug hat Verspätung.**"

3 **Beispiellösung:**

Der Schaffner hat uns mitgeteilt, dass der Regional Express aus Aachen zur Weiterfahrt nach Hamm mal wieder Verspätung hat. Er sagte, der Zug würde heute in den Bereichen A bis C halten. Deshalb bat er die Fahrgäste, die sich in dem Abschnitt D befinden, in die Abschnitte A bis C zu kommen. Dadurch würden wir helfen, weitere Verzögerungen zu vermeiden.

4 **Beispiellösung:**

Die Beraterin hat gesagt, dass die Reise 5 Tage dauert. Sie meinte, dass wir am Gründonnerstag starten und dann am Ostermontag gegen 18.00 Uhr zurück in Augsburg wären. Sie erklärte, dass wir auf dem Weg nach Marburg einen Zwischenstopp in Weimar machen. Weiterhin sagte sie, dass wir, bevor wir nach Marburg fahren, noch die Möglichkeit haben das Bauhaus in Dessau und die Innenstadt von Weimar zu besichtigen. Außerdem meinte sie, dass wir nach einer Übernachtung in Weimar Karfreitag gegen Mittag in Marburg ankommen.
Sie versprach, dass unser Hotel direkt an der Lahn liegt und wir von dort aus jeden Tag verschiedene Ausflüge starten. Sie gab mir eine Broschüre und sagte, dass es sich bei den Ausflügen um freiwillige Angebote handelt. Zudem meinte Sie, dass der Besuch des Schlosses und eine Radtour ins schöne Hinterland sehr zu empfehlen sind und eine Kanutour auf der Lahn von Marburg nach Wetzlar etwas ganz besonderes ist.

8 Negation

1

Interview mit Alice Schwarzer

von Roger Köppel

*Frau Schwarzer, was machen Sie eigentlich, wenn Sie sich **nicht** mit der Sache der Frau beschäftigen?*
Alice Schwarzer: Mich interessiert alles **außer** Sport. Diesen Luxus leiste ich mir. Dann gibt es Domänen, in denen ich bedauerlicherweise **nicht** ausreichend gebildet bin, Naturwissenschaften zum Beispiel. Sonst bin ich in Bewegung. Mich interessiert, was aufkommt.

Stören Sie Etiketten wie ‚Feministin' und ‚Emanze'?
Schwarzer: **Nein**, aber ich habe **noch nie** ein Abzeichen getragen. Mir ist jede Etikettierung fremd. Selbst in den Hoch-Zeiten des Feminismus habe ich **keine** Frauenzeichen getragen …

*Heute stimmen Ihnen doch alle zu. Es traut sich fast **niemand** mehr, Sie zu kritisieren.*
Schwarzer: Tatsächlich? Und wenn, es stört mich **nicht**, recht zu haben.

*Sie haben Ihre Karriere aus dem **Nichts** aufgebaut. Beneiden Sie Leute, die aus besseren Startpositionen heraus ihr Leben gestalten konnten?*
Schwarzer: Beneiden? Warum sollte ich? Eine Karriere scheint mir **keinesfalls** erstrebenswert um ihrer selbst willen. Ich bin Schritt für Schritt vorangegangen, weil mich die Art der Tätigkeit und die Inhalte interessierten. Ich habe lediglich darauf geachtet, finanziell **un**abhängig zu bleiben. [...]

Wie war Deutschland, als Sie Mitte der siebziger Jahre aus Frankreich zurückkamen?
Schwarzer: Das war ein Kulturschock für mich. Am **un**erträglichsten war das Schwarzweißdenken. Vor allem in Berlin herrschte eine hochneurotische Stimmung. Es gab nur links oder rechts, richtig oder falsch, Parole. Man musste sich in einem bestimmten Vokabular ausdrücken, sonst gehörte man **nicht** dazu.

Hat die Frauenbewegung die Männer verändert?
Schwarzer: Es brauchte ja **nicht** die Frauenbewegung, um die Männer etwas menschlicher werden zu lassen. Mein Großvater, der ein sehr liebenswerter Mann war, hätte doch als junger Mann **nie** einen Kinderwagen angefasst – mich, seine Enkelin, aber hat er gewickelt und gefüttert. In den Vierzigern. In den letzten dreißig Jahren nun hat eine wahre Kulturrevolution stattgefunden. Frauen und Männer haben sich verändert. Alle Untersuchungen, auch die von EMMA in Auftrag gegebene aktuelle Umfrage, zeigen: Zwei Drittel der Männer sind auf unserer Seite, finden die Emanzipation gut. Nur ein Drittel bleibt hart dagegen.

*Die Männer sind weiblicher und damit für Frauen **un**attraktiver geworden?*
Schwarzer: Ach was? Weil sie mal Ohrringe oder Kajal tragen? **Nein**, im Ernst, natürlich sind Zeiten der Veränderung Zeiten der Verunsicherung. Aber wir schaffen das schon.
[...]

2

Zu sensibel für diese Welt?

Der Nachbar hat mich eben im Treppenhaus **nicht** gegrüßt – was hat der auf einmal gegen mich?
Helen hat **noch nicht** auf meine Mail geantwortet – meine Einladung ist für sie wohl **nicht** wichtig genug.
Und was reden die beiden in der Ecke da? Ich hoffe, die tratschen **nicht** über mich.
Mein Chef hat mich seit Monaten **nicht mehr** gelobt, ich glaube, da stimmt etwas **nicht**.
Ich habe mir auf der Betriebsfeier wirklich Mühe gegeben, aber Martin hat **nicht** mich, sondern diese neue Sekretärin zum Tanzen aufgefordert.
Die finnischen Vokabeln kann ich mir überhaupt **nicht** merken, bin ich zu doof fürs Sprachen lernen?
Ich versuche schon seit 20 Minuten, einen Kaffee zu bestellen, aber der Kellner sieht mich **nicht** – bin ich etwa Luft für die anderen Menschen?

3 Die Stellung von *nicht*

- Bei der **Satznegation** negiert *nicht* die Aussage des ganzen Satzes, es steht eher am Ende des Satzes, nach Akkusativ- und Dativergänzung:
  ... aber der Kellner sieht mich nicht ...
- Die Satznegation mit *nicht* steht bei mehrteiligen Verbformen immer vor dem zweiten, dem infiniten Verbteil (Partizip II, Infinitiv oder Präfix eines trennbaren Verbs):
  Der Nachbar hat mich eben im Treppenhaus nicht gegrüßt ...
- *Nicht* steht auch immer vor Konstruktionen von *sein / werden / bleiben* + Adjektiv / Nomen (Prädikativ-Ergänzung):
  ... meine Einladung ist für sie wohl nicht wichtig genug.
- *Nicht* steht meist vor Präpositionalergänzungen:
  Helen hat noch nicht auf meine Mail geantwortet ...

4.1 › Satzteilnegation

4.2
  ☒ **Sie ist hübsch.**   ☐ Sie ist wunderschön.
  ☐ Sie hat Komplexe.   ☒ **Sie ist nicht arrogant.**

5.1

> ➤ Es sollten **keine** Grillroste verwendet werden, an denen alte Essensreste kleben, nach jedem Grillvorgang muss der Rost gut gereinigt werden.
>
> ➤ Auf **keinen** Fall sollte man den Grill mit Spiritus anzünden. Durch die plötzlich hochschießende Stichflamme passieren jedes Jahr rund 4000 Unfälle in Deutschland.
>
> ➤ Geschmacklich macht es **keinen** Unterschied, ob man mit Gas, Strom oder Kohle grillt.
>
> ➤ Man darf **kein** Nadelholz zum Grillen verwenden, es enthält zu viel Harz, das ungesund ist und den Geschmack des Fleisches zerstört. Wenn schon Holz, dann Laubholz wie Birke oder Buche.
>
> ➤ Gepökeltes Fleisch wie Kassler, Schinken oder Wiener Würstchen darf **nicht** gegrillt werden, beim Grillen dieser Produkte können krebserregende Stoffe entstehen.
>
> ➤ Die Gabel ist **kein** geeignetes Grillbesteck – Würstchen und Steaks sollten **nicht** mit der Gabel gewendet werden. Mit einer Zange werden die Poren des Fleisches **nicht** verletzt und es trocknet **nicht** so schnell aus.
>
> ➤ Das Fleisch sollte **nicht** zu oft gewendet werden. Besser ist es, das Fleisch von beiden Seiten anzubraten und es dann abseits der direkten Glut fertig zu garen.
>
> ➤ Der Kartoffelsalat oder Nudelsalat sollte **nicht** in fetter Mayonnaise schwimmen, eine Salatsoße aus Olivenöl und Essig ist besser für die Verdauung.
>
> ➤ Sie haben noch **keine** Idee für den Nachtisch? Probieren Sie doch mal gegrillte Äpfel, Birnen oder Bananen.

5.2 Beispiellösung:

b „Es gibt keine Holzkohle mehr, benutzen wir doch das Fichtenholz."
Nein, zum Grillen benutzt man kein Nadelholz, sondern Laubholz.
c „Werfen wir doch noch die Wiener Würstchen auf den Grill."
Zum Grillen nimmt man kein gepökeltes, sondern nur rohes Fleisch.
d „Ich benutze immer eine Gabel zum Drehen der Steaks."
Du solltest keine Gabel benutzen, sondern eine Grillzange.
e „Ich habe einen Kartoffelsalat mit leckerer Mayonnaise mitgebracht, wie findest du den?"
Nun, den finde ich nicht so gesund. Ich mache meinen Kartoffelsalat nicht mit Mayo, sondern mit Essig und Öl.

5.3   **Beispiellösung:**

Keine Angst vor klaren Regeln, aber man sollte seinem Kind nicht zu viel verbieten.
Erklären Sie Ihrem Kind die Regeln, warum es z.B. abends nicht länger aufbleiben darf.
Sie können sowieso nicht alles kontrollieren – also vertrauen Sie Ihrem Kind.
Nicht immer gleich laut werden!

Wenn Sie keine Ahnung vom Kochen haben, versuchen Sie sich zuerst an einfachen Gerichten, sonst haben Sie keine Erfolgserlebnisse und sind schnell frustriert.
Nicht gleich ein 5-Gänge-Menü planen. Überraschen Sie stattdessen Ihre Familie mit einem Hauptgericht und einem unkomplizierten Dessert.
Suchen Sie am Anfang Rezepte aus, bei denen Sie die Zutaten kennen, damit Sie nicht zu viel Zeit damit verbringen, ein exotisches Gewürz im Fachhandel zu finden.

6

Auf den Brockengipfel – eine einsame Bergwanderung 1925

Das Frühstück in meiner Pension war wunderbar, ich habe **nichts** auf dem Teller gelassen. Satt und zufrieden machte ich mich auf den Weg zum Brocken, noch **nie** habe ich mich so auf ein Wanderziel gefreut, wie auf diesen wilden Hexentanzplatz! Ich ging schnell voran und erreichte so auch recht bald Torfhaus, wo es zu regnen und zu schneien begann. Ab Torfhaus wurde es sehr einsam um mich herum – **niemand** begegnete mir auf den Bergpfaden, **nirgends** sah ich Spuren im tiefen Schnee. Je höher man den Berg hinaufsteigt, desto kürzer, zwergenhafter werden die Tannen, sie scheinen immer mehr zu schrumpfen, bis nur noch Heidelbeersträucher und Bergkräuter übrig bleiben. Es tut so gut, allein durch die stille Natur zu wandern und mit **niemandem** ,Konversation' machen zu müssen, **niemals** werde ich mich an den Lärm der Großstadt gewöhnen ... So erstieg ich ohne wesentliche Schwierigkeiten den höchsten Berg des Harzes und staunte sehr über die Windgeschwindigkeiten, die dort auf dem Gipfel zustande kommen ...

7   **Beispiellösung:**

Ich wollte schon immer mal ganz allein zelten, am besten in einer dünn besiedelten Region, zum Beispiel in Nordschweden. Tagelang mit niemandem sprechen – herrlich. Nirgendwo muss ich mir Gedanken machen, wie ich aussehe, ich muss mich nicht unterhalten, nichts kann mich aus der Ruhe bringen. Nach einem Samstagnachmittag in der vollen Innenstadt kann ich mir nichts Schöneres vorstellen.

8.1

Ich bin Mitte 20 und habe es in meinem Leben zu nichts, aber auch gar nichts gebracht. Andere Frauen haben einen Mann, eine Familie, ein Haus und sind glücklich. Die große Liebe habe ich nie gefunden, ich bin einsam, erfolg**los**, arbeits**los**, **un**zufrieden und **un**fähig. Meine trost**lose** Situation **miss**fällt mir, aber ich kann sie nicht verbessern. Ich brauche einen totalen Neuanfang, helfen Sie einer **un**glücklichen Frau und sagen Sie mir, was ich tun soll!

8.3   **Beispiellösung:**

Nadine,
09.12.2012 20:45

Liebe Claudi,
Sie klingen wirklich sehr unglücklich. Dabei sind Sie doch jung und ungebunden, Ihnen stehen so viele Möglichkeiten offen.
Lassen Sie sich von kleinen Misserfolgen nicht unterkriegen! Statt sich einzureden, dass Sie unfähig und erfolglos sind, sollten Sie etwas optimistischer auf Ihr Leben blicken. Sie sind alles andere als chancenlos!

9.1   **Ich gehe ziemlich / recht / ganz gern shoppen.**

9.2   **Beispiellösung:**

Ich mag es überhaupt nicht, wenn mir distanzlose Verkäuferinnen etwas aufschwatzen wollen. Wenn ich sage, dass ich mich erst einmal umschauen möchte und sie mir trotzdem ständig etwas empfehlen, bin ich auch nicht gerade freundlich.
Oft verstehen die Verkäufer auch nicht, was mir gefällt und was mir nicht gefällt.

10   **Der Prinz hat Emilia mit Gefallen angesehen.**

9   Textstruktur und Textaufbau

9.1   Erzählen

9.1.1   Literarische Erzählung: „Es war einmal . . .“

1.1   a   Es gibt auf Rhodos **eine alte Geschichte, die im Volk erzählt wird, wie die Insel entstanden ist.**

b   Vor einiger Zeit kehrte spät abends im „Goldenen Löwen" zu Kassel **ein elegant, aber nachlässig gekleideter Fremder** ein, der augenscheinlich eine längere Fußtourgemacht hatte.

c   Vor langer, langer Zeit **lebte die Schneiderin Anna.**

d   Unter meinen Jugendbekannten war **der Sohn eines berühmten Arztes, dessen Geschichte so tragisch ist, dass sie nicht in Vergessenheit geraten darf.**

1.2   a   **die Enstehungsgeschichteder Insel Rhodos**   c   **die Schneiderin Anna**

b   **der Fremde und sein Besuch im Gasthaus „Goldener Löwe"**   d   **der Schulfreund des Erzählers**

2.1

> ### Wilhelm Busch: Eine Nachtgeschichte
>
> Vor einiger Zeit kehrte spät abends im „Goldenen Löwen" zu Kassel **ein elegant, aber nachlässig gekleideter Fremder** ein, der augenscheinlich eine längere Fußtour gemacht hatte. Aus **seinen schmerzlichen Zügen** sprach eine stille Verzweiflung, ein heimlicher Kummer musste **seine Seele** belasten. **Er** aß nur äußerst wenig und ließ sich bald **sein Schlafzimmer** anweisen.
>
> Es mochte wohl eine Viertelstunde später und nahezu Mitternacht sein, als der Kellner an Nr. 6, dem Zimmer **des Fremden**, vorüberkam. Ein lautes, herzzerreißendes Ächzen und Stöhnen drang daraus hervor. Dem erschrockenen Kellner erstarrte das Blut in den Adern. Irgendetwas Entsetzliches musste da vorgehen. Schleunige Hilfe tat Not; der Kellner stürzt zur Polizei.
>
> Unterdessen hat die Regierungsrätin v.Z., welche in Nr. 7 schläft, dieselbe schreckliche Entdeckung gemacht und bereits das ganze Wirtshaus in Alarm gebracht, als der Kellner mit der Polizei zurückkommt. Man dringt nun sofort in das Zimmer **des Fremden**. Aber leider kam die Hilfe zu spät, denn **derselbe** hatte bereits in Ermangelung eines anderen Instrumentes mit eigener Hand unter Schmerzen und Wehklagen **seine engen Stiefel** ausgezogen.

2.2

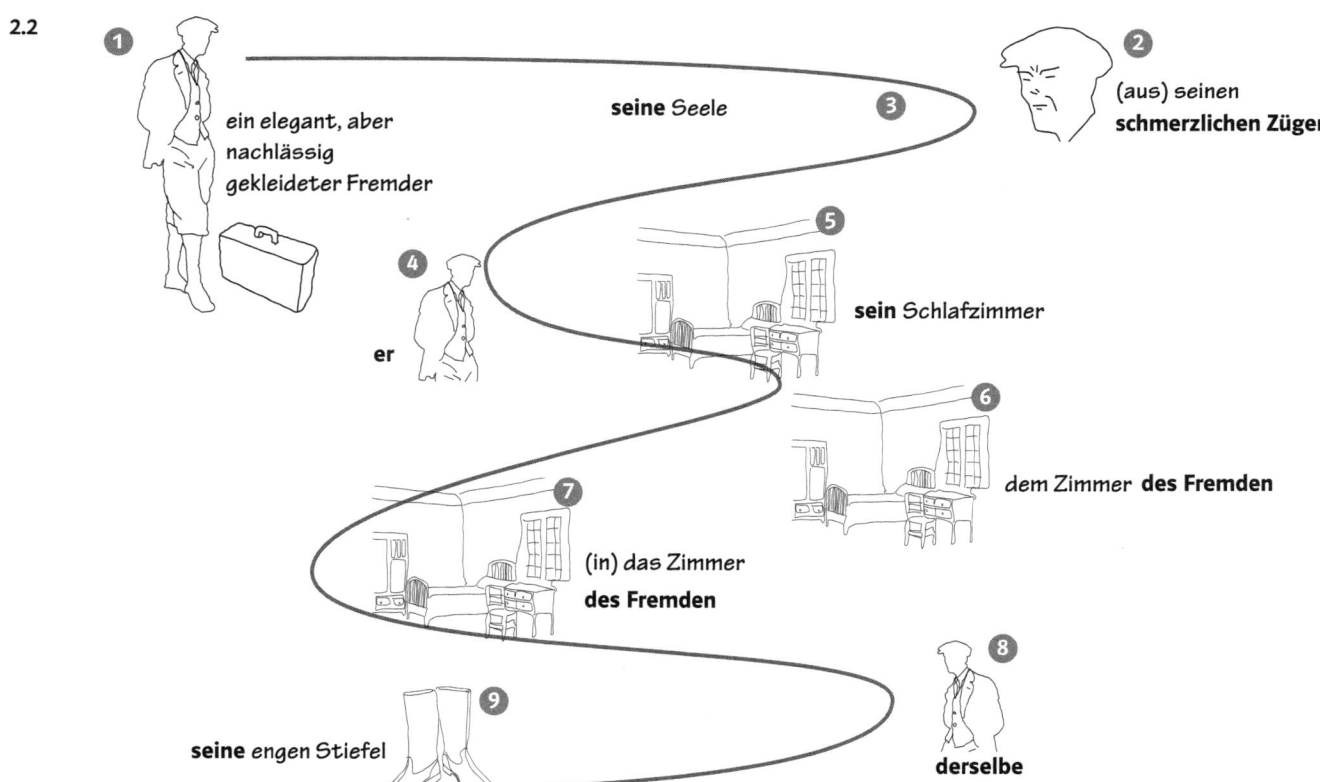

① ein elegant, aber nachlässig gekleideter Fremder

seine Seele   ③

② (aus) seinen **schmerzlichen Zügen**

④ **er**

⑤ **sein** Schlafzimmer

⑥ dem Zimmer **des Fremden**

⑦ (in) das Zimmer **des Fremden**

⑧ **derselbe**

⑨ **seine** engen Stiefel

2.3 **1** Der Protagonist ist noch nicht bekannt, das signalisiert der unbestimmte Artikel *ein*. Hier wird der Protagonist eingeführt und ein wenig beschrieben.

2 Wir haben den Protagonisten schon in **1** kennen gelernt, jetzt wird über einen Teil von ihm gesprochen, *seine Züge* (=die Mimik). *Seine* zeigt, dass es immer noch um die gleiche Person geht. Wir haben unsere Aufmerksamkeit schon auf ihn gerichtet und sollen diese Orientierung beibehalten.

3 **Es wird wieder über einen anderen Teil des Protagonisten gesprochen, *seine Seele*. *Seine* zeigt wiederum, dass die Orientierung beibehalten werden soll.**

4 **Noch immer geht es um den Protagonisten. *Er* zeigt an, dass wir die Orientierung beibehalten sollen.**

5 **Nun geht es um das Zimmer, in dem der Protagonist übernachtet. *Sein* zeigt an, dass die Orientierung auf ihn immer noch gültig ist.**

6 Um wen geht es nun? Wir kennen die Person schon, deshalb steht der bestimmte Artikel *des*. Von einem *Fremden* war schon am Textanfang die Rede, es ist also wieder der Protagonist. Wir sollen uns also erneut auf ihn orientieren, nachdem eben die Aufmerksamkeit auf den Kellner gerichtet war.

7 **Wie schon in **6** haben wir unsere Aufmerksamkeit auf andere Personen gerichtet (den Kellner, die Regierungsrätin) und sollen uns nun wieder dem Protagonisten zuwenden. Er wird mit dem bestimmten Artikel *des* und dem schon eingeführten Nomen *Fremden* benannt.**

8 Es ist viel Bewegung in die Geschichte gekommen, das ganze Wirtshaus ist aufgeregt, die Polizei ist auch schon da. In dieser Szene soll die Aufmerksamkeit wieder auf den Protagonisten gerichtet werden: *derselbe* zeigt an, dass wir eine Person fokussieren sollen, die schon vorkam: den Fremden.

9 Schließlich geht es um ein Kleidungsstück des Protagonisten, *seine Stiefel*. Wie schon mehrmals zuvor, zeigt *seine* an, dass die Orientierung auf ihn noch immer besteht.

2.4 daraus: **aus dem Zimmer**
dieselbe schreckliche Entdeckung: **die Entdeckung, dass aus dem Zimmer Ächzen und Stöhnen dringt**

3.2

	Originalgeschichte	Veränderte Version
Andeutungen oder Vermutungen?	**Man weiß nicht, was mit dem Fremden los ist. Es gibt aber mehrere Andeutungen und Vermutungen, dass etwas nicht stimmt:** Am Anfang der Geschichte spekuliert der Erzähler darüber, wie es dem Fremden geht: „eine stille Verzweiflung, ein heimlicher Kummer musste seine Seele belasten". Er verwendet das Modalverb *müssen* in subjektiver Bedeutung und drückt dadurch aus, dass er sich mit seiner Vermutung sehr sicher ist. Als Leser wird man nun neugierig: Warum ist der Fremde wohl verzweifelt oder traurig? Als der Kellner an dem Zimmer vorbeigeht, vermutet er, dass etwas Schreckliches passiert: „Irgendetwas Entsetzliches musste da vorgehen." Man möchte wissen, was da passiert ist, erhält aber immer noch keine Lösung.	**Die Andeutungen und Vermutungen fehlen. Deshalb hat man als Leser gar nicht erst den Eindruck, dass mit dem Fremden etwas nicht stimmt und ist auch nicht neugierig.**
Adjektive zur Beschreibung von Gefühlen und Stimmung?	**Der Fremde wird durch zahlreiche Adjektive beschrieben: Dadurch wird betont, dass es etwas Geheimnisvolles an ihm gibt („ein heimlicher Kummer"). Durch die Adjektive wird ein Gegensatz zwischen seinem Äußerem und Inneren deutlich. Als der Kellner an dem Zimmer vorübergeht, lassen vor allem die verwendeten Adjektive an etwas Furchtbares denken („herzzereißend", „schrecklich").**	**Es werden kaum Adjektive verwendet.**
Tempus?	**Es gibt einen Wechsel vom Präteritum ins Präsens: „Schleunige Hilfe tat Not; der Kellner stürzt zur Polizei." Der Wechsel vom Präteritum zum Präsens erhöht die Spannung, da er eine größere Nähe herstellt: Man hat als Leser das Gefühl, die Handlung mitzuerleben.**	**Es wird vor allem Präteritum verwendet.**
Verzögerung?	**Die Auflösung der Geschichte wird durch die Regierungsrätin verzögert, die das gleiche wie der Kellner hört. Man muss dadurch länger auf die Erklärung warten. Auch im letzten Satz steht die lang erwartete Information erst ganz am Ende - viele Satzglieder sind dazwischen geschoben.**	**Das Rätsel wird ebenfalls erst am Ende gelöst, aber es gibt keine zusätzlichen Verzögerungen.**

9.1.2 Alltägliche Erzählung: „Jetzt muss ich noch was erzählen ..."

1.3 / 1.4

1 und ich dann im Nachthemd | und dann mein Mantel über |

2 und dann auf der Tour meiner Tochter | und fahr dann den Neustädter Berg hoch |

3 kommt da denn die Polizei. Fahrzeugkontrolle. |

4 und nu wollten sie mal gucken | und da hab ich gesagt „was". | Ja äh und dann haben sie gesagt „Und hier ist also/ die TÜV-Plakette ist abgelaufen." | und ich sag "Was? Wie viele Polizisten stehen hier? Eins, zwei drei" Ich war geladen. | „Na, wenn sie damit ihr Geld verdienen!" hab ich gesagt, "na dann also wirklich, Halleluja." |

5 Weißt du, was das gekostet hat? Hundertzwanzig Mark!

> Die Handlungen werden in chronologischer Reihenfolge genannt und häufig durch und bzw. und dann miteinander verknüpft. Auffällig ist außerdem, dass das Verb manchmal an erster Stelle steht und das Subjekt ich weggelassen wird. Das ist typisch für mündliches Erzählen.

2

Erzählerin	Zuhörer
– *Die Erzählerin sagt mehrmals „weißt du". Damit stellt sie sicher, dass die Zuhörer ihrer Geschichte folgen können.* – *Sie greift den Kommentar von Ulla auf („und dann war ich noch frech")*	– *Lachen:* An manchen Stellen lachen die Zuhörer. Sie zeigen damit, dass sie Elkes Erzählung folgen und sie verstehen. – *Zuweisung der Erzählerrolle:* **Hartmut gibt Elke das „Ticket" zum Erzählen, indem er sie daran erinnert, was „sie damals alles zu fassen hatten" (=akzeptieren mussten).** – *Kommentare:* **Ulla kommentiert die Geschichte an zwei Stellen, am Ende folgen mehrere Bewertungen der anderen Zuhörer.** – *Übernahme der Erzählerrolle:* **Als Ulla fertig ist, möchte Hartmut offensichtlich etwas erzählen. Er leitet seine Erzählung dadurch ein, dass er etwas Ähnliches erlebt habe: „So ist das, ich kenn das auch, ich bin ..."**

3.1 **Sie zählt die Polizisten (*eins, zwei, drei*)**
Dadurch nimmt man die Szene so wahr, wie die die Erzählerin damals wahrgenommen hat: Sie sieht die Polizisten nacheinander.

Sie verwendet zweimal den Ausdruck *hier*.
Dadurch lenkt sie die Aufmerksamkeit auf etwas, das damals in ihrer Nähe war (TÜV-Plakette, Polizisten). Man nimmt den Ort so wahr, wie sie ihn damals wahrgenommen hat.

Sie zitiert, was die Personen gesagt haben.
Dadurch fühlt man sich, als ob man die Personen selbst sprechen hört.

Die Erzählerin wechselt ins Präsens.
Dadurch markiert sie das unerwartete Ereignis (die Polizeikontrolle mitten in der Nacht). Es ist so nah, als ob es jetzt gerade passiert.

4 ☒ **Elke findet es nicht richtig, dass sie so viel Geld wegen der Plakette zahlen musste.**
☒ **Ulla meint, dass die Strafe so hoch war, weil Elke so reagiert hat.**
☐ Klaus denkt, dass die Strafe angemessen war.
☒ **Hartmut kennt die Geschichte schon und sagt nicht, was er davon hält.**
> Klaus kommentiert nicht die Geschichte, sondern Ullas Bemerkung.

9.2　Beschreiben in Sach- und Fachtexten

9.2.1　Wörterbucheinträge: „im Strafraum verhängter Strafstoß"

1.1　› Das Bild zeigt einen Wörterbucheintrag. Man verwendet Wörter, um sich inhaltlich oder formal
　　　über einen Begriff zu informieren.

1.2

> **Funktion von Wörterbucheinträgen**
>
> - In einem Wörterbuch gibt es **viele** Wörterbucheinträge.
> - Die Einträge sind eine **erste einführende Erklärung** zu einem bestimmten Begriff. Sie
>   enthalten **Grundinformationen**, die als gesichert und relevant gelten.
> - Die Begriffe werden sowohl grammatisch-formal als auch **semantisch-inhaltlich** beschrieben.

4.1

> **Elf'me•ter** [a] (m.; -s, -; [b] Sp. [c]) *im Fußball nach schweren Regel-widrigkeiten
> (Foul, Handspiel) im Strafraum verhängter Strafstoß, bei dem von der Elfmeter-
> marke aus direkt auf das Tor geschossen wird* [d]; *Sy* Strafstoß, Penalty [e]

nicht beschrieben werden: **f, g**

4.2　- In welchem Sportbereich wird der Strafstoß verhängt? **im Fußball**
　　　- Wann / in welchen Fällen wird der Strafstoß verhängt? **nach schweren Regelwidrigkeiten (Foul,
　　　　Handspiel)**
　　　- Wo wird der Strafstoß verhängt? **im Strafraum**
　　　- Wohin und wie wird geschossen? **direkt auf das Tor**
　　　- Von wo aus wird geschossen? **von der Elfmetermarke aus**

4.3　- „X ist ein/eine …" wird weggelassen (= Ellipse), denn es ist allgemein bekannt, dass in einem
　　　　Lexioneintrag etwas definiert wird.
　　　- Handelnde Akteure werden nicht erwähnt. Im Wörterbucheintrag wird aus Platzgründen nur das
　　　　Wichtigste mitgeteilt. Es kann auch als bekannt vorausgesetzt werden, dass Schiedsrichter
　　　　Strafstöße bzw. Elfmeter verhängen, und dass Fußballspieler Elfmeter schießen. Die Verwendung
　　　　der nominalen Gruppe mit Partizipialattribut („verhängter Strafstoß") als Passiversatzform
　　　　(Strafstoß wird verhängt) und der Passivform „wird geschossen" drücken eine allgemeine Gültigkeit
　　　　aus.
　　　- Ein Elfmeter ist ein im Fußball vom Schiedsrichter nach schweren Regelwidrigkeiten (Foul, Handspiel)
　　　　im Strafraum verhängter Strafstoß, bei dem von der Elfmetermarke aus von einem Fußballspieler
　　　　direkt auf das Tor geschossen wird.
　　　- Das Thema ist Elfmeter und steht ganz am Anfang. Das Relativpronomen dem stellt einen
　　　　Rückbezug zu Elfmeter her.
　　　　Das bekannte Thema steht vorn, die neue Information eher hinten.

4.4　im Fußball nach schweren Regelwidrigkeiten (Foul, Handspiel) im Strafraum verhängter **Strafstoß**,
　　　bei dem von der Elfmetermarke aus direkt auf das Tor geschossen wird

5.1

> Elfmeter der <-s. -> SPORT **im Fußball** │*der*│ **als Bestrafung für ein Foul vollzogene,**
> **direkte** │*Schuss*│ **auf das Tor** Der Schiedsrichter müsste einen Elfmeter geben., Wer hat
> den Elfmeter geschossen / verwandelt?, Der Torwart hat den Elfmeter gehalten.

› Kern der nominalen Gruppe: der (…) Schuss

5.2 Foul: **jmd. (Akk.) foulen**
vollzogene: **etwas (Akk.) vollziehen**
Schuss: **etwas (Akk.) schießen**

5.3

> **Elfmeter, der** Ein Schiedsrichter bestraft eine Mannschaft mit einem Elfmeter, wenn ein Spieler dieser Mannschaft einen Spieler der anderen Mannschaft foult. Den Elfmeter vollzieht ein Spieler der Mannschaft, deren Spieler gefoult wurde, indem er direkt auf das Tor schießt.

> Durch die Verwendung von Nominalstil wird der Text kürzer, mit Verben kann man die Informationen nicht so kurz ausdrücken. Man bräuchte viel mehr Platz, den man im Wörterbuch nicht hat.

6 im Strafraum verhängter Strafstoß: **Schiedsrichter**
als Bestrafung für ein Foul vollzogener, direkter Schuss auf das Tor: **Fußballspieler**

---

7 Sprachliche Mittel in Wörterbuch- und Lexikonartikeln

Wörterbucheinträge enthalten kurze Informationen in schriftlicher Form. Es gibt sprachliche Mittel, die Platz sparen und eine hohe Informationsdichte erlauben:

- Das Kopulaverb (meistens *sein*) in der Definition wird *oft* weggelassen.
- Passiv ist ein **typisches** sprachliches Mittel in Lexikonartikeln, Handelnde werden meist nicht genannt.
- Es gibt **viele** Attribute vor (Adjektive, Partizipialattribute) und nach dem Kern der Nominalphrase (z. B. Präpositionalattribute, Genitivattribute, Relativsätze, Appositionen).
- Wörterbuchartikel zeichnen sich durch Nominalstil aus. Es gibt **viele** Substantivkomposita und Substantive, die aus Verben gebildet wurden.
- Die bekannte Information (das Thema) steht **meist** vorn, die neue Information (das Rhema) eher hinten im Satz.
- **Oft** gibt es mehrere Bedeutungen, oft mit Zahlen gekennzeichnet.

9.2.2 Audioguides: „... auf der Rampe in der großen Glaskuppel mit einem Spiegeldings in der Mitte"

1.3

	Audioguide 🎧	Wörterbucheintrag 📑
Enthält wichtige Informationen.	x	x
Man betrachtet etwas und hört gleichzeitig einen Text dazu.	x	
Es gibt viele Kennzeichen der Mündlichkeit (z. B. deiktische Lokalangaben wie *hier* oder *da oben*).	x	
Der Hörer wird direkt einbezogen (z. B. durch *Schauen Sie* ... oder *Wir sehen* ...).	x	
Es gibt viele Beschreibungen.	x	x

1.4

	Erwachsene	Kinder
Abschnitt a	x	
Abschnitt b	x	
Abschnitt c		x
Abschnitt d		x

1.5

	Anfang	Mitte	Ende
Abschnitt a	x		
Abschnitt b		x	
Abschnitt c		x	
Abschnitt d			x

2.1 **Kinder: Kuppelöffnung**
Erwachsene: Spiegel / Fotovoltaikanlage

2.2 „Wenn Sie nach oben sehen, werden Sie erkennen, dass die Kuppel offen ist."
indirekte Aufforderung

„Auf dem Dach des Reichstagsgebäudes dient eine Fotovoltaikanlage von mehr als 300 m² Fläche als emissionsfreie Stromquelle. Haben Sie eigentlich in der Mitte der Kuppel die Spiegel entdeckt?" **direkte Frage**

Briegel der Busch: „Wisst ihr eigentlich, warum die Kuppel offen ist?"
Chili das Schaf: „Vergessen zuzumachen?"
Bernd das Brot: „Keine Ahnung, mir egal?"
Kind: „Wieso offen?" **Gespräch mit Fragen**

„Hallo du da am Kopfhörer, schau mal in die Mitte der Kuppel. Hast du da die vielen Spiegel bemerkt?" **direkte Frage**

3 **Audioguide für Erwachsene**

Merkmal	Beispiel
Zuhörer wird höflich gesiezt (mit „Sie" angesprochen)	Wenn Sie ..
Zuhörer wird über Fragen direkt angesprochen und auf etwas hingewiesen, was dann thematisiert wird.	**Haben Sie eigentlich in der Mitte der Kuppel die Spiegel entdeckt?**
Aufforderungen sind eher indirekt.	**Wenn Sie nach oben sehen . . .**
Zuhörer wird durch Lokaladverbien räumlich orientiert.	**nach oben**
Es wird eher sachlich-nominal beschrieben.	**Im Licht- und Ablufttrichter ist zudem eine Wärmerückgewinnungsanlage verborgen.**

Audioguide für Kinder

Merkmal	Beispiel
Zuhörer wird geduzt (mit „du" angesprochen)	**Hallo du . . .**
Zuhörer wird über Fragen direkt angesprochen und auf etwas hingewiesen, was dann thematisiert wird.	**Hast du da die vielen Spiegel bemerkt?**
Zuhörer wird als Teil einer Gesprächsgruppe angesprochen und geführt.	**So, da sind wir.**
Aufforderungen sind eher direkt.	**Schau mal runter!**
Aufforderungen werden auch in rhythmisch-musikalischer Form gegeben.	**Geh die Rampe entlang!** (gesungen)
Zuhörer wird durch Lokaladverbien räumlich orientiert.	**unter uns**
Es werden Späße gemacht.	**Jetzt geht es nur noch bergab, wie immer in meinem Leben.**

4 (Die Öffnung) dient zur Entlüftung des Plenarsaals – **der Trichter ist zur Belüftung gedacht**
   Die Wärme der Abluft wird zur Beheizung des Gebäudes genutzt – **die warme Luft aus dem Saal wird dann**
   zum Heizen genutzt
   Bei Regen und Schnee – **Wenn es regnet**
   (Die Kuppel) hat einen Durchmesser von 40 Metern – **Die ist 40 Meter im Durchmesser.**
   Die Kuppel wiegt 1.200 Tonnen, wovon 900 Tonnen auf die Stahlkonstruktion entfallen. – **(Die Kuppel) wiegt**
   1200 Tonnen, so viel wie etwa 480 Elefanten.

> Die Äußerungen in der Kinderversion sind weniger komplex und orientieren sich an der Erfahrungswelt von Kindern.
Um z.B. die Schwere der Kuppel verständlich zu machen, wird das Gewicht nicht nur in der Standardeinheit Tonne
angegeben, sondern auch in Elefanten (1 200 Tonnen, so viel wie etwa 480 Elefanten).

5.1

die Öffnung	öffnen
die **Entlüftung**	**entlüften**
das **Gebäude**	**bauen**
die **Konstruktion**	**konstruieren**
die **Beheizung**	**heizen**

5.2

Kompositum	Teil, der Grundwort näher bestimmt	Fugen-element	Grundwort
der **Plenarsaal**	plenar *(von lat. ‚plenus' = voll)*		**der Saal**
das Straßenniveau	die Straße	-n-	das Niveau
die **Kuppelöffnung**	**die Kuppel**		**die Öffnung**
der **Licht- und Ablufttrichter**	das Licht und die Abluft		**der Trichter**
die **Wärmerückgewinnungsanlage**	die Wärmerückgewinnung	-s-	**die Anlage**
die **Stahlkonstruktion**	**der Stahl**		**die Konstruktion**
das **Reichstagsgebäude**	**der Reichstag**	-s-	**das Gebäude**
die **Glaskonstruktion**	**das Glas**		**die Konstruktion**
das **Staatswesen**	**der Staat**	-s-	**das Wesen**

Fachbereiche: **Bauwesen / Architektur, Physik, Politik**

5.3
Wenn Sie nach oben sehen, werden Sie erkennen, dass die Kuppel offen ist. Die Öffnung hat einen
Durchmesser von 10 Metern und dient zur Entlüftung **des Plenarsaals**. Vom Straßenniveau bis zur
Kuppelöffnung beträgt die Höhe **des Gebäudes** insgesamt 54 Meter. Die Kuppel ist von der Höhe der
Terrasse bis zu ihrer Öffnung 23 ein halb Meter hoch. Bei Regen und Schnee fällt das Wasser durch die
Kuppelöffnung in den Trichter. Es wird innerhalb der Konstruktion aufgefangen und abgeleitet, sodass kein
Regen in den Plenarsaal eindringen kann. Im Licht- und Ablufttrichter ist zudem eine
Wärmerückgewinnungsanlage verborgen. Die Wärme der Abluft wird zur Beheizung **des Gebäudes** genutzt.
Die Kuppel wiegt 1.200 Tonnen, wovon 900 Tonnen auf die Stahlkonstruktion entfallen. Sie ist mit 3.000 qm
Glas gedeckt und hat einen Durchmesser von 40 Metern. Insgesamt hat Norman Foster 40.000 qm Glas im
Reichstagsgebäude verbaut. Die Transparenz **der Glaskonstruktion** soll zugleich auf die Transparenz **unseres**
demokratischen Staatswesens hinweisen.

6.1 Diese: **die Spiegel**   da: **in der Mitte der Kuppel**
     Hierdurch: **Lichtumlenkung des Tageslichts**   die Spiegel: **die vielen Spiegel in der Mitte der Kuppel**
     das: **das Sonnensegel**   Das: **das Metallsegel**

> Die Wörter stehen eher vorn im Satz. Sie nehmen bekannte Information (das Thema) wieder auf.
  Schon bekannte Elemente stehen meist vorn.

6.2

Haben Sie eigentlich **in der Mitte der Kuppel** die Spiegel entdeckt?
Diese 360 Spiegel lenken als so genannte Lichtumlenkelemente Tageslicht in den Plenarsaal. Hierdurch wird weniger Strom zur künstlichen Beleuchtung des Plenarsaals benötigt. Mithilfe des Sonnensegels, das **entlang der Rampe parallel zur Sonne** mitläuft und das Licht bricht, wird verhindert, dass es zu Blendeffekten kommen kann.

BERND DAS BROT:	Hallo du **da am Kopfhörer**, schau mal **in die Mitte der Kuppel**. Hast du **da** die vielen Spiegel bemerkt?
BRIEGEL DER BUSCH:	Klaro, die Spiegel leiten Tageslicht in den Plenarsaal. Seht ihr das Metallsegel? Das dreht sich immer mit der Sonne als Jalousien, damit die Abgeordneten nicht geblendet werden.

6.3   Hallo du da am Kopfhörer: **Zuhörer wird direkt angesprochen**
Schau mal: **Zuhörer wird direkt aufgefordert**
Hast du da die vielen Spiegel bemerkt?: **Der Zuhörer wird direkt gefragt und mit** *da* **wird auf eine bestimmte Position in seiner Umgebung verwiesen.**
Klaro: **umgangssprachliche zustimmende Antwort**
Seht ihr das Metallsegel?: **Zuhörer wird als Teil einer Gruppe direkt gefragt**

7   **Audioguides: Funktion und typische sprachliche Mittel**

- Audioguides vermitteln dem interessierten Zuhörer **neues Wissen** zu einer bestimmten Sache.
- Der Zuhörende wird direkt **beim Betrachten** einer Sache geführt, d.h. während er / sie ein Bild oder einen Teil eines Gebäudes ansieht, hört er / sie dazu passende Informationen.
- Durch die Verwendung von Lokalangaben wird der Zuhörende **räumlich orientiert**.
- Der Zuhörende wird oft **direkt angesprochen** und etwas gefragt oder zu etwas aufgefordert.
- Durch die Verwendung von Personendeixis (*wir, Sie, du, uns*) wirken Audioguides oft wie **Dialoge**.

9.2.3   Grafiken: „Das Kreisdiagramm zeigt . . .“

1.3   **Grafikbeschreibung: Gliederung**

Grafikbeschreibungen sind meist in folgende Teile gegliedert:
- Nennung des Themas (Abschnitt: **1**)
- Angabe der Quelle (Abschnitt: **2**)
- Darstellung und Vergleich der Daten (Abschnitt: **3**)
- Schlussfolgerung / Fazit (Abschnitt: **4**)

2   **Beispiellösung:**

> Die Grafiken 1 und 2 haben unterschiedliche Formen. Beide Grafiken thematisieren erneuerbare Energien. Grafik 1 zeigt, wie hoch 2008 der Anteil der erneuerbaren Energien am Gesamtenergieverbrauch in den 27 EU-Mitgliedsländern ist und wie hoch der geplante Anteil für das Jahr 2020 ist. Grafik 2 zeigt die Verteilung der verschiedenen Energieträger in Österreich im Jahr 2009.
> Beide Grafikbeschreibungen sind sehr ähnlich aufgebaut: Zunächst werden das Thema und die Quelle der Grafik genannt. Es folgen eine Darstellung und der Vergleich der Daten. Am Ende wird ein Fazit gezogen.

3.1 Grafiken: Formen und Funktionen

- Eine Grafik ist eine bildliche Darstellung von Informationen, Grafiken zeigen **Daten und Fakten**.
- Grafiken **beantworten** oft die Fragen: „Wie viele?" und „Wann?"
- Anhand von Grafiken kann man Daten vergleichen, Entwicklungen aufzeigen und **Zusammenhänge herstellen**.
- Grafiken werden nach ihrer Form unterschieden, es gibt z.B. Kreisdiagramme, Liniendiagramme, Säulendiagramme oder Balkendiagramme.
- **Kreisdiagramme** stellen meist die einzelnen Anteile eines Ganzen (100%) dar.
- **Linien-, Säulen- und Balkendiagramme** zeigen oft zeitliche Entwicklungen.
- Daten werden statt in Grafiken auch oft als Tabellen präsentiert.

3.2 Grafik 1: **Säulendiagramm** Grafik 2: **Kreisdiagramm**

4.1

Das Diagramm zeigt die jeweiligen Anteile der erneuerbaren Energien am gesamten Energieverbrauch der 27 EU-Mitgliedsländer (EU27) im Jahr 2008 mit besonderem Fokus auf Österreich. **In der Abbildung sind** außerdem die Zielsetzungen bis zum Jahr 2020 **angezeigt**.[1] **Die Daten stammen vom** Statistischen Amt der Europäischen Union.[2] **Der Durchschnitt** der 27 EU-Länder **liegt bei ca**. 10%. **Mit fast** 30% **liegt** Österreich im EU-Vergleich **an** vierter Stelle. **Während** innerhalb der EU **der Anteil** der erneuerbaren Energieträger am gesamten Energieverbrauch **bis zum** Jahr 2020 **um** 10 % **auf** 20 % **steigen soll**, **strebt** Österreich **dagegen bis** 2020 **einen Anstieg um** 5,5 % **auf** 34 % **an**.[3] **Damit zeigt die Grafik, dass** Österreich im Bereich der erneuerbaren Energien im europäischen Vergleich eine Spitzenposition einnimmt.[4]

Das Kreisdiagramm, **das von** der Bundesanstalt Statistik Austria **stammt**, **gibt Auskunft über** die Anteile der verschiedenen Energieträger in Österreich im Jahre 2009. **Der Anteil** des Öls **beträgt rund** 40 %, es **nimmt** damit **den größten Anteil** bei der Energieversorgung **ein**. **An zweiter Stelle folgen mit** fast 30 % die erneuerbaren Energien. **Der Anteil** des Energieträgers Gas **beläuft sich auf** etwas mehr als ein Fünftel (22 %). Kohle deckt **nur rund** 10 % des Energiebedarfs in Österreich. **Zwar ist der Anteil** der erneuerbaren Energien in Österreich **mit fast** 30 % **deutlich höher als** in den meisten anderen Ländern der EU (vgl. Grafik 1), **trotzdem werden** 70,5 % des Energiebedarfs durch die Energieträger Öl, Kohle und Gas **gedeckt**. **Somit macht das Diagramm auch deutlich**, **dass** in Österreich der Anteil der fossilen Energieträger noch immer bei rund 70 % liegt.

4.2 Sprachliche Mittel der Grafikbeschreibung I

Thema	– **Das Diagram zeigt** … – In der Abbildung sind … angezeigt. – **Das Kreisdiagramm** … **gibt Auskunft über** …
Angabe der Quelle	– **Die Daten stammen vom statistischen Amt der EU.** – Das Kreisdiagramm, das von … stammt.
Vergleiche und Entwicklungen	– **Der Durchschnitt** der 27 EU-Länder liegt bei ca. 10 %. – Mit fast 30 % **liegt** Österreich im EU-Vergleich **an** vierter Stelle. – **Während** innerhalb der EU **der Anteil** der erneuerbaren Energieträger am gesamten Energieverbrauch **bis zum** Jahr 2020 **um** 10 % **auf** 20 % **steigen soll**, **strebt** Österreich dagegen **bis** 2020 **einen Anstieg von** 5,5 % **auf** 34 % **an**. – **Der Anteil** des Öls beträgt rund 40 %, es **nimmt** damit **den größten Anteil** bei der Energieversorgung **ein**. – **An zweiter Stelle folgen** mit fast 30 % die erneuerbaren Energien. – Der Anteil des Energieträgers Gas **beläuft sich auf** etwas mehr als ein Fünftel (22 %). – Kohle deckt **nur rund 10%** des Energiebedarfs in Österreich. – **Zwar ist der Anteil der** erneuerbaren Energien in Österreich mit fast 30 % **deutlich höher als** in den meisten anderen Ländern der EU (vgl. Grafik 1), **trotzdem werden** 70,5 % des Energiebedarfs durch die Energieträger Öl, Kohle und Gas **gedeckt**.
Fazit / Schlussfolgerung	– **Damit zeigt die Grafik, dass** … – Somit macht das Diagramm auch deutlich, dass …

4.3　> Grafikbeschreibungen sind oft im Nominalstil formuliert und enthalten in komprimierter Form viele Informationen. Ein typisches Kennzeichen dieses Nominalstils ist die Verwendung von Genitivattributen:

die jeweiligen Anteile <u>der erneuerbaren Energien</u> am gesamten Energieverbrauch <u>der 27 EU-Mitgliedsländer</u>

5.1

die Hälfte	1/2	50%
ein Drittel	1/3	**ca. 33%**
ein Viertel	1/4	25 %
ein Fünftel	1/5	20 %

5.2　Beispiellösung:

10,2 % **rund 10%**　　24,6 % **fast ein Viertel**　　32 % **knapp ein Drittel**
76 %　**ca. drei Viertel**　51,3 % **gut die Hälfte**

6.1　Thema der Grafik:

Das Schaubild **gibt Auskunft über** öffentliche Ausgaben je Schüler an allgemeinbildenden Schulen pro Bundesland.
Die Tabelle **stellt** den Anteil an Betriebsräten in größeren und kleineren Unternehmen **dar**.
Die von der Hans-Böckler-Stiftung im Jahr 2011 herausgegebene Grafik **hat** die Entwicklung des Verhältnisses der Deutschen zum Beruf des Politikers **zum Gegenstand**.

Quelle oder Details der Darstellung:

In der Grafik **wird** zwischen Betrieben mit 5 bis 50 Mitarbeitern und Betrieben mit mehr als 500 Mitarbeitern in Ost- und Westdeutschland **differenziert**.
Die Daten **stammen vom** Statistischen Bundesamt.
Die Tabelle **macht Aussagen** zu den Bundesländern Thüringen, Hamburg, Nordrhein-Westfalen und Gesamtdeutschland.

Beschreibung (Vergleich und Entwicklung):

Während im Jahre 1978 noch 24% der befragten Deutschen angaben, den Beruf des Politikers zu achten, **gaben** dies 2008 nur noch 6% der Befragten **an**.
Während die Anzahl der kleineren Betriebe mit Betriebsräten mit 6 Prozent relativ gering ist, **liegt** der Anteil der größeren Betriebe mit Betriebsrat deutlich **höher**, nämlich bei fast 100 Prozent – und zwar in den neuen und den alten Bundesländern.
Im Vergleich fällt auf, dass in Nordrhein-Westfalen die Ausgaben mit 4.900 € pro Schüler deutlich unter dem bundesdeutschen Durchschnitt von 5.600 € **liegen**. Dagegen wird in Thüringen pro Schüler mit 7.100 € wesentlich mehr als im Durchschnitt ausgegeben.

Schlussfolgerung:

Die Grafik **lässt erkennen**, dass die Achtung der Deutschen dem Beruf des Politikers gegenüber in den letzten Jahren kontinuierlich gesunken ist.
Somit sind Betriebsräte in größeren Unternehmen fast immer die Regel, wohingegen sie in kleineren Betrieben eher die Ausnahme darstellen.
Es wird ersichtlich, dass einzelne Bundesländer unterschiedlich viel in Bildung investieren.

6.2　BILDUNG – Grafik 1:

Das Schaubild gibt Auskunft über öffentliche Ausgaben je Schüler an allgemeinbildenden Schulen pro Bundesland. Die Daten stammen vom Statistischen Bundesamt. Die Tabelle macht Aussagen zu den Bundesländern Thüringen, Hamburg, Nordrhein-Westfalen und Gesamtdeutschland. Im Vergleich fällt auf, dass in Nordrhein-Westfalen die Ausgaben mit 4.900 € pro Schüler deutlich unter dem bundesdeutschen Durchschnitt von 5.600 € liegen. Dagegen wird in Thüringen pro Schüler mit 7.100 € wesentlich mehr als im Durchschnitt ausgegeben.
Es wird ersichtlich, dass einzelne Bundesländer unterschiedlich viel in Bildung investieren.

MITBESTIMMUNG – Grafik 2:

Die Tabelle stellt den Anteil an Betriebsräten in größeren und kleineren Unternehmen dar. In der Grafik wird zwischen Betrieben mit 5 bis 50 Mitarbeitern und Betrieben mit mehr als 500 Mitarbeitern in Ost- und Westdeutschland differenziert. Während die Anzahl der kleineren Betriebe mit Betriebsräten mit 6 Prozent relativ gering ist, liegt der Anteil der größeren Betriebe mit Betriebsrat deutlich höher, nämlich bei fast 100 Prozent – und zwar in den neuen und den alten Bundesländern.

Somit sind Betriebsräte in größeren Unternehmen fast immer die Regel, wohingegen sie in kleineren Betrieben eher die Ausnahme darstellen.

DEMOKRATIE – Grafik 3:

Die von der Hans-Böckler-Stiftung im Jahr 2011 herausgegebene Grafik hat die Entwicklung des Verhältnisses der Deutschen zum Beruf des Politikers zum Gegenstand. Während im Jahre 1978 noch 24% der befragten Deutschen angaben, den Beruf des Politikers zu achten, gaben dies 2008 nur noch 6% der Befragten an.

Die Grafik lässt erkennen, dass die Achtung der Deutschen dem Beruf des Politikers gegenüber in den letzten Jahren kontinuierlich gesunken ist.

7   Die Statistik gibt Auskunft **über** die voraussichtliche Entwicklung **der** Erwerbspersonen-zahlen **von** 2005 **bis** 2030. **Während es** im Jahr 2005 bundesweit **noch ca.** 43 Mill. Erwerbspersonen **gab**, wird diese Zahl **bis zum** Jahr 2020 **um ca.** 3,1 Mill. Erwerbspersonen **zurück gehen**, d.h. im Jahr 2020 wird die Erwerbspersonzahl rund 40 Mill. **betragen.** Zwischen 2020 und 2030 könnte die Erwerbspersonenzahl bundesweit **nochmals um** 4,5 Mill. **auf dann nur noch** 35 Mill. **sinken. Das heißt**, dass im Jahr 2030 **ungefähr** 8 Millionen Erwerbspersonen **weniger** auf dem Arbeitsmarkt zur Verfügung stehen werden.

8   Beispiellösung:

Die Grafik gibt Auskunft über die Entwicklung der Anzahl ausländischer Ärzte in Deutschland zwischen 1991 und 2010. Während im Jahre 1991 nur etwa 10.000 ausländische Ärzte in Deutschland arbeiteten, stieg die Zahl im Laufe der Jahre bis 2010 auf mehr als das Doppelte, nämlich ca. 25.000 an. Das Schaubild zeigt somit einen kontinuierlichen Anstieg der in Deutschland arbeitenden ausländischen Ärzte.

9.3 Argumentieren und Diskutieren

9.3.1 Argumentieren: „Es steht außer Frage, dass ..."

1.1   Es geht um das Problem: **Fachkräftemangel in Deutschland**

1.2   ☐ Die demographische Entwicklung in Deutschland geht zurück.
      ☐ Es fehlen Fachkräfte in Deutschland.
      ☒ **Soll sich Deutschland um die Zuwanderung ausländischer Fachkräfte bemühen?**
      ☐ Sollten Ältere und Frauen besser in den Arbeitsmarkt integriert werden?

1.3   Argumentieren: Zentrale Kategorien

- Beim Argumentieren wird eine **These** (eine bestimmte Meinung oder Behauptung) zu einer **Steitfrage** geäußert.
- Um eine **These** zu begründen oder einer These zu widersprechen werden **Argumente, Beispiele** und **Belege** vorgetragen.
- Um eine These zu bekräftigen, verweist man oft auch auf eine **Autorität** (z.B. eine anerkannte Person oder Institution).
- Die Streitfrage, die Thesen und Argumente bleiben oft **impliziert**, man muss sie selbst erschließen. Es ist auch nicht immer **eindeutig**, ob etwas eine These, ein Argument oder ein Beispiel ist.

1.4

Problem:	Fachkräftemangel in Deutschland	
Streitfrage	Soll sich Deutschland um die Zuwanderung ausländischer Fachkräfte bemühen?	
Argumente	Pro:	Contra:
	Ja, Deutschland soll sich um die Zuwanderung ausländischer Fachkräfte bemühen.	Nein, Deutschland soll sich nicht um die Zuwanderung ausländischer Fachkräfte bemühen.
Wer vertritt die These?	**Bundesagentur für Arbeit**	**CSU-Politiker**
Thesen	Durch die demographische Entwicklung werden in den nächsten Jahren immer mehr Fachkräfte fehlen. In südeuropäischen Staaten sind viele junge Hochschulabsolventen arbeitslos	Die Zahl der Arbeitslosen in Deutschland muss abgebaut werden, anstatt Arbeitskräfte aus dem Ausland zu holen.
Beispiele / Belege	Rückgang der Erwerbstätigen bis 2025 um 6,5 Millionen	Über 7 Prozent Arbeitslose in Deutschland
Autoritäten	Wissenschaft (wissenschaftliche Prognosen)	Statistisches Bundesamt

1.5　**Um** einen Fachkräftemangel **zu** verhindern, will die BA Ingenieure, Pfleger und Ärzte in den südeuropäischen Krisenländern Griechenland, Spanien und Italien suchen. Ihr **Argument**: die demographische Entwicklung. **Ohne Zweifel** werden in ein paar Jahren **viel weniger** erwerbsfähige Menschen in Deutschland zur Verfügung stehen, **wenn wir so weitermachen wie bisher**. Aber die BA kann doch als Lösung **nicht ernsthaft** ausländische Fachkräfte vorschlagen, **während** die Arbeitslosenquote in Deutschland **laut** Statistischem Bundesamt bei über sieben Prozent liegt. **Deswegen ist eines klar**: wir brauchen mehr Qualifizierung innerhalb Deutschlands. **Anstatt** sich also im Ausland umzuschauen, **sollten** sich die Jobcenter auf den Abbau der Arbeitslosigkeit hierzulande konzentrieren.

1.6　Beispiellösung:

Die Frage, ob sich Deutschland um die Zuwanderung ausländischer Fachkräfte bemühen muss, ist eindeutig zu beantworten: Ja. Um einen Fachkräftemangel zu verhindern, sollten wir gezielt Ingenieure, Pfleger und Ärzte in den südeuropäischen Krisenländern Spanien, Portugal und Griechenland suchen. Wissenschaftliche Prognosen belegen, dass die Zahl der erwerbsfähigen Menschen hierzulande bis 2025 um 6,5 Millionen zurückgehen wird, wenn niemand zuwandert. Deshalb steht außer Frage, dass sich Deutschland um ausländische Fachkräfte bemühen muss. Sicher, um Engpässe zu vermeiden, ist ein Bündel von Maßnahmen notwendig. So müssen auch Schulabbrecher gefördert werden, auch sollten Ältere und Frauen stärker in den Arbeitsmarkt integriert werden. Weil das aber nicht ausreicht, brauchen wir zusätzlich Zuwanderung.

9.3.2　Stellung nehmen im Leserbrief: „Her mit den Griechen!"

1.1

Her mit den Griechen! In Deutschland fehlen Fachkräfte, während sie in Südeuropa arbeitslos sind.

Die Zahlen sind alarmierend. **2030 werden in Deutschland mehr als sechs Millionen Arbeitskräfte fehlen, die Hälfte davon Akademiker. Aus eigener Kraft**, mit Arbeitslosen, **lässt sich diese Lücke nicht schließen. Allein schaffen wir das nicht. Es ist doch ein Trugschluss zu glauben, dass** das ohne qualifizierte Zuwanderer ginge. **Schon heute fehlen tausende Ingenieure und Ärzte, sehr bald werden es auch tausende Facharbeiter und Pflegerinnen sein. Als Abteilungsleiter eines großen Technologiebetriebes weiß ich, wovon die Rede ist: seit mehr als zwei Monaten suchen wir verzweifelt nach Ingenieuren und finden sie nicht**.

Die Bundesagentur für Arbeit hat die Zeichen der Zeit erkannt: Sie will junge Fachkräfte im Ausland anwerben. **Davon profitieren nicht nur wir Deutsche, sondern auch die jungen Menschen ohne Job in Griechenland, Spanien und Portugal. Und was spricht dagegen** Fachkräfte auch aus Indien, Brasilien oder Ägypten **zu** holen? **In der Politik ist man wohl aber noch nicht so weit**. So fordern CSU-Mitglieder von der Bundesagentur, sie **möge** sich doch lieber um Jobs für Langzeitarbeitslose, Ältere und Frauen bemühen. **Zuwanderer scheinen bei uns offenbar noch immer als Bedrohung, und nicht als Bereicherung zu gelten.**

› Herr Baumeister argumentiert <u>für</u> die Zuwanderung von Fachkräften nach Deutschland.

1.2

Autor fordert Zuwanderung:	Her mit den Griechen!
Einleitung einer These: Einleitung einer These:	Es ist doch **ein Trugschluss zu glauben, dass** … Und was **spricht dagegen**, … zu …?
Argumente – fehlende Fachkräfte in Deutschland und arbeitslose Fachkräfte in Südeuropa: – Vorteil für beide Seiten:	In Deutschland fehlen Fachkräfte – während sie in Südeuropa arbeitslos sind. Davon **profitieren nicht nur wir Deutsche, sondern auch die jungen Menschen ohne Job in Griechenland, Spanien und Portugal.**
Fachkräftemangel wird mit „alarmierenden" Zahlen untermauert – heute fehlen: – in Zukunft fehlen:	2030 werden in Deutschland mehr als sechs Millionen Arbeitskräfte fehlen, die Hälfte davon Akademiker. **tausende Ingenieure und Ärzte** **auch tausende Facharbeiter und Pflegerinnen**
angegebene Zahlen sind vage und nach oben offen	**mehr als** sechs Millionen; tausende (2x)
Verwendung von Futur, um Zukunft zu prognostizieren	**werden fehlen; werden sein**
Autor nennt zwei Standpunkte: – er bewertet den Standpunkt der Arbeitsagentur – und im Gegensatz dazu den Standpunkt der Politik	hat die Zeichen der Zeit erkannt in der Politik **ist man wohl aber noch nicht so weit**
Autor verallgemeinert durch unpersönliche Ausdrücke	lässt sich; man (in der Politik)
Autor distanziert sich von anderen Meinungen	ginge; **möge**; scheinen … offenbar … zu gelten
Autor berichtet eigene Erfahrung aus der Arbeitswelt als Beispiel für Fachkräftemangel	**Als Abteilungsleiter eines großen Technologiebetriebes weiß ich, wovon die Rede ist: seit mehr als zwei Monaten suchen wir verzweifelt nach Ingenieuren und finden sie nicht.**
Autor versteht sich als Teil der deutschen Gesellschaft	aus eigener Kraft; **allein schaffen wir das nicht**; wir Deutsche; **bei uns**

1.3

Ein Leser möchte einen Artikel kommentieren. Er [*ein Leser*] schreibt einen Leserbrief.	aufnehmendes Personalpronomen [Rückbezug] → altes Thema wird aufgenommen
Dieser Leserbrief [*der Leserbrief*] nimmt Bezug auf einen Zeitungsartikel, in dem [**der Zeitungsartikel**] die Zuwanderung nach Deutschland thematisiert wird.	fokussierendes Artikelwort [**Rückbezug**] fokussierendes Relativpronomen [**Rückbezug**] → Aufmerksamkeit wird auf ein anderes Thema gelenkt
Der Leser ist davon [**Zuwanderung ist nötig**] überzeugt, dass Zuwanderung notwendig ist.	fokussierendes Adverb [**Vorausweisen**] → *da-* bezieht sich auf einen noch folgenden Textteil, die Präposition *von* ist vom Verb gefordert (überzeugt sein von)
Darüber [**die Zuwanderung nach Deutschland**] wird nicht nur in den Medien diskutiert.	aufnehmendes und fokussierendes Adverb [**Rückbezug**] → Thema wird wieder aufgenommen und fokussiert, *da-* bezieht sich auf einen vorher erwähnten Textteil, die Präposition *über* ist vom Verb gefordert (diskutieren über)
Auch in der Politik ist eines [**es fehlen Fachkräfte**] klar: Es fehlen Fachkräfte.	indefiniter Artikel [**Vorausweisen**] → Aufmerksamkeit wird auf einen folgenden Textteil gelenkt
Deshalb [**weil Fachkräfte fehlen**] wird auch hier [**in der Politik**] über Vor- und Nachteile der Zuwanderung nach Deutschland diskutiert. Damit steht fest, dass Zuwanderung ein viel diskutiertes Thema ist.	fokussierendes Adverb [**Rückbezug**] fokussierendes Adverb [**Rückbezug**] fokussierendes Adverb [**Rückbezug**] → Aufmerksamkeit wird auf zuvor erwähnte Textteile gelenkt

1.4 <u>sie</u> (Überschrift): **Rückbezug auf „Fachkräfte"**

Die Hälfte <u>davon</u> (Zeile 5): **Rückbezug auf „mehr als sechs Millionen Arbeitskräfte"**

<u>Diese</u> Lücke (Zeile 6): **Rückbezug auf „mehr als sechs Millionen Arbeitskräfte"**

…, dass <u>das</u> ohne Zuwanderung ginge (Zeile 8): **Rückbezug auf Schließen der Lücke**

<u>Sie</u> will junge Fachkräfte anwerben (Zeile 17): **Rückbezug auf „die Bundesagentur für Arbeit"**

<u>Davon</u> profitieren (Zeile 18): **Rückbezug auf Anwerben junger Fachkräfte im Ausland**

<u>Die</u> jungen Menschen (Zeile 19): **Rückbezug auf junge Fachkräfte im Ausland**

9.3.3 Erörterung: „Insgesamt bin ich der Meinung, dass . . .“

Einleitung   **Hauptteil**

1.1/2/3

Im Zuge der Bologna-Hochschulreform wurden in Deutschland Bachelor- und Masterstudiengänge eingeführt. Ein Bachelorstudium dauert meistens drei Jahre. Für einen Masterabschluss muss man zusätzlich noch einmal ein bis zwei Jahre studieren. Viele Studierende stehen vor der Frage: Soll ich nach dem Bachelor noch einen Masterstudiengang absolvieren?

Zur Beantwortung dieser Frage möchte ich im Folgenden auf die Aspekte Zeit, Aufgabenbereiche und Gehalt eingehen. Zunächst zum Zeitfaktor: Einen Bachelorabschluss hat man gewöhnlich schon nach drei Jahren, für einen Masterabschluss braucht man insgesamt wenigstens vier bis fünf Jahre. Will man also möglichst schnell in den Beruf einsteigen, reicht dafür ein Bachelorabschluss völlig aus.[1] Andererseits ist jedoch auch zu bedenken, dass man mit einem Bachelor zwar relativ schnell eine Berufsqualifikation erwirbt, aber dafür im Job später meist weniger Verantwortung trägt. Statistiken zeigen, dass Mitarbeiter mit einem Masterabschluss etwa in Projektteams häufig mehr Verantwortung tragen als Mitarbeiter mit einem Bachelorabschluss.[2] Eng damit verbunden ist das Gehalt. Mehr Verantwortung wird honoriert. Auch hier belegen Studien, wie sie etwa von IW-Personaltrends durchgeführt wurden, dass Masterabsolventen insbesondere aus den Betriebs- und Naturwissen-

schaften im Vergleich zu den Bachelorabsolventen deutlich mehr verdienen.[3] Ein weiterer Aspekt ist die Spezialisierung. Ein Master bietet die Chance, sich zu spezialisieren. Seit Langem ist bekannt, dass in Deutschland Fachkräfte fehlen. Mit dem fundierten Wissen aus einem Master hat man sicher auch mehr Chancen auf dem Arbeitsmarkt.[4]

In meinem Heimatland Ägypten gibt es abgesehen von ein paar Ausnahmen keine Masterstudiengänge. Wir haben einen vierjährigen Bachelor. Nur die besten Absolventen können dann ein Magisterstudium anschließen, das vor allem auf die Tätigkeit an einer Universität vorbereitet. Dort kann man also nicht selbst entscheiden, ob man weiterstudiert. Nur mit den allerbesten Noten hat man überhaupt die Möglichkeit dazu.[5] Insgesamt bin ich der Meinung, dass jeder für sich selbst entscheiden sollte, ob er/sie einen Master machen will oder ob ein Bachelor ausreicht. Letztlich ist es doch auch eine Typfrage. Nicht jede/r hat Lust über Jahre hinweg zu studieren. Mit einem Bachelorabschluss kann man schnell und bei ganz gutem Gehalt einen Beruf ausüben. Wer allerdings gern studiert, gern mehr Verantwortung im Beruf übernimmt und dies auch finanziell honoriert sehen will, der sollte sich auf jeden Fall für ein Masterstudium entscheiden.[6]

Belege aus Statistiken   **Schluss**

1. Aspekt: Dauer des Studiums   3. Aspekt: Gehalt   5. eigene Erfahrung/Bezug zum Heimatland
2. Aspekt: Aufgabenbereiche im Berufsleben   4. weiterer Aspekt: Spezialisierung   6. Fazit / eigener Standpunkt

1.4   Einleitung der Frage bzw. These der Erörterung: Viele Studierende **stehen vor der Frage:**
Einleitung des Hauptteils: Zur **Beantwortung dieser Frage möchte ich im Folgenden** ...
Ausdrücke, die Argumente miteinander verbinden und die Erörterung gliedern:
Zunächst zum Zeitfaktor; Andererseits ist jedoch auch zu bedenken, dass..., aber dafür ...;
Eng damit verbunden ist das Gehalt; ein **weiterer Aspekt ist** die Spezialisierung
Wie werden Belege angeführt? Statistiken zeigen, dass ...; **Auch hier belegen Studien** ...
Einleitung des Bezugs zum Heimatland: **In meinem Heimatland** ...
Einleitung des Schlussteils: **Insgesamt bin ich der Meinung, dass** ...

1.5 | **Erörterung: Formulierungshilfen**

- Thema der Erörterung für die Einleitung formulieren:
  Wir stehen vor der Frage: ...?; **Es stellt sich die Frage: ...?**
- Hauptteil einleiten:
  Um die Frage beantworten zu können, ...; **Zur Beantwortung dieser Frage** ...
- Belege und Beispiele anführen:
  Statistiken belegen, dass ..., **Als Beispiel für ... kann ... angeführt werden**
  Aus eigener Erfahrung weiß ich, dass ..., **Studien bestätigen, dass** ...
- Schlussteil einleiten:
  Insgesamt lässt sich zusammenfassen: ...
  Meines Erachtens ..., **Ich denke** ...
  Schließlich / Letztlich ..., **Ich bin der Meinung, dass** ...

2 Beispiellösung:

In der Diskussion um Klimawandel und die Begrenztheit von Rohstoffen spielt auch die Nutzung von Ökostrom immer weder eine Rolle. Es geht um die Frage: Soll die gesamte Stromversorgung mit Ökostrom abgedeckt werden?

Auf den ersten Blick liegen die Vorteile von Ökostrom auf der Hand: Ökostrom wird meist aus nachwachsenden, erneuerbaren Rohstoffen wie Holz, Getreide oder aus Abfällen der Land- und Forstwirtschaft hergestellt. Im Falle von Erdwärme, Wind- oder Sonnenenergie kommt Ökostrom sogar ganz ohne den Verbrauch von Ressourcen aus. Ökostrom wird also CO_2-frei erzeugt und leistet somit einen wichtigen Beitrag zur Erreichung der klimapolitischen Zielsetzungen.

Gerade daher wächst in der Europäischen Union der Anteil erneuerbarer Energien am gesamten Energieverbrauch, auch in Zukunft soll der Anteil weiter steigen. Auf diese Weise wird man unabhängiger von Energie-Importen aus dem Ausland. Gleichzeitig schafft der Ausbau von Ökostromanlagen neue Arbeitsplätze.

Allerdings darf man die sozialen Aspekte nicht vergessen: Ökostrom ist trotz finanzieller Förderung nach wie vor teurer als herkömmlicher Strom – nicht jeder, der Ökostrom befürwortet, kann ihn sich auch leisten. Außerdem werden teilweise Ökostromanlagen errichtet, die die Umwelt zerstören. Hinzu kommen die globalen Auswirkungen. Rohstoffe wie Holz oder Getreide sind nur begrenzt verfügbar. Auf den Anbauflächen, auf denen Pflanzen zur Stromgewinnung wachsen, können keine Nahrungsmittel für Menschen und Tiere angebaut werden. Dadurch steigen die Preise für Nahrungsmittel.

Meiner Meinung nach ist Ökostrom grundsätzlich positiv zu bewerten und leistet einen wichtigen Beitrag zum Klimaschutz. Selbstverständlich darf aber Ökostrom nicht auf Kosten von lebenswichtigen Nahrungsmitteln produziert werden.

3 Beispiellösung:

Statistiken belegen: Facebook hat im Jahr 2011 580 Mio. Nutzer und jeder Facebooknutzer hat im Durchschnitt 130 Freunde.

Es stellt sich die Frage: Was bedeutet „Freunde"? Geht durch die Nutzung von sozialen Netzwerken wie Facebook oder Myspace „reale Freundschaft" und Intimität verloren?

Um diese Frage beantworten zu können, muss zunächst geklärt werden, was Freundschaft und Intimität eigentlich bedeuten. Freundschaft ist eine Beziehung zwischen Menschen, die Sympathie und Vertrauen füreinander haben. Intimität bedeutet große Vertrautheit im persönlichen Bereich. All dies wird durch Facebook nicht prinzipiell untergraben. Erstens hängt es von einem selbst ab, welche Freunde man einlädt und welche man „zulässt". Zweitens kann man auch selbst entscheiden, welche persönlichen Daten man von sich preisgeben möchte. Drittens dienen die Netzwerke auch als reine Informations- und Werbequellen. So kann man sich dort als Künstler etwa mit seinen Werken präsentieren und Interessierte haben die Möglichkeit sich Lieder, Bilder oder Fotos anzuschauen.

Sicherlich hat man in der digitalen Netzwerkwelt viel mehr Freunde als in der realen Welt, aber auch Internet-Freunde findet man doch aufgrund gemeinsamer Interessen irgendwie sympathisch. Auf jeden Fall verbinden einen Themen, über die man gemeinsam reden kann. Und genau das zeichnet Freundschaft aus. Hinzu kommt, dass niemand zu Kommunikationen im persönlichen Bereich Zugang hat, der nicht explizit eingeladen ist. Gewiss, viele Daten, die auch außerhalb des persönlichen Bereichs veröffentlicht werden, sind sehr intim und können zu Missbrauch führen. Allerdings muss man diese Daten ja nicht preisgeben.

Letztlich hängt es davon ab, wie man solche Netzwerke nutzt. Nicht jedem muss man Zugang gewähren und man sollte auch nicht sein ganzes Leben öffentlich machen. Ich bin der Meinung, ein vernünftiger Umgang mit den Netzwerken schafft also weder Freundschaft noch Intimität ab, sondern kann diese sogar fördern.

9.3.4 Diskussionsforum: „Stimmt genau!"

1.1

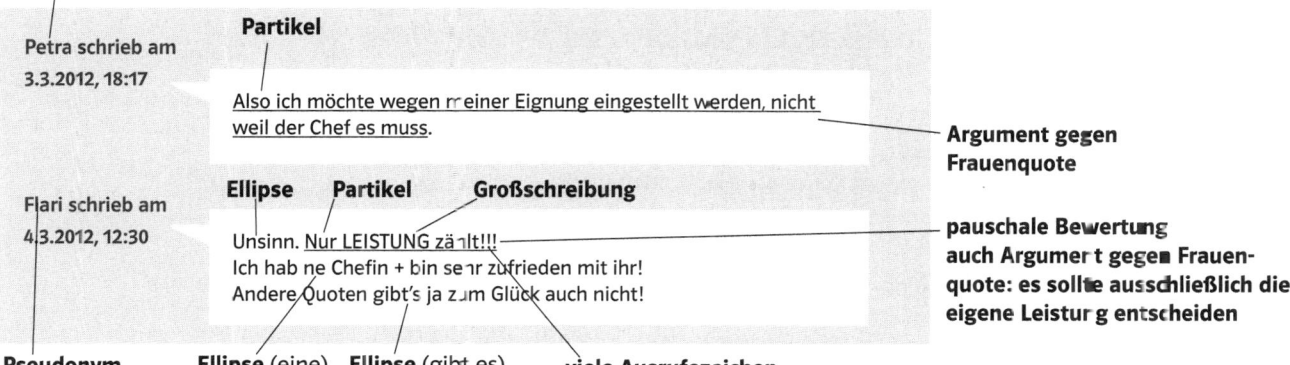

Vorname, richtiger Name könnte ganz anders lauten

Petra schrieb am 3.3.2012, 18:17

Partikel

Also ich möchte wegen meiner Eignung eingestellt werden, nicht weil der Chef es muss.

Argument gegen Frauenquote

Flari schrieb am 4.3.2012, 12:30

Ellipse   **Partikel**   **Großschreibung**

Unsinn. Nur LEISTUNG zählt!!!
Ich hab ne Chefin + bin sehr zufrieden mit ihr!
Andere Quoten gibt's ja zum Glück auch nicht!

pauschale Bewertung auch Argument gegen Frauenquote: es sollte ausschließlich die eigene Leistung entscheiden

Pseudonym   **Ellipse** (eine)   **Ellipse** (gibt es)   **viele Ausrufezeichen**

1.2 Das Thema lautet so oder ähnlich:
Brauchen wir eine Frauenquote?
Wir brauchen eine Frauenquote!

1.3 Beispiellösung:
Eine Frauenquote einzuführen wäre meiner Meinung nach Unsinn.

2.1 Die Frage lautet so oder ähnlich:
Brauchen wir eine Einheitsdenkmal?
Ist ein Einheitsdenkmal sinnvoll?

2.2 **2**

  a **für** und **gegen** das Denkmal → **Ich bin froh** über dieses Denkmal. / **Ich frage mich aber schon** ...
  b Bewertung: Die Symbolik ist sehr gut.
  c Pro-Argument: Es **zeigt anschaulich, dass man** zusammen etwas bewegen kann.
  d Einschränkung, neuer Aspekt: Ich **frage mich aber schon, ob das nicht eher** ein Demokratiedenkmal ist.

3

  a **für** und **gegen** das Denkmal → Wenn schon, **dann so**.
  b Bewertung: ..., dass ein MitMach-Mal die **beste** Variante ist
  c Pro-Argument: ..., denn **was ist** zum Nachdenken und Erinnern **besser als** Kunst, an der man selbst Teil hat?
  d Einschränkung, Kontra-Argument: **Sicherlich brauchen wir** in Berlin **nicht noch mehr** „Denkmäler", in einer Stadt, die **schon so** geschichts- und denkmallastig ist.

4

  a **gegen** das Denkmal → Ich bin **dagegen**.
  b Bewertung: **Nicht zu fassen; Unglaublich!**
  c Kontra-Argumente: dafür ist Geld da (Geld besser anders zu gebrauchen); weit weg von den Bürgern
  d Kontra-Argumente als neue Aspekte: **dafür** ist Geld **da, aber für** Steuersenkungen und kaputte Straßen **nicht**; alles **weit weg** in Berlin **und weit weg von uns**, den Bürgern.

2.3 **Steuerzahler schrieb am 01.07.2011 / 23:17**
Haben wir zuviel Geld? Berlin ist **pleite**, Deutschland ist so gut wie pleite, wesentliche Beiträge zur Erhaltung der Infrastruktur können nicht mehr geleistet werden, die Schere zwischen Arm und Reich klafft immer weiter auseinander. Und jetzt werden **mal eben** 10 Millionen Euro für ein Denkmal verschwendet, das **nun wirklich** keiner braucht und das auch noch nur wenige tausend Meter entfernt ist vom **größten** Symbol der deutschen Einheit, dem Brandenburger Tor. **Was soll der Schwachsinn**? Daran sieht man **mal wieder**, wie unsere Steuergelder **verschleudert** werden. Das ist **ein Schlag ins Gesicht** all derer, bei denen scheinheilig gekürzt wird. **Weg damit!** Bringt lieber unsere **schlecht** ausgestatteten Schulen und Kindergärten in Schuss**!**

S.B. schrieb am 01.07.2011 / 23:58
Stimmt genau Ich **schließe mich** dem Beitrag von Steuerzahler zu 100 % **an**!

2.4 Haben wir zuviel Geld? Berlin ist pleite, Deutschland ist so gut wie pleite, **wesentliche Beiträge zur Erhaltung der Infrastruktur können nicht mehr geleistet werden**, die Schere zwischen Arm und Reich klafft immer weiter auseinander. **Und jetzt werden mal eben 10 Millionen Euro für ein Denkmal verschwendet**, das nun wirklich keiner braucht und das auch noch nur wenige tausend Meter entfernt ist vom größten Symbol der deutschen Einheit, dem Brandenburger Tor. Was soll der Schwachsinn? Daran sieht man mal wieder, **wie unsere Steuergelder verschleudert werden**. Das ist ein Schlag ins Gesicht all derer, **bei denen scheinheilig gekürzt wird**. Weg damit! Bringt lieber unsere schlecht ausgestatteten Schulen und Kindergärten in Schuss!

9.3.5 Radiodiskussion: „Nee, ganz und gar nicht."

1.1 Beispiellösung:

Es geht um die Frage, ob eine Frauenquote notwendig ist oder nicht.

1.2

MODERATOR (M):	Unser Thema heute ist die gesetzlich vorgeschriebene Frauenquote. Seit langem wird die ja von den einen vehement gefordert, von den anderen entschieden abgelehnt. In der Sendung werden wir verschiedene Meinungen hören und am Ende vielleicht genauer sagen können, ob wir eine Frauenquote brauchen oder nicht? Ich begrüße meine drei Gäste, da sind Frau Kersting, freie Mitarbeiterin in einer Frauenberatungsstelle. Herr Schmiedel ist Pilot. Und dann ist da noch Frau Naumann. Frau Naumann ist in der Führungsetage eines großen Unternehmens tätig und Mutter von zwei Kindern. Frau Naumann, Sie haben zwei Kinder und arbeiten von früh bis spät. Wie kriegen Sie das hin?
FRAU NAUMANN:	Ach, das ist gar nicht so schwer, schließlich ist mein Mann ja auch noch da und die Zeiten, da Frauen nur Mütter und für den Haushalt zuständig sind, sind ja zum Glück vorbei.
M:	Sind sie für oder gegen eine Frauenquote?
FRAU NAUMANN:	Ich bin dagegen, ich finde, nur Leistung sollte zählen. Und ich wehre mich dagegen, Frauen in unserem System zur Minderheit zu machen, aber genau das würde eine gesetzlich vorgeschriebene Frauenquote bedeuten.
FRAU KERSTING:	**Nee, ganz und gar nicht. Das sehe ich ganz anders.**
M:	Frau Kersting, was ist Ihre Meinung?
FRAU KERSTING:	Leistung **ist das eine**, die – **da gebe ich Ihnen recht** – bringen Frauen ganz genauso wie Männer. **Ich denke, darüber sind wir uns einig. Aber genau hier liegt das Problem**. Trotz der Leistung sitzen Frauen viel seltener in höheren Positionen. Warum? Man stellt sie nicht ein, weil nun mal Frauen die Kinder kriegen, rein biologisch kann das ja der Mann noch nicht, und dann ist da die Angst bei Arbeitgebern, dass die Frauen dann erst mal ausfallen. Dann muss Ersatz her. Und in verantwortlichen Positionen ist das nicht immer so einfach. Logisch, dass man da Männer vorzieht. Aber die Frauen können ja nichts dafür, dass sie beides können, also arbeiten und Kinder kriegen. Um die Fähigkeiten von Frauen zu fördern, muss der Staat her. Deshalb brauchen wir die Frauenquote.
M:	Jetzt Herr Schmiedel. Wie sehen Sie das? Würde es mit der Frauenquote mehr Pilotinnen geben?
HERR SCHMIEDEL:	**Also ich weiß nicht, sicher ist es richtig, dass** Frauen beides können, und … **Ja klar**, Frauen bringen auch Leistung, aber ich glaube, man muss unterscheiden zwischen Intelligenz und körperlicher Anstrengung. Männer sind doch robuster und halten rein körperlich oft mehr aus. Das sehe ich täglich in meinem Beruf als Pilot, wie anstrengend das ist. Und ich glaube, dass Frauen oft viel emotionaler reagieren und mit Stresssituationen einfach nicht so gut zurechtkommen. Wenn –
FRAU KERSTING:	**Wie bitte? Das kann doch nicht Ihr Ernst sein**! Das sind doch reine Vorurteile. Also da
HERR SCHMIEDEL:	**Nun lassen sie mich doch ausreden**. Warum bewerben sich dann nur ca. 5 Prozent Frauen für den Beruf des Piloten? Das ist vielen Frauen einfach zu anstrengend.
FRAU KERSTING:	**Wenn ich dazu mal was sagen darf**?
M:	Ja bitte!
FRAU KERSTING:	**Es geht doch hier um** die Frauenquote, mit anderen Worten also um eine Einmischung des Staates in …

1.3 Frau Kersting: **ist für die Frauenquote, weil sie die Fähigkeiten von Frauen unterstützt**
Frau Naumann: **ist gegen die Frauenquote, weil allein Leistung zählt**
Herr Schmiedel: **äußert sich nicht dazu**

9.4 Auffordern

9.4.1 Mahnung: „Sicherlich haben Sie nur übersehen, die Prämie zu entrichten."

1.1   > Herr Hettinger hat die fällige Prämie für die Haushaltsversicherung in Höhe von 148,75 € nicht
         bezahlt.
      > Er wird aufgefordert, diese Prämie in den nächsten zwei Wochen zu überweisen (oder seine
         Einwilligung für den Einzug der Prämien im Lastschriftverfahren zu erteilen).
      > Wenn er die Aufforderung nicht befolgt, verliert er den Versicherungsschutz.

1.2

Aufforderung	Kundenbeziehung
Imperativ, z. B.: **Bitte überweisen Sie ...** **Bitte verwenden Sie ...**	Sicherlich haben Sie nur übersehen ...: Der Autor zeigt Verständnis für den Kunden: Er denkt nicht, dass dieser absichtlich nicht gezahlt hat. „Sicherlich" hat er es „nur" vergessen – das kann jedem passieren.
Betreffzeile: **Zahlungserinnerung** Nennung einer Frist: **innerhalb von zwei Wochen** Nennung der negativen Konsequenz: **Verlust des Versicherungsschutzes**	**..., damit sie Ihren Versicherungsschutz nicht verlieren.** Der Autor zeigt, dass er sich um den Kunden sorgt und negative Konsequenzen für ihn vermeiden möchte. Der Autor wählt das Verb **verlieren** bei dem kein Täter genannt wird, um die möglichen Konsequenzen als eine Art Automatismus darzustellen. Um Ihnen den Weg zum Kreditinstitut sowie Überweisungsmöglichkeiten zu ersparen, können Sie die fälligen Prämien im Lastschriftverfahren einziehen lassen. **Der Autor signalisiert, dass er es dem Kunden möglichst einfach machen möchte.** Das Modalverb **können** zeigt an, dass es sich um eine Handlungsmöglichkeit handelt. Unterschrift: Durch die handschriftliche Unterschrift wirkt der Brief persönlicher. dreimalige Verwendung von *bitte*: **Höflichkeitssignale, gerade bei den direkten Aufforderungen.**

2   > „Zahlungserinnerung" klingt höflicher als „Mahnung": Man möchte den Kunden an etwas erinnern, das
       er nur vergessen hat. Wenn man hingegen das Wort „Mahnung" verwendet, signalisiert man eine
       weniger positive Einstellung: Vielleicht hat der Kunde ja absichtlich nicht gezahlt?
       Außerdem ist „Mahnung" auch ein juristischer Begriff - wer trotz Mahnung nicht zahlt, muss mit
       Konsequenzen rechnen.

3

AKA VERSICHERUNGS-AG
Postfach 73 56 86
51643 Gummersbach

AK Insurance

Briefkopf

Anschrift des Absenders

Herrn
Manfred Hettinger
Heinrich-Heine-Straße 50
28211 Bremen

Anschrift des Empfängers

Versicherungs-Nr.: 56789324
Kunden-Nr.: 1974238

Gummersbach, 28.04.2011

Datum

Zahlungserinnerung: Haushaltversicherung

Prämie für den Zeitraum von 01.04.2011 bis 01.04.2012, Einlösebetrag: 148,75 €

Betreffzeile

Sehr geehrter Herr Hettinger,

Anrede

sicherlich haben Sie nur übersehen, die fällige Prämie an uns zu entrichten. Bitte überweisen Sie den angeforderten Betrag innerhalb von zwei Wochen, damit Sie Ihren Versicherungsschutz nicht verlieren.

Um Ihnen den Weg zum Kreditinstitut sowie Überweisungsgebühren zu ersparen, können Sie die fälligen Prämien im Lastschriftverfahren einziehen lassen. Bitte verwenden Sie hierfür das als Anlage beigefügte Formular.

Sollte die Zahlung in den letzten Tagen bereits erfolgt sein, betrachten Sie dieses Schreiben bitte als gegenstandslos.

Mit freundlichen Grüßen

AKA VERSICHERUNGS-AG

Dr. Schütte

Grußformel

Unterschrift

AKA-Kundenservice Mo. - Fr. 7-20 Uhr   Telefon: 01803 / 543354*   Telefax: 01803 / 543354-99*

9.4.2 Verbots- und Warnschilder: „Rauchen verboten"

1.1

 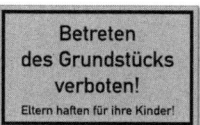  

Was darf man nicht tun?	Man darf nicht mit dem Handy telefonieren.	**Man darf das Grundstück nicht betreten.**	**Man darf keine Fahrzeuge parken bzw. abstellen.**	**Man darf nicht rauchen.**
Wo könnte das Schild stehen?	z. B. in einer Bibliothek, in einer Schule oder in einem Krankenhaus	**z.B. auf einer Baustelle oder einem Privatgrundstück**	z. B. vor einem Krankenhaus oder einer Feuerwehr	**z.B. in einem öffentlichen Gebäude oder im Zug**

1.2 Beispiellösung:

- Satzzeichen: Auf den zwei Schildern mit Text werden Ausrufezeichen verwendet. Hinter „Parken verboten" wird kein Satzzeichen verwendet.
- Inhalt: Auf keinem der Schilder steht, von wem das Verbot stammt. Auf einem Schild wir zumindest darauf hingewiesen, dass Eltern für ihre Kinder zur Verantwortung gezogen werden. Es handelt sich bei allen Schildern um Verbote, es wird nicht erklärt, warum etwas verboten ist.
- Bild/Text: Die Texte auf den zwei Schildern mit Sprache sind kurz, zwei Schilder kommen ganz ohne Sprache aus, das Verbot wird dort durch Bilder bzw. Piktogramme ausgedrückt.

2.1

Verbote: **2, 4**   Warnungen: **1, 3, 5**

2.2 Beispiellösung:

 Achtung – Gefährliche Stoffe!
Vorsicht Gift!

 Fotografieren ist hier nicht gestattet.
Fotografieren verboten

 Schwimmen verboten
Hier nicht schwimmen!

 Achtung Wildwechsel!
Vorsicht: Diese Straße wird von Wild überquert.

 Warnung vor dem Hund
Achtung, Wachhund!

9.4.3 Kochrezept: „Die Äpfel nach Belieben schälen"

1.2
- **Kochrezepte** bestehen normalerweise aus einem Zutatenteil und einem Zubereitungsteil.
- **Die Zutaten** werden im Zubereitungsteil mit dem bestimmten Artikel eingeführt, weil sie im Zutatenteil schon aufgeführt sind. Sie werden nur selten durch Pronomen ersetzt.
- **Die Handlungen** werden in der Reihenfolge genannt, in der sie ausgeführt werden sollen.
- **Die Verben** haben zum Teil eine sehr spezifische Bedeutung (*unterrühren, unterziehen, ausstechen*). **Mengenangaben** stehen im Zutatenteil, damit man vor dem Backen schnell überprüfen kann, ob man genug Mehl, Zucker, Eier usw. hat.

1.3 Beispiellösung:

Zuerst wird das Eigelb mit dem Zucker schaumig gerührt, dann wird der Quark mit Zitrusschale und Zimt untergerührt. Das Eiweiß wird mit einer Prise Salz cremig geschlagen und mit Milch und Mehl unter die Eigelb-Quark-Masse gezogen. Die Äpfel können nach Belieben geschält werden, die Kerngehäuse sind zu entfernen und die Äpfel sollten dann in etwa 5 mm dicke Scheiben geschnitten werden. Als nächstes wird etwas Öl in einer Pfanne erhitzt und die Apfelscheiben werden hineingelegt. Auf die Mitte jeder Scheibe werden je 1 bis 2 Esslöffel Teig gegeben. Wenn die Küchlein von der Unterseite goldbraun sind, müssen sie vorsichtig gewendet werden. Nach Belieben kann ein Stich Butter zugegeben werden. Mit Zimt-Zucker bestreut werden sie auf ein Backblech gelegt und im vorgeheizten Backofen bei 180 Grad 5 Minuten gebacken.

2 > Bei dieser Aufgabe gibt es keine umfassende Lösung. Liefländer-Koistinen (1993) hat deutsche und finnische Kochrezepte verglichen und dabei z.B. folgende Unterschiede festgestellt:
– Layout: Das Layout der deutschen und finnischen Kochrezepte ist ähnlich.
– Aufbau: In deutschen Rezepten werden die Zutaten im Zutatenteil genau in der Reihenfolge genannt, wie sie im Zubereitungsteil vorkommen. Für finnische Kochrezepte gilt das nicht so streng.
– Form der Aufforderung: Im Deutschen wird der Infinitiv verwendet (also: Äpfel schneiden), im Finnischen der Imperativ (Schneide die Äpfel!).
> Wichtig ist zu erwähnen, dass auch deutsche Kochrezepte nicht immer gleich aufgebaut sind. Wenn Sie sich zum Beispiel ein altes Kochbuch anschauen, wird dort häufig noch der Konjunktiv I verwendet („Man nehme 100 g Butter …"). Dies ist heute nicht mehr üblich. Unterschiede gibt es aber auch zwischen Kochrezepten in Büchern, in Zeitschriften und im Internet.

9.4.4 Medikamentenbeipackzettel: „Fragen Sie Ihren Apotheker."

1 > Wenn man das Wort in seine Bestandteile zerlegt, kann man seine Bedeutung verstehen: Ein Medikamentenbeipackzettel ist ein Zettel, der zu einem Medikament gepackt wird oder einem Medikament beigepackt ist.

2.1 > Der Beipackzettel ist in Fragen und Antworten gegliedert. So kann man schnell Informationen zu einer bestimmten Frage finden kann (z.B. Dosierung). Fragen sind für den Leser verständlicher als Überschriften.
> Die Zahlen vor den einzelnen Fragen sind sehr groß. Dadurch kann man die Fragen sehr schnell finden.
> Wichtige Informationen sind fett gedruckt oder unterstrichen.

2.2 **Beispiellösung:**

Der Text enthält viele Imperative. Das ist eine sehr direkte Form der Aufforderung.
Lesen Sie die gesamte Packungsbeilage sorgfältig durch, …
Heben Sie die Packungsbeilage auf …
Fragen Sie ihren Apotheker, …
Er wirkt aber nicht unhöflich, denn es wird deutlich, dass die Beziehung zum Kunden wichtig ist und die Sorge um seine Gesundheit im Vordergrund steht.
Sie (die Packungsbeilage) enthält wichtige Informationen für Sie.
Vielleicht möchten Sie sie später noch einmal lesen.
Außerdem ist der Beipackzettel (im Vergleich zum Kochrezept) viel persönlicher, der Kunde wird immer wieder direkt angesprochen.
…, informieren Sie bitte Ihren Arzt oder Apotheker.

2.3 **Beispiellösung:**

Bedingung	Folge
Ich benötige weitere Informationen / einen Rat.	Ich soll einen Arzt oder Apotheker fragen.
Symptome verschlimmern sich / es tritt keine Besserung ein	**Ich muss auf jeden Fall einen Arzt aufsuchen.**
Eine der aufgeführten Nebenwirkungen beeinträchtigt mich stark oder ich habe eine Nebenwirkung, die nicht angegeben ist.	**Ich soll einen Arzt oder Apotheker informieren.**
Ich bin nicht ganz sicher, wie ich ACC akut einnehmen soll.	**Ich soll einen Arzt oder Apotheker fragen.**
Das Medikament wurde nicht anders verordnet.	**Die übliche Dosis ist …**
Das Krankheitsbild verschlimmert sich oder nach 4-5 Tagen tritt keine Verbesserung auf.	**Ich sollte einen Arzt aufsuchen.**
Ich habe das Gefühl, dass die Wirkung von ACC akut zu stark oder zu schwach ist.	**Ich soll mit einem Arzt oder Apotheker sprechen.**
Ich habe das Medikament überdosiert.	**Es können Reizerscheinungen im Magen-Darm-Bereich auftreten.**
Ich habe den Verdacht, das Medikament überdosiert zu haben.	**Ich soll auf jeden Fall einen Arzt oder Apotheker aufsuchen.**
Ich habe einmal vergessen, ACC akut einzunehmen oder habe zu wenig eingenommen.	**Ich soll die Einnahme beim nächsten Mal wie gewohnt fortsetzen.**
Ich habe weitere Fragen zur Anwendung des Arzneimittels.	**Ich soll einen Arzt oder Apotheker fragen.**

2.4 Beispiellösung:

a Für den Fall, dass **Sie weitere Informationen oder einen Rat benötigen**, fragen Sie bitte einen Arzt oder Apotheker.

b Verschlimmern sich die Symptome **oder tritt keine Besserung ein, müssen Sie auf jeden Fall einen Arzt oder Apotheker aufsuchen.**

c Bei **Unsicherheit bezüglich der richtgen Einnahme** fragen Sie einen Arzt oder Apotheker.

d **Die übliche Dosis ist eine Tablette**, es sei denn, **das Medikamrnt wurde anders verordnet.**

e Im Falle einer Überdosierung **können Reizerscheinungen im Magen-Darm-Bereich auftreten.**

f Bei **weiteren Fragen zur Anwendung des Arzneimittels fragen Sie einer Arzt oder Apotheker.**

9.5 Kontaktieren

9.5.1 Glückwunschkarte: „Alles Gute zur Hochzeit"

1 Wer: **Bernd und Klara** Wer: **Barbara und Ulrich mit ihren Kindern Jan und Christine)**
  Wem: **Petra und Michael** Wem: **dem Brautpaar**
  Wozu: **Hochzeit** Wozu: **Hochzeit**

2.1

> Liebe Maya,
> nun bist du schon 12 Jahre alt. Zu d**einem**
> **Geburtstag** möchten **wir** d**ir** ganz herzlich
> gratulieren. Für das neue Lebensjahr w**ünschen**
> **wir** dir alles Gute. Feiere schön!
> Oma und Opa

> Liebe Heidi, lieber Christof,
> Kinder bedeuten nicht viel und nicht wenig, sie bedeuten alles (P. Rosegger).
> **Zur Geburt** eurer Tochter möchten wir euch **ganz herzlich** gratulieren
> und **euch** alles nur erdenklich **Gute** wünschen. Möge sie wachsen und
> gedeihen!

2.2 a Zu unserer Hochzeit möchten wir euch **ganz / sehr** herzlich einladen. Wir hoffen **sehr**, dass ihr kommen könnt.

b Zum Geburtstag wünsche ich dir **alles** Gute: **Viel** Glück und Gesundheit.

c Zur Geburt eures Kindes gratulieren wir **ganz / sehr** herzlich und wünschen euch **alles** nur erdenklich Gute für das Leben zu dritt.

d Für die Glückwünsche möchten wir uns **ganz / sehr** herzlich bedanken.

e Es hat uns **sehr** gefreut, dass ihr bei unserer Hochzeit wart.

f **Viel**en Dank für die Blumen!

9.5.2 SMS-Kommunikation: „Lust auf nen Kaffee?"

1 Sophie und Johannes:
Johannes hat eine Prüfung und Sophie drückt ihm die Daumen. Die Prüfung läuft nicht so gut, aber er besteht sie. Danach geht er (mit anderen Leuten) in das Café Schroeder's. Er fragt Sophie, ob sie mitkommen möchte und sie sagt zu.

Kerstin und Anke:
Kerstin fährt übers Wochenende zu Anke, ihr Zug ist sehr voll und es ist sehr heiß. Als sie in Leipzig ankommt, ist sie sehr kaputt. Anke hat etwas für sie gekocht.

2 Das Subjekt „ich" oder „wir" wird meistens weggelassen (Beispiele: drück dir die Daumen, gehen jetzt ...,
mach mich ...). Andere Subjektpronomen werden hingegen nicht weggelassen (du, ihr, er, sie, es).
Manchmal fehlt auch der Artikel (Beispiel: Prüfung).
Es gibt Sätze ohne Verb (Beispiel: und das bei der Affenhitze).
Wenn man sich regelmäßig schreibt, werden Anrede oder Grußformeln weggelassen. Bei unregelmäßigem
Kontakt verwendet man sie meistens nur bei der ersten SMS.
Die Situation wird oft nicht erklärt, man schreibt beispielsweise einfach „Drück dir die Daumen" – der
Empfänger weiß ja, dass er gleich eine Prüfung hat.
Man benutzt Imperative, um den anderen um Kontaktaufnahme zu bitten (Beispiele: meld dich mal; sag mal
Bescheid). Das ist kürzer als Formulierungen wie „Ich würde mich freuen, wenn du dich melden würdest." oder
„Es wäre nett, wenn du Bescheid geben könntest, ..."
Emoticons können ganze Sätze ersetzen (Beispiel: ☹).
Abkürzungen (WE, L) stehen für Wörter.

3 ☺ **Juchu! (Ich freue mich.)**

; -) **Wie schade! (Ich bin traurig)**

WE **Nimm es nicht so schwer!**

LG **Liebe Grüße**

4   Petra, 27.05., 15:30   Susanne, 27.05., 18:42   Petra, 28.05., 09:30   Susanne, 28.05, 14:24:17

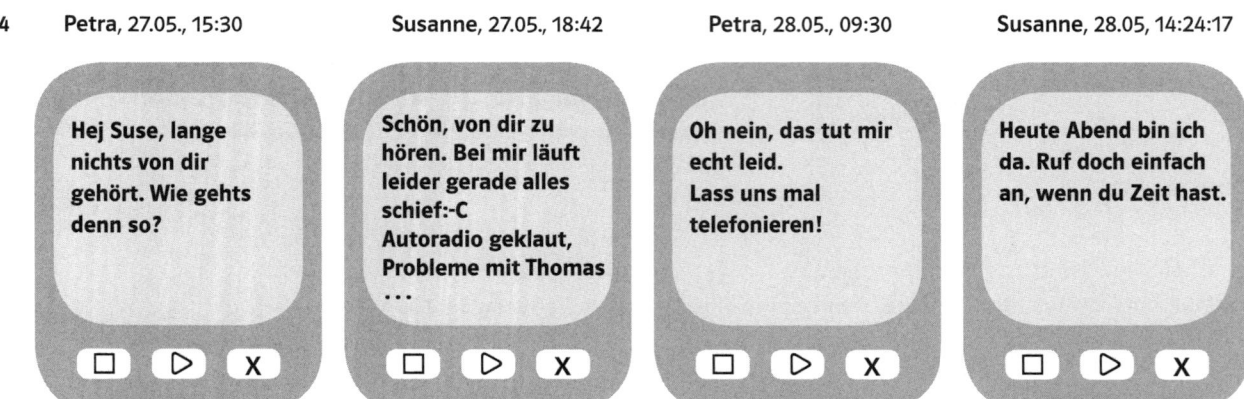

5 Beispiellösung:

Bernt, 26.10, 12:14   Arne, 27.10, 09:32

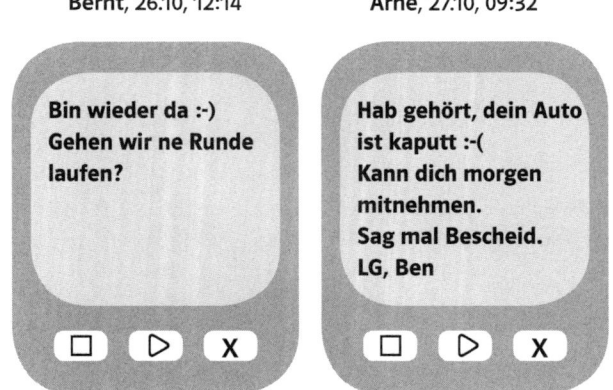

9.5.3 Alltagsgespräche: „Bei dem Wetter ...“

1.1 ☒ **Der Kunde und die Friseurin haben keine private Beziehung, sondern eine typische „Servicebeziehung“. Nach der Besprechung des Haarschnittes könnten sie schweigen. Da dies dem Kunden möglicherweise unangenehm wäre, beginnt die Friseurin ein Gespräch über das Wetter.**

  ☒ **Das Gespräch ist ein typisches Beispiel für einen Smalltalk: Man wählt ein Thema wie das Wetter, bei dem ein Konflikt unwahrscheinlich ist, weil die meisten Menschen ähnliche Erfahrungen haben. Man verwendet Signale wie ne, weil man möchte, dass der Gesprächspartner zustimmt.**

1.2 1: *Der Kunde hustet.*
  2: **Wetter als Grund für Erkältung**
  3: **Erkältungsanfälligkeit beider Sprecher allgemein**
  4: **Wetterumschwung**
  5: *die verfrühten Eisheiligen*

2

1	ja ↗ „Ich bin überrascht und will sichergehen, dass ich das auch richtig verstanden habe“	Ja, echt? **Ja, meinen Sie?**
2	ja → „Ich weiß, dass etwas offensichtlich oder schon bekannt ist. Daher nehme ich an, dass der Hörer es auch weiß.“	Bei dem Wetter ist es ja auch kein Wunder, dass man sich erkältet, **... da wird's ja endlich wärmer ...** **... da hab ich ja auch Urlaub.** **... das sind ja die verfrühten Eisheiligen, ne?**
3	ja → „Ich möchte weitersprechen oder jetzt anfangen zu sprechen. Das, was ich sagen möchte, passt zu dem, was gerade gesagt wurde.“	Ja, ich hab im Allgemeinen auch nichts damit zu tun. **Ja, es muss ja auch endlich mal wärmer werden ...** **Ja, die kommen normalerweise erst Mitte Mai ...**

3 da (Zeile 3): *bei dem Wetter*
  damit (Zeile 5): **mit Erkältungen**
  da (Zeile 9): **ab Donnerstag**
  das hier (Zeile 11): **das Wetter zum Zeitpunkt des Gesprächs**
  die (Zeile 16): **die Eisheiligen**

4 z.B. Urlaub, Neuigkeiten in der Satdt, Prominentenklatsch, Hobbys